Towards the Monitoring of Dumped Munitions Threat (MODUM)

NATO Science for Peace and Security Series

This Series presents the results of scientific meetings supported under the NATO Programme: Science for Peace and Security (SPS).

The NATO SPS Programme supports meetings in the following Key Priority areas: (1) Defence Against Terrorism; (2) Countering other Threats to Security and (3) NATO, Partner and Mediterranean Dialogue Country Priorities. The types of meetings supported are generally "Advanced Study Institutes" and "Advanced Research Workshops". The NATO SPS Series collects together the results of these meetings. The meetings are co-organized by scientists from NATO countries and scientists from NATO's "Partner" or "Mediterranean Dialogue" countries. The observations and recommendations made at the meetings, as well as the contents of the volumes in the Series, reflect those of participants and contributors only; they should not necessarily be regarded as reflecting NATO views or policy.

Advanced Study Institutes (ASI) are high-level tutorial courses to convey the latest developments in a subject to an advanced-level audience.

Advanced Research Workshops (ARW) are expert meetings where an intense but informal exchange of views at the frontiers of a subject aims at identifying directions for future action.

Following a transformation of the programme in 2006, the Series has been re-named and re-organised. Recent volumes on topics not related to security, which result from meetings supported under the programme earlier, may be found in the NATO Science Series.

The Series is published by IOS Press, Amsterdam, and Springer, Dordrecht, in conjunction with the NATO Emerging Security Challenges Division.

Sub-Series

A.	Chemistry and Biology	Springer
B.	Physics and Biophysics	Springer
C.	Environmental Security	Springer
D.	Information and Communication Security	IOS Press
E.	Human and Societal Dynamics	IOS Press

http://www.nato.int/science
http://www.springer.com
http://www.iospress.nl

Series C: Environmental Security

Towards the Monitoring of Dumped Munitions Threat (MODUM)

A Study of Chemical Munitions Dumpsites in the Baltic Sea

edited by

Jacek Bełdowski
Department Marine Chemistry & Geochemistry, Institute of Oceanology PAS
Sopot, Poland

and

Robert Been
Centre for Maritime Research and Experimentation (STO - CMRE), Science and Technology Organisation, La Spezia, Italy

and

Eyup Kuntay Turmus
Emerging Security Challenges Division, Science for Peace & Security (SPS) Programme, Brussels, Brussels Hoofdst.ge., Belgium

Springer

Published in Cooperation with NATO Emerging Security Challenges Division

Results of the multi-year NATO SPS project
Towards the Monitoring of Dumped Munitions Threat (MODUM)
2013–2016

Library of Congress Control Number: 2017949359

ISBN 978-94-024-1161-4 (PB)
ISBN 978-94-024-1152-2 (HB)
ISBN 978-94-024-1153-9 (e-book)
DOI 10.1007/978-94-024-1153-9

Published by Springer,
P.O. Box 17, 3300 AA Dordrecht, The Netherlands.

www.springer.com

Printed on acid-free paper

MODUM

TOWARDS THE MONITORING
OF DUMPED MUNITIONS THREAT

NATO
OTAN

This publication is supported by:

The NATO **Science** for **Peace** and **Security** Programme

Preface: The NATO Science for Peace and Security Perspective

The Science for Peace and Security (SPS) Programme is an established brand for NATO based on three pillars – science, partnership and security – and has been contributing to the core goals of the Alliance for many decades. In the spirit of cooperative security, SPS provides concrete, practical opportunities for cooperation to NATO's wide network of partner countries based on security-related civil science, technology and innovation. The origins of the SPS Programme date back to the 1950s, and its outlook has been adapted to the changing security environment over the decades. The programme provides funding and expert advice for security-relevant activities in the form of workshops, training courses or multi-year research projects and thus promotes dialogue among partners.

All activities funded under the SPS Programme address one or more of the programme's key priorities and have a link to security. The SPS key priorities are based on NATO's Strategic Concept, agreed by Allies in the 2010 Lisbon Summit, and the strategic objectives of NATO's partner relations agreed in Berlin in 2011. According to these priorities, the programme promotes cooperation, scientific research and innovation to address contemporary security challenges, including terrorism; defence against chemical, biological, radiological and nuclear agents; cyber defence; energy security; and environmental concerns, as well as human and social aspects of security such as the implementation of United Nations Security Council Resolution 1325 on Women, Peace and Security.

Dumped chemical weapons pose an actual environmental and security hazard in the Baltic Sea Region. Official and unofficial dumpsites cover vast areas of the Baltic Sea bottom, with at least 50,000 tons of CW having been dumped, containing approximately 15,000 tons of CWA. Their actual position is unknown, and pollution originating from corroded munitions is only roughly estimated. The risk created by the dumpsites is twofold: They pose important risks for users, such as fishermen and maritime entrepreneurs, and they create a diffuse source of contaminants to surrounding sediments and near bottom waters, hence threatening marine biota. With more and more activities being performed in the Baltic Sea area, the threat level rises constantly and could potentially create chemical weapon danger for the Baltic States population. In this regard, the multi-year NATO SPS project *Towards the*

Monitoring of Dumped Munitions Threat (MODUM) was initiated in 2013 in order to address environmental and security challenge created by dumping operations. Building on the studies of two EU projects, *Modelling of Ecological Risks Related to Sea-Dumped Chemical Weapons* (MERCW) and *Chemical Munitions Search and Assessment* (CHEMSEA), dealing especially with the dumpsites located in the Bornholm and Gotland Deep, the SPS project complemented EU efforts and actively included specialized EU agencies. While abovementioned projects depended on ship-based measurement platforms for detection of dumped munitions, supplemented by ship-based samplers for sediment collection video identification by remotely operated vehicles, NATO activities complemented EU research and focused on the broadening of knowledge about the risks posed by sea-dumped munitions and the creation of monitoring guidelines for the dumpsite areas using autonomous underwater vehicles (AUVs) and remotely operated underwater vehicles (ROVs) and utilizing existing research vessels of partner institutions as launching platforms. The project thereby concentrated on three representative areas: habitat status evaluation, fish health studies and modelling of possible threats to adjacent areas and performed field measurement campaigns.

The SPS project established standard operational procedures to monitor the breakdown of munitions and release of agents into sediments and water, to be able to spot a mass breakout, potentially creating chemical weapon danger for the Baltic States population, in advance. These procedures include a continuous survey procedure, based on the best available technology, a procedure for the identification of dumped munitions, based on ROV video and acoustic camera imagery, and a procedure for establishing continuous risk assessment, including sediment pollution, effect on biota and habitat damage. The success of the project provides a scalable solution for the monitoring of Baltic Sea chemical weapon dumpsites, and information contained will enable states to proclaim potentially dangerous areas for fishermen and offshore industry. Capacity for the creation of permanent monitoring of the dumpsites, including equipment and human potential, is built in the Baltic Sea area and is currently used as a basis for the Baltic Sea Region EU project Decision Aid for Marine Munitions (DAIMON). Project DAIMON aims at creating decision support tool for maritime administration of Baltic countries, in regard to munition dumpsite management. It develops further risk assessment methods created during the MODUM project and prepares algorithms for their interpretation.

By connecting scientists, experts, government representatives and civil society on key issues of security, the SPS Programme is able to make a significant positive impact upon society and achieve tangible and lasting results. The aim of SPS activities is to set a working model often by providing an initial operational capability that could then be expanded through national and/or other resources. Under the umbrella of the MODUM project, young scientists training in summer schools in Canada and the United States, as well as publications and conference presentations, successfully achieved a knowledge transfer to the global scientific community, with this book providing another valuable contribution to the research on dumped munition. The selection of NATO and partner countries represented a regional approach, and the linkage with the International Dialogue on Underwater Munitions (IDUM) ensured the global dimension, making a transfer of project results to other affected areas possible, to increase human security worldwide.

The programme further looks at engaging partners in civil science, technology, innovation and beyond. Interested parties are therefore encouraged to submit an application for funding. Proposed projects must be led by project directors from at least one Allied and one partner country as well as address the SPS key priorities that have a clear link to security. The developed collaborative activity must fit within one of the SPS grant mechanisms, that is, multi-year projects, training courses or workshops, as mentioned above. Applications received by the SPS Programme will undergo a comprehensive evaluation and approval process, taking into account expert, scientific and political guidance. For more information and the latest news about the SPS Programme, please visit our website (www.nato.int/science) where you will also be able to find detailed application guidelines and forms. Alternatively, you can e-mail us under sps.info@hq.nato.int.

Eyup Kuntay Turmus

Science for Peace & Security (SPS) Programme,
Emerging Security Challenges Division
J-202, NATO Headquarters CB, 1700,
Brussels, Belgium
turmus.eyup@hq.nato.int

Acknowledgements

Studies described in this book were funded by the NATO Science for Peace project #984589 (MODUM) and polish science funds for years 2014–2016 for the implementation of international co-funded project 3055/NATO/2014/0.

Some of included results were obtained in the frame of DAIMON project – in which the research work was fund by the European Union (European Regional Development Fund) under the Interreg Baltic Sea Region Programme 2014–2020, project #R013 DAIMON, cofunded by the Ministry of Science and Higher Education (Poland) from the 2016–2019 science funding allocated for the implementation of international co-financed project.

Contents

Editors and Contributors

Editors

Jacek Bełdowski Institute of Oceanology, Polish Academy of Sciences, Sopot, Poland

Robert Been Science and Technology Organisation – Centre for Maritime Research and Experimentation (STO - CMRE), La Spezia, SP, Italy

Eyup Kuntay Turmus Science for Peace & Security (SPS) Programme, Emerging Security Challenges Division, Brussels, Belgium

Contributors

Robert Been Science and Technology Organisation – Centre for Maritime Research and Experimentation (STO - CMRE), La Spezia, SP, Italy

Jacek Bełdowski Institute of Oceanology, Polish Academy of Sciences, Sopot, Poland

Federico Cernich Science and Technology Organisation – Centre for Maritime Research and Experimentation, La Spezia, SP, Italy

Ilias Christensen Department of Environmental Science, Aarhus University, Roskilde, Denmark

Department of Environmental Engineering, Technical University of Denmark, Lyngby, Denmark

Michał Czub Institute of Oceanology, Polish Academy of Sciences, Sopot, Poland

Patrik Fauser Department of Environmental Science, Aarhus University, Roskilde, Denmark

Stefano Fioravanti Science and Technology Organisation – Centre for Maritime Research and Experimentation, La Spezia, SP, Italy

Maria Golenko Atlantic Branch of the P.P. Shirshov Institute of Oceanology, Russian Academy of Sciences, Kaliningrad, Russian Federation

Miłosz Grabowski Institute of Oceanology, Polish Academy of Sciences, Sopot, Poland

Katarzyna Grzelak Institute of Oceanology, Polish Academy of Sciences, Sopot, Poland

Jaromir Jakacki Institute of Oceanology, Polish Academy of Sciences, Sopot, Poland

Piia Jõul Tallinn University of Technology, Tallin, Estonia

Mihkel Kaljurand Tallinn University of Technology, Tallin, Estonia

Zygmunt Klusek Institute of Oceanology, Polish Academy of Sciences, Sopot, Poland

Lech Kotwicki Institute of Oceanology, Polish Academy of Sciences, Sopot, Poland

Thomas Lang Thünen-Institut für Fischereiökologie, Cuxhaven, Germany

Heidi Lees Tallinn University of Technology, Tallin, Estonia

Terrence Long International Dialogue of Underwater Munitions, Leitches Creek, NS, Canada

Roger Magnusson Swedish Defence Research Agency, Umeå, Sweden

Vitalijus Malejevas The Environmental Protection Agency, Vilnius, Lithuania

Anders Östin Swedish Defence Research Agency, Umeå, Sweden

Vadim Paka Atlantic Branch of the P.P. Shirshov Institute of Oceanology, Russian Academy of Sciences, Kaliningrad, Russian Federation

Erik Amos Pedersen Department of Environmental Science, Aarhus University, Roskilde, Denmark

Department of Environmental Engineering, University of Southern Denmark, Odense, Denmark

Johanna Qvarnström Swedish Defence Research Agency, Umeå, Sweden

Daniel Rak Institute of Oceanology, Polish Academy of Sciences, Sopot, Poland

Jenny Rattfelt-Nyholm Swedish Defence Research Agency, Umeå, Sweden

Hans Sanderson Department of Environmental Science, Aarhus University, Roskilde, Denmark

Martin Söderström Finnish Institute for Verification of the Chemical Weapons Convention, University of Helsinki Finland, Helsinki, Finland

Morten Swayne Storgaard Department of Environmental Science, Aarhus University, Roskilde, Denmark

Katharina Straumer Thünen-Institut für Fischereiökologie, Cuxhaven, Germany

Marta Szubska Institute of Oceanology, Polish Academy of Sciences, Sopot, Poland

Eyup Kuntay Turmus Science for Peace & Security (SPS) Programme, Emerging Security Challenges Division, Brussels, Belgium

Merike Vaher Tallinn University of Technology, Tallin, Estonia

Paula Vanninen Finnish Institute for Verification of the Chemical Weapons Convention, University of Helsinki, Helsinki, Finland

Kela Weber Environmental Sciences Group, Department of Chemistry and Chemical Engineering, Royal Military College of Canada, Kingston, Ontario

Lina Weirup Thünen-Institut für Fischereiökologie, Cuxhaven, Germany

Victor Zhurbas Atlantic Branch of the P.P. Shirshov Institute of Oceanology, Russian Academy of Sciences, Kaliningrad, Russian Federation

Chapter 1
Introduction

Jacek Bełdowski, Terrence Long, and Martin Söderström

Abstract Dumped chemical weapons (CW) pose an actual environmental and security hazard in the Baltic Sea Region. Their actual position is unknown, and pollution originating from corroded munitions is only roughly estimated. One of possible low-cost solution is the creation of monitoring network, providing info about exact location and environmental threat posed by sea dumped CW.

MODUM project aimed at the establishment of the monitoring network observing Chemical Weapons dumpsites in the Baltic Sea, using autonomous underwater vehicles (AUV's) and remotely operated underwater vehicles (ROV's), and utilizing existing research vessels of partner institutions as launching platforms. Project consisted of the test phase, which helped choosing best available solutions for the difficult Baltic Sea environment, Survey phase, which located actual objects of concern, and monitoring phase, which concentrated on the collection of environmental data close to the objects of concern. The result of the project is a set of methods and procedures that are applicable for the dumpsites monitoring, ranging from sea bottom survey, object identification to actual environmental parameters collection, biota exposure and portable chemical warfare agents (CWA) analysis. It uses such data for the calculation of environmental risk assessment in the dumpsites.

J. Bełdowski (✉)
Institute of Oceanology, Polish Academy of Sciences,
ul. Powstańców Warszawy 55, 81-712 Sopot, Poland
e-mail: hyron@iopan.gda.pl

T. Long
International Dialogue of Underwater Munitions, Leitches Creek, NS B2A 3Z7, Canada
e-mail: tplong@eastlink.ca

M. Söderström
Finnish Institute for Verification of the Chemical Weapons Convention, University of Helsinki
Finland, A. I. Virtasen aukio 1, P.O. Box 55, Helsinki 00014, Finland
e-mail: martin.soderstrom@helsinki.fi

J. Bełdowski et al. (eds.), *Towards the Monitoring of Dumped Munitions Threat (MODUM)*, NATO Science for Peace and Security Series C: Environmental Security, DOI 10.1007/978-94-024-1153-9_1

1.1 Introduction

The first modern use of CW took place in 1915 during the First World War (WWI). Later in the war all sides started producing and using CWA. The widespread use of CW in WWI resulted in hundreds of thousands of casualties on the battlefield. The horrors, that came after their use during the War motivated diplomats to negotiate the 1925 Geneva Protocol, a multilateral treaty banning the use of chemical weapons in armed conflict. As a result, some of the first major dumping of chemical weapons, took place in the English Channel in the 1920's that, many believed, led to the contamination of mussels and shellfish in the Channel. In years following the First World War many countries including US, UK, USSR, Germany, France and Canada, manufactured massive quantities of chemical weapons and agents. This build-up of chemical stockpiles continued throughout much of the twentieth century.

Sea disposal of chemical munitions or containers with CWA was a common practice until 1970's, and dumping operations were performed by as many as 40 countries (Carton and Jagusiewicz 2009). Dumping operations in US started in 1919, after the 1st world war, and included mostly mustard gas. According to US regulations, dumping was required to take place in marine waters, of depth not less than 30 m (updated later to 55 m) and at a distance of minimum 65 km from the shore. However, other countries did not follow similar regulations, and chemical munitions were frequently dumped in the coastal waters, sometimes in very shallow places, and in some cases dumping was conducted also in freshwater environments, such as lakes and even rivers (Walker 2012). Until 1970's, dumping of obsolete or outdated chemical munitions was conducted mostly by USA, France, Great Britain, Japan and Russia and many others. It is estimated that about 1 million tons of chemical munitions, originating from both world wars and post-war period, is dumped into seas and oceans. One hundred and twenty seven documented dumping areas are known, while over 300 total dumpsites are suspected to exist (James Martin Center, WWW). Dumpsites are located in Atlantic, Pacific and Indian Oceans, at east and west Canadian and US coasts, in Gulf of Mexico, Australian, New Zealand, India, Philippines, Japan, Great Britain and Irish coasts as well as in Caribbean, Black, Red, Baltic, Mediterranean and North Seas.

At the end of World War II, The Potsdam Agreement, was signed between the three Allied Powers, United Kingdom, United States, Union of Soviet Socialist Republics, for the military occupation and reconstruction of Germany. The Potsdam's agreement led to the disposal of massive quantities of captured, damaged, and obsolete chemical and conventional weapons, by dumping them into the seas and ocean. Under the Potsdam Agreement shells and bombs were jettisoned from sea going vessels, but more often were loaded as cargo into ships, that were sunk by opening seacocks or by artillery fire, explosives or torpedoes. Most dumping operations were carefully undertaken, including the keeping of records of where the dumping occurred, a listing of the material that was dumped, and the quantities of dumped material.

The Potsdam Agreement, was the catalyst for global dumping of munitions, that was carried out, up until the 1970's, by most Nations of the World as a cheap means of disposal. Today the Potsdam Agreement still keeps a secret the location of the underwater sites, and is due to be open to the public in 2017.

In Europe, chemical munition identified dumpsites exist in coastal waters of the North Sea (Belgium, Holland, Germany), deep sites in the Irish Sea and Biscay Bay, and selected sites in the semi enclosed Adriatic and Baltic Seas (Arison 2013). Munitions there consist mostly of Mustard Gas, and smaller quantities of arsenic based agents, such as Clark I and II, Adamsite and Lewisite, in some sites nerve agents were dumped as well (Arison 2013). North Sea sites originate from WWI, while Baltic, Adriatic and Irish Sea contain warfare material mostly from the WWII (Tobias Knobloch et al. 2013). Altogether, the amount of chemical munitions in European Waters is estimated to be 684 thousand tons (Arison 2013).

In the Baltic Sea chemical weapons originating from captured German arsenals were dumped in the Baltic Proper and Skagerrak Strait on the orders of the British, Russian and American occupation authorities, while some of the weapons were dumped also by Wehrmacht at the last days of WWII (Tobias Knobloch et al. 2013). At least 170,000 t of CW was dumped in the Skagerrak, mainly in the Norwegian trench and in the eastern Skagerrak, off the Swedish coast. In most of the dumping operations in the Skagerrak, complete ships were sunk with their cargo. In the Baltic Sea, at least 50,000 t of CW were dumped – it is assumed that these contained roughly 15,000 t of CWA; the most important dumpsites here are located in the Lille Belt, near the island of Bornholm, and in the Gotland basin. In most cases, the CW was thrown overboard, either loose (bombs, shells) or in containers, but some ships were also sunk. In most cases those dumped materials contained explosives (bursters for the CW), in some cases dumping of conventional munitions was commenced in the same locations as CW. There are strong indications that part of the CW was thrown overboard during transport to the Baltic dumpsites; how many tonnes were thus dumped is not known.

There is, though never verified, information that chemical munitions were dumped in the Baltic Sea for many years after the year 1948 by the army of the German Democratic Republic and the Soviet Army, most probably still in the 1980s. Since those suspected operations were unofficial, little is known on the types of munitions or containers dumped (Neffe et al. 2011).

1.2 Study Area

The following section describes areas of concern because of documented, possibly detected or suspected presence of dumped chemical munitions. Location of designated and suspected dumping areas are presented at the picture below (Fig.1.1).

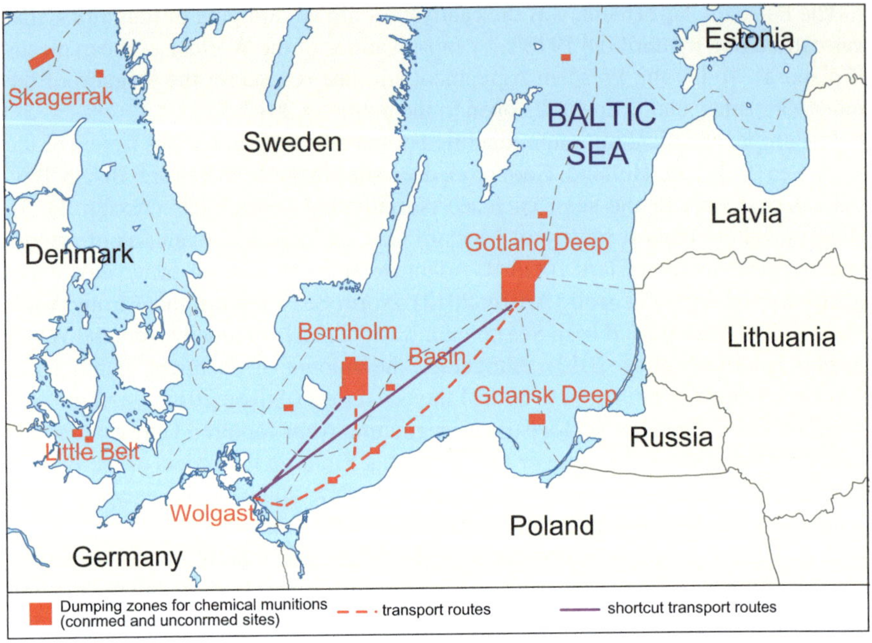

Fig. 1.1 Overview map of known and suspected dumpsites of chemical warfare materials (Reproduced with permission from the CHEMSEA project)

1.2.1 Gotland Basin

It has been alleged that, during WWII, 2000 t of chemical warfare materials (consisting of approximately 1000 t of chemical warfare agent payload) were transported from the loading port of Wolgast, and dumped there by item-by-item disposal in southern part of the Gotland Basin. There is suspicion that, chemical warfare materials were thrown overboard while the ships were en route, as it took place, during dumping runs to the Bornholm Basin. During the CHEMSEA Project, all targets were categorized into five classes based on the size of highlights and acoustic shadows of detected objects. Figure 1.2 represents the ROV operations conducted by Swedish Maritime Administration (SMA) in Gotland Deep during the CHEMSEA Project, showing the correctness of target classification. Targets classified primarily as Class 1 (targets within dimensions of 1.8 × 0.5 m – the most probable munition pieces) and visually confirmed.

Gotland Deep in located in the middle of southern Baltic and is in Latvian, Lithuanian and Swedish EEZs. It also borders Russian (Kaliningrad) EEZ. In general, the water depth in the large area marked on sea charts as 'explosives dumping ground' ranges between 93 m and more than 120 m. The thickness of the sediments reaches 4–6 m, with clayey muds prevailing. The sedimentation rate is also low, varying between 0.5 and 1 mm per year and resulting in an additional layer of up to 6 cm since 1948 (MERCW 2006). The overall observation can be made that the

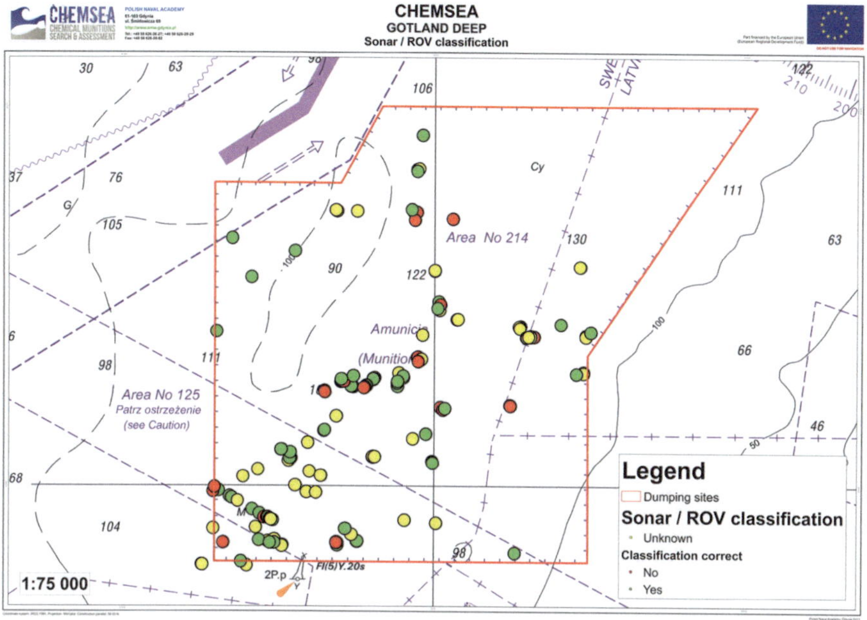

Fig. 1.2 Map of the Gotland Dumpsite Including target classification and verification by ROV missions

distribution of munitions in the Gotland Basin is broader than earlier estimated, that it is not a primary dumpsite, but objects are quite evenly distributed in the area. Dumping of ca. 2000 t of CW material on behalf of the Soviet Authorities took place in 1947. The dumped items contained ca. 930 t of agents. Most of this was sulfur mustard (580 t) and arsenicals (294 t).

In Gotland Deep dumping of sulfur mustard, Clark, arsine oil, Adamsite has been confirmed by documents and chemical analysis. In addition to these agents records on dumping of CN have been found. Dumping of tabun has been suspected as it has been stored in Wolgast harbor, where the ships sails for dumpsites.

1.2.2 Bornholm Deep

Bornholm Deep is the largest of the dump sites and it is situated east of Bornholm island within the Danish extended economic zone (EEZ). Some dumping operations were carried out by British Military Administration in Germany already in 1945–1946. The major dumping took place in 1947–1948 by Soviet Military Administration in Germany. Some additional dumping was carried out in 1959–1967 by former GDR authorities. (Tobias Knobloch et al. 2013) In total ca. 32,000 t of CW material was dumped under Soviet control and further 60 t under GDR control. The amount of actual CWA chemicals in the Soviet controlled dumping operarations was ca. 10,700 t. Amount of sulphur mustard was ca. 6700 t and arsenicals was ca. 3400 t.

Dumping of German sulfur mustard, Clark, arsine oil, Adamsite and CN have been confirmed by documents and chemical analysis. (Tobias Knobloch et al. 2013). As for other areas, dumping of Tabun is suspected. Phosgene was included in the GDR dumping operations. Lewisite II-related chemicals (present in Lewisite I) have been found in chemical analysis of sediment samples from Bornholm dumpsite area, (Nord Stream Project 2010; 2011) but their origin remains unknown.

The dumping area – sometimes referred commonly as the 'primary dumpsite' in the Baltic Sea – was located around a point with coordinates 55° 20′ N, 15° 37′ E (WGS84). The northern part of the area is currently marked as 'larger explosives dumping ground' on sea charts. A variety of different chemical warfare materials were dumped there. While wooden or metallic encasements were also used, more than 80% of the dumped chemical warfare materials were munitions containing explosives, with large KC50 and KC250 aircraft bombs forming the largest contingent. Dumping was carried out by item-by-item disposal and the wooden crates were often seen drifting before sinking. (Sanderson et al. 2008). It was reported that in other cases, the bombs were taken out of the crates, disposed of and the crates brought back to the originating harbour of Wolgast for further use (Bruchmann 1953). Because of these facts, research covered the area of 'primary dumpsite' – the high priority area, as well like 'extended dumpsite' area and possible transports routes – medium priority area depicted at the Fig. 1.3, below.

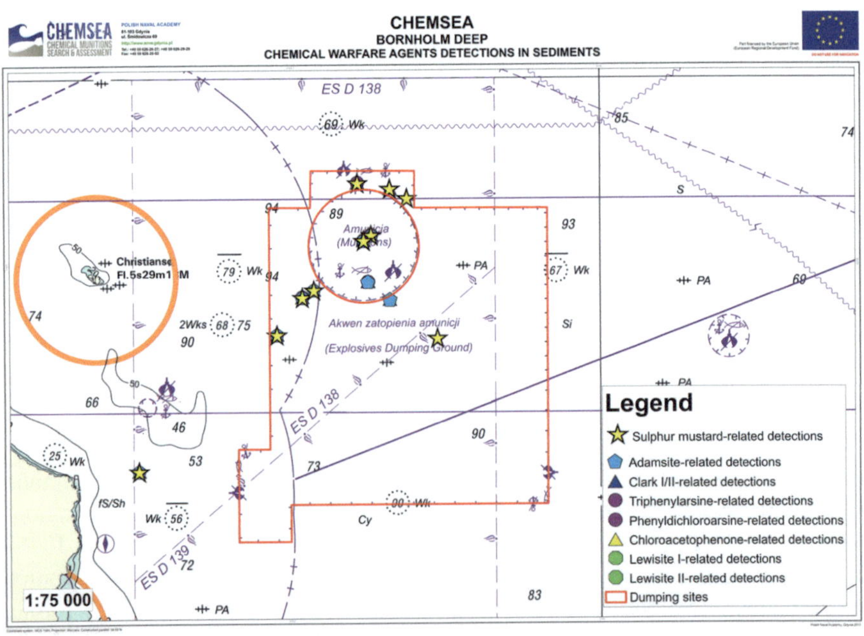

Fig. 1.3 Map of Bornholm dumpsite showing the findings of the previous CHEMSEA project. See Fig. 1.1 for the overview map (Reproduced with permission from the CHEMSEA project)

1.2.3 *Little Belt*

The southern entrance to the Little Belt is an area that was used by German side for munitions disposal at the end of WWII (Jäckel 1969, BSH 1993, Wichert 2012). Towards the end of April 1945, two barges, loaded with approximately 1250 t of chemical grenades and crates containing 250 t of powder, were tugged on order of the German Navy to a designated area located at the southern entrance to the Little Belt and scuttled. There are suppositions, that during that operation, about 5000 t of chemical bombs and shells were dumped in the area by item-by-item disposal. The participants of the disposal runs reported that manual item-by-item disposal was also conducted en route from Flensburg to the dumping area. It is assumed that a cargo of at least 1200 t of chemical warfare materials was dumped in the Flensburg Fjord, which is very important place for the survey on the basis of above facts. It was reported that 600 t (approximately 250 items) of missiles or their parts had been dumped during an additional run along the cost of the Flensburg Fjord. Figure 1.4 represents suspected dumpsite area with designed transport route.

In general, the water in the area is around 25–31 m deep and the seafloor is in most places characterized by a thick layer of soft and muddy sediment (0–8 m). The rate of sedimentation in this area can be estimated up to 1–2 mm/year, resulting in a muddy sediment layer of up to 13 cm since 1945. The dumpsite is included on sea charts as an area encompassing around 4180 ha and is designated as 'foul (explosives)' and dangerous for anchoring and fishing.

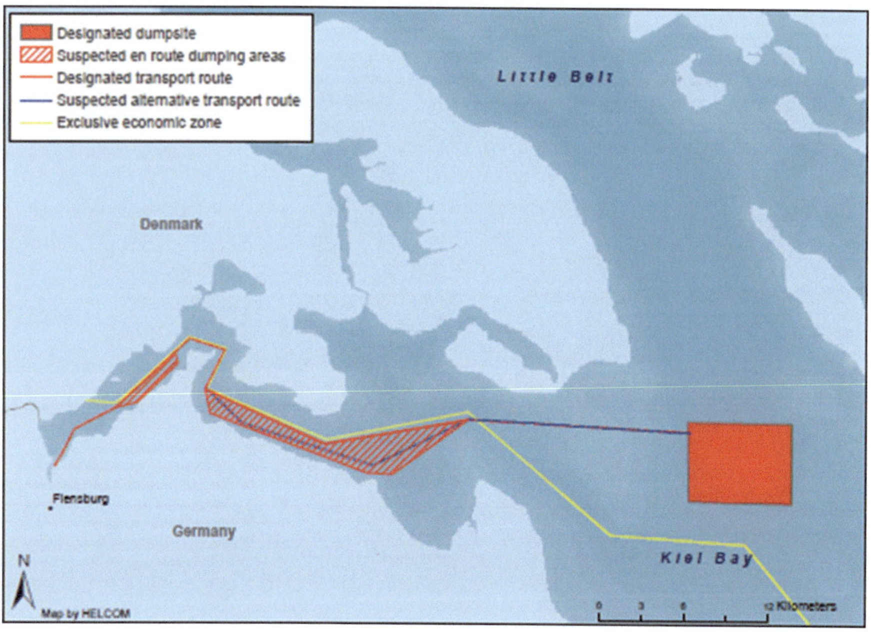

Fig. 1.4 Map of the designated and suspected dumpsite and transport routes at the south of Little Belt (www.helcom.fi)

1.2.4 Gdansk Basin

The Gdańsk Deep is probably the relatively insignificant dumpsite, with only several tonnes of chemical munitions dumped alongside conventional munitions (Fig. 1.5). Based on incidents in 1950s involving sulfur mustard, it was suspected that dumping operations had taken place also in Gdańsk Deep in the Gulf of Gdańsk on the Polish coast and within the Polish EEZ. A chemical munitions dumpsite is suspected to exist in the southern part of the Deep (Andrulewicz 1996), an area with a depth ranging between 80 and 110 m. A small circular area with a diameter of 0.62 nautical miles designated as a formerly used explosives dumping grounds marked on navigational charts at position 54° 45′N, 19° 10′E. The Gdańsk Deep is covered with mud and clayey mud; the thickness of the unconsolidated layer can reach 10 m. The sedimentation rate in the area is estimated to be 1.8 mm/year, meaning the objects in question could be covered by at least 11 cm of sediments (Bełdowski and Pempkowiak 2007). Bottom water is usually anoxic, with episodic inflows of North Sea water, oxygenating the area during the medium and major inflows. The extent of sediment pollution is currently unknown. Based on results from the CHEMSEA project, at least sulphur mustard and arsine oil containing munitions have been dumped in Gdańsk Deep. Degradation products for Clark were found, but these could originate either from arsine oil in winter-grade mustard or from Clark-containing munitions.

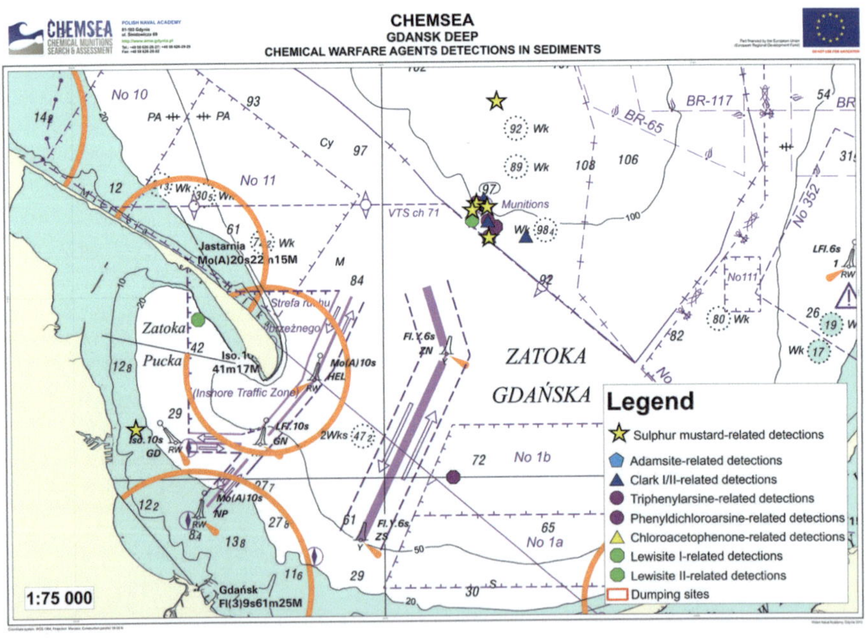

Fig. 1.5 Map of Gdańsk dumpsite showing the findings of the previous CHEMSEA project. See Fig. 1.1 for the overview map (Reproduced with permission from the CHEMSEA project)

Chemical weapons sites, have created a public health hazard. In areas of substantial dumping, such as off the coast of United States, Japan and in the Baltic and Adriatic Seas, a number of injuries have resulted from exposures to accidentally recovered chemical weapons. In most cases, CW materials are ensnared in fishing nets or have been accidentally disturbed during dredging operations in Singapore, Vietnam, and Philippians.

1.3 Environmental Fate of Munitions and Chemicals

Chemical munitions dumped at sea create a hazard for environment, if the warfare agents contained in them are released to sediments and overlying water. Release of constituents can occur either through explosion of burster charge or corrosion of casings. Given the fact, that deliberately dumped munitions were not armed, the latter seem more likely. According to existing models (Korotenko 2000) in the Baltic conditions, corrosion rate of the bombshells ranges from 0.05 to 0.575 mm/year depending upon shell type. This process may be enhanced up to four-fold by even slow water flow, which is a normal situation in most areas of the Baltic. Other processes, that may speed up corrosion rate include pH changes and abundance of chemotrophic bacteria (Silva and Chock 2016). According to models, the highest rate of release of most abundant compound, sulphur mustard would occur after 125 years from dumping over 60 year period. The amount was calculated to be in the order of >6000 t by Sanderson et al. (2008).

Existing models represent average situation, since environmental condition in bottom sediments are characterized by considerable patchiness, resulting in variable pH, content of organic matter and bacteria abundance. This has been documented in the Baltic ie. based on mercury concentration in sediemnts, which shows about 20% variability in areas less than 1 km apart (Bełdowski et al. 2014). Examination of munitions in the Baltic dumpsites reveal both completetly corroded objects, and almost completely intact.

CWA mixtures and explosives are chemicals that may have reacted with other materials in the container or with themselves. Effectively, this aging process may have changed the properties of the chemical contents. With regard to CWA, compounds with less pronounced or without warfare capabilities may have emerged.

When sea water comes into contact with these chemicals, it may act as solvent or suspension agent. Consequently, the chemicals will leak into the environment, first spread locally, possibly enter a sediment sorption/desorption equilibrium process, and will, with time, be distributed on a larger scale by hydrological processes and anthropogenic activities.

Once under the influence of environmental factors, chemicals may also undergo changes by abiotic (e.g. reactions with sea water and its components like dissolved oxygen or hydrogen sulfide, or closer to the surface, sunlight-mediated degradation) or biotic processes (e.g. bacteria-mediated biotransformation).

The propensity to undergo chemical transformations and the pathways and modes of environmental distribution, taken together can be defined as the *environmental fate* of a chemical. The environmental fate depends on the nature of the chemical (e.g. reactivity, polarity) and on the prevailing ambient conditions (e.g. temperature, reaction partners, bacterial population). Resulting from these transformations are chemicals which may have properties similar to or quite unlike the parent compounds.

Some parent or transformation chemicals will undergo fast reactions, in other cases transformations will occur only very slowly. The latter chemicals are persistent in the environment and, given suitable hydrophobic (fat-soluble) properties, have the potential to bioaccumulate in living organisms via food webs (food chains). Persistent organic pollutants (POPs) is one of the principal issues of environmental pollution.

With regard to organic chemicals, the highest possible stage of chemical breakdown is mineralization – conversion to e.g. carbon dioxide, ammonia, water and hydrogen sulfide. In the case of organometallic (e.g. organoarsenic-based CWA) or inorganic (e.g. metals from containers or the primary explosive lead (II) azide or mercury fulminate from detonators) chemical warfare materials, transformations will lead to inorganic species of heavy metals which can be converted to different organometallic species by biotic processes. These latter inorganic and organometallic species do occur naturally and their toxic properties depend on the chemical "wrapping", oxidation state and nature of the metal atom and may either be pronounced or even negligible (e.g. arsenobetaine). Nonetheless, since the amounts of bioavailable heavy metals introduced by anthropogenic activities is considerable in comparison to the naturally bioavailable amounts, discharge of heavy metals into the environment is one of the principal issues of environmental pollution.

Speed of corrosion and subsequent start of release of all chemical contents is strongly dependent on the local environment a given chemical warfare material container rests. In general, the presence of oxygen and engulfing currents will promote corrosion of a container, while burial in sediment and a low oxygen environment will preserve its original state. Even if the outer hull is still pristine, the chemical contents of a bulk container or, even more likely due to its more complex composition, of a munition may have changed with time.

Details of the degradation pathways of dumped warfare agents are given in Chap. 4.

1.4 Risks and Impacts on the Marine Environment

Scarcity of data results in gaps in information regarding the effects on environment. Both completely corroded munitions and intact pieces were found in the dumpsites. However corrosion rates, their dependence on environmental factors and fluxes of warfare agents and explosive degradation products to the environment were only briefly described, so no definite answer exists. Chemical Warfare Agents (CWA) degradation products were observed in the sediments on distances up to 40 m from objects, but in some instances the range was considerably smaller. Arsenic based agents seem to spread further than mustard degradation products – analysis of samples collected at distances 0.5 m and 25 m containing both types of agents, show steeper decline in mustard concentration than As based (Fig. 1.6). Model studies show, that the penetration of CWA and their degradation products should be limited

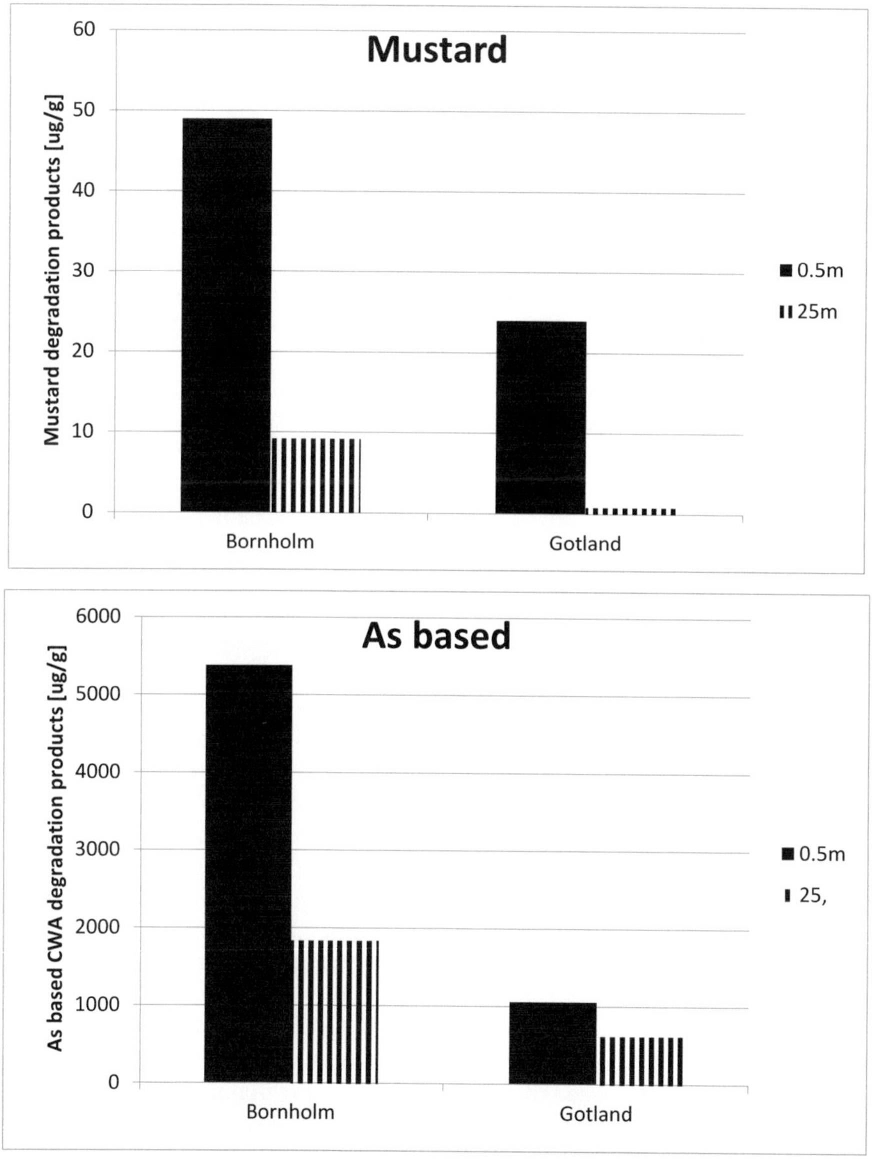

Fig. 1.6 Mustard degradation products and as based agents degradation products concentration next to the object and at 25 m distance in two points at Bornholm Deep and Gotland Deep

to meters – such discrepancy shows that further studies are needed in an area of CWA degradation and transport.

During the CHEMSEA and MODUM projects and Mediterranean studies biomarker effects of environmental stress were observed in organisms close to munition dumpsites, but no clear link could be provided to CWA, because biomarkers used were too general. Lack of specific, CWA oriented biomarkers, and toxicity studies is another gap in the current state of the art.

The substances which are dealt with by the environmental models assessing the environmental risk from munitions dumping sites include mainly mustard gas, agents based on arsenic oil (Clark I and Clark II, adamsite and lewisite) and less frequently tabun.

Because of their high toxicity, extremely slow hydrolysis and the equally toxic breakdown products, the greatest environmental threat is generally believed to be posed by Clark I and Clark II, as well as nitrogen mustard gas and sulphur mustard gas.

Mustard gas is extremely toxic to all species, and its degradation products cause a variety of adverse effects towards microorganisms, changing both their composition and abundance (Medvedeva et al. 2009).

In the case of the Baltic dumps, the extent of dispersion in the seabed sediments through sub-surface leakage, the quantity of toxic agents which may penetrate to the near-bottom water and the duration of contamination are the important aspects of assessment of the risk related to possible leakage of chemical warfare agents. Moreover, there is a potentially significant overlap between the dump site, fertile fishing grounds and the breeding grounds of cod (*Gadus morhua*) east of Bornholm, suggesting that this economically and ecologically important fish species to the Baltic Sea might be particularly at risk from dumped CWAs (Sanderson et al. 2008).

Based on the measured concentrations of parent CWA compounds in the sediments of the Bornholm deep, there are no concern for direct acute risks towards the fish community. The potential chronic and indirect ecosystem effects are however not clarified (Sanderson et al. 2010). CWA conditions, levels, and risks elsewhere in the Baltic Sea are less well analysed.

According to Russian studies during MERCW project, enhanced numbers of mustard resistant bacteria were found in the nearbottom water at all Baltic dumpsites, also decreased biodiversity of bacteria was observed close to the identified objects. This indicates a probable leakage of CWA into the nearbottom waters and pollution of both surface sediments and water. Microorganisms from those areas were characterized with enhanced resistance to CWA degradation products and they were able to use some of those compounds as a carbon source, which could explain low spreading of those substances to the adjacent areas (Medvedeva et al. 2009).

Both original CHEMU report and the 2010 thematic report on hazardous substances in the Baltic Sea (HELCOM Baltic Sea Environment Proceedings no. 120B) concluded that dissolved warfare agents do not pose a wide-scale threat to the marine environment. The main reason is that these compounds cannot occur in higher concentrations in the water.

This can also be concluded from the MERCW project that studied bottom samples from within a primary dump site. No intact warfare agent chemicals were found.

However, the project documented findings of degradation products of chemical warfare agents, such as sulphur mustard, Tabun, ά-chloroacetophenone, Adamsite, Clark I, Clark II or phenyldichloroarsine.

In the recent years, there have been carried out a significant number of ecotoxicological tests (Sanderson et al. 2008, 2009; Emelyanov et al. 2010) in the light of which it can be concluded that the considered chemical warfare agents do not accumulate in living organisms or the tendencies for bioaccumulation are low. Such threat cannot be ruled out, however, in particular in the case of arsenic CWAs. The compounds, after breakdown to non-organic compounds, due to possible accumulation in the organisms of fish, constitute a threat also to people. In the opinion of experts, further research is needed in that scope, in particular with respect to poorly soluble compounds, such as viscous mustard gas mixed with organic arsenic compounds.

As indicated by the research performed in CWA contaminated area of Bari, Italy, new ecotoxicological tools, as biomarkers, can point at threats to the marine ecosystem posed by CWA, which were not identified by studying the concentrations of CWA compounds alone. It is mostly a threat to demersal fish, which are especially vulnerable for sediment associated contaminants. The ammunitions dumped in the Bari area are similar in composition to that dumped in the Baltic Sea. It contains artillery shells and aerial bombs filled with mustard and arsenic based CWAs – lewisite and CLARK. The As content in eels from this area was elevated as compared to fish from the control area, while in other fish, with no apparent differences in arsenic concentrations in tissues, adverse effects of CWA were visible by means of biomarkers assays. Fish from the dumping area had disturbed enzymatic system, DNA damage and damaged internal organs (AMATO et al. 2006).

There are of course gaps in our knowledge concerning the properties of chemical warfare agents at 50–100 m depth in the Baltic. The investigations so far reveal a further need to refine the chemical analytical methods. However, with the state-of-the-art analysis performed by Finnish research institute, VERIFIN, the detection frequency of parent CWA compounds is very low. Degradation products have been found more frequently in the primary dump site of the Bornholm Deep.

1.5 Management Requirements

Dumped chemical munitions are no doubt point sources of contaminants to the benthic ecosystems and they are also a threat to commercial operations on the seabottom. Due to unknown degree of corrosion and decomposition of chemical warfare agents contained in the munitions, it is hard to predict the magnitude of fluxes released into the environment, and possible temporal trends in the contaminants release. Therefore, a number of management options was suggested in the past (Duursma and Surikov 1999).

The most simple and environmentally friendly option is the monitoring of dumpsites, in order to control possible spreading of contaminants outside of dumpsite area. Such monitoring includes condition of munitions, levels of CWA degradation products in the sediments and nearbottom water, but also environmental parameters responsible for water and suspended matter movements, corrosion rate and transformation of toxic chemicals – especially turbulence, currents speed and direction, oxygen concentration and parameters related to organic matter decomposition.

Other options can create potential adverse effects in the environment, therefore they are recommended only in cases where the impact of dumped munitions on the environment is well proven, and exceeds threats created by remediation activity. Those options can be divided into in-situ methods and removal.

In-situ methods include hydrolysis of munition constituents in underwater domes and various sediment capping options (Duursma and Surikov 1999), which could either transform toxicants into less toxic compounds or separate them form the bottom water for long enough to be buried by sediment layers or for the natural depuration processes to complete. Disadvantage of such option is the elimination of selected seabottom areas from the ecosystem, in terms of habitat, and other ecosystem roles and services they normally provide. Such methods were successfully demonstrated in the Black Sea, where post soviet chemical munitions were sarcophaged by concrete sediment capping (Korendovych 2012).

Removal options include various means of retrieval, although most environmentally friendly option seem to be in-situ overpacking partially corroded munitions into hermetic containers. Retrieval operation creates a risk of resuspension of contaminated sediments and release of toxic agents into the water, however such risks may be minimized by careful operation of divers and underwater robots. Retrieved munitions may be neutralized by various means, including shipboard installations and land based facilities. Various options include plasma ovens, detonation chambers or combustion chambers, with specialized off gas treatment (Tobias Knobloch et al. 2013).

Due to the fact, that all the options are costly, they should be preceded with careful risk assessment, both for potential munition spreading and consequences of different management options. To this end, EU has funded an integrated project Decision aid for munition Management (DAIMON), which is aimed at the development of risk assessment methods and remediation option selection.

NATO SPS project "Towards the Monitoring of Dumped Munition Threat" (MODUM) addresses the uncertainty of the impact of dumped chemical munition on the Baltic environment. Very limited datasets that exist nowadays show that the degradation products of conventional and chemical munitions are present in the sediments close to sunk objects (Knobloch et al. 2013), and adverse chronic effects on fish are not excluded. Therefore MODUM project aimed at creation of cost effective monitoring and survey methods for explosives and chemical munitions dumping grounds, in order to control the magnitude of the leakage of pollutants associated with warfare material, and their impact on the Baltic ecosystem.

1.6 Conclusions

Dumped chemical munitions are no doubt point sources of contaminants to the benthic ecosystems and they are also a threat to commercial operations on the seabottom. Due to unknown degree of corrosion and decomposition of chemical warfare agents contained in the munitions, it is hard to predict the magnitude of fluxes released into the environment, and possible temporal trends in the contaminants release. Therefore, a number of management options was suggested in the past (Duursma and Surikov 1999).

The simplest and environmentally friendly option is the monitoring of dumpsites, in order to control possible spreading of contaminants outside of dumpsite area. Such monitoring includes condition of munitions, levels of CWA degradation products in the sediments and nearbottom water, but also environmental parameters responsible for water and suspended matter movements, corrosion rate and transformation of toxic chemicals – especially turbulence, currents speed and direction, oxygen concentration and parameters related to organic matter decomposition.

Methods developed within the MODUM project provide a complex system of tools that enable holistic estimation of the impact of dumped ammunition on environment. Survey methods based on AUV carried sonars seem to be very effective in locating dumped warfare material, both conventional and chemical. Usage of towed magnetometer greatly reduces the number of false positives, and reduces the amount of time consuming ROV missions. Sampling strategies tested in the project are valid for all kinds of point source associated pollutants, but are excellent for the estimation of warfare agent degradation products, especially hydrophobic mustard degradation products, which tend to contaminate very limited areas. Chemical detection methods developed within the MODUM project are aimed specifically for CWA, therefore they can provide fast answer if given dumped object is a chemical or conventional munition.

Bioindicator and biomarker approach developed in MODUM are nonspecific, therefore they could be used for monitoring of both CWA degradation products, and toxic degradation products of conventional munitions, such as e.g. DNT. Their usage within the MODUM project indicate, that in the Baltic we may observe an impact of dumped munition on biota, however not on fish consumers at the present.

Efficient monitoring of the dumpsites is possible using the tools developed within the MODUM project, however several steps have to be taken. First of all, funding has to be secured for multi year monitoring program, in which time series would be obtained, that enable prediction of future trends. Secondly, methods developed in the Baltic need to be transferred to other areas – which will require further studies there. As for the application of MODUM methods to conventional munitions, the only element that is missing are the mobile detection methods.

References

Amato E, Alcaro L, Corsi I, Della Torre C, Farchi C, Focardi S, Marino G, Tursi A (2006) An integrated ecotoxicological approach to assess the effects of pollutants released by unexploded chemical ordnance dumped in the southern Adriatic (Mediterranean Sea). Springer, Heidelberg

Andrulewicz E (1996) War gases and ammunition in the polish economic zone of the Balic Sea. In: Kafka AV (ed) Sea-dumped chemical weapons: aspects, problems and solutions, NATO ASI Series, vol 7. Kluwer Academic Publishers, Dordrecht, pp 9–15

Arison LH (2013) European disposal operations: the sea disposal of chemical weapons create space independent publishing platform

Bełdowski J, Pempkowiak J (2007) Mercury transformations in marine coastal sediments as derived from mercury concentration and speciation changes along source/sink transport pathway (Southern Baltic). Estuar Coast Shelf Sci 72(1–2):370–378

Bełdowski J, Miotk M, Beldowska M, Pempkowiak J (2014) Total, methyl and organic mercury in sediments of the southern Baltic Sea. Mar Pollut Bull 87(1–2):388–395. doi:10.1016/j.marpolbul.2014.07.001

Bruchmann (1953) Report about the investigations concerning areas of chemical warfare agent dumping in the Baltic Sea. Report of the regional office of the People's Police Rostock to (i.a.) the Ministry of Interior of the German Democratic Republic

BSH (1993) Chemical munitions in the southern and western Baltic Sea – report by a federal/Länder government working group. (in German) Federal Maritime and Hydrographic Agency (Bundesamt für Seeschifffahrt und Hydrographie, BSH) (1993) and cited references, Hamburg, Germany

Carton G, Jagusiewicz A (2009) Historic disposal of munitions in U.S. and European coastal waters, how historic information can be used in characterizing and managing risk. (w:) J Marine Tech Soc 43(4):16–32

Duursma EK, Surikov BT (1999) Dumped chemical weapons in the sea : options: a synopsis. A.H. Heineken Foundation for the Environment, Amsterdam, Netherlands

Emelyanov E, Kravtsov V, Savin Y, Paka V, Khalikov I (2010) Influence of chemical weapons and warfare agents on the metal contents in sediments in the Bornholm Basin, the Baltic Sea. Baltica 23(2):77–90

Jäckel K (1969) Investigation report about the dumping of munitions stocks (warfare agent munitions) of the German Wehrmacht after the Second World War in the sea territories of Bornholm, Gotland and off the south exit of the Little Belt. File number 57-50-10 VS-NfD (in German), report to the Federal Ministry of Defense (1969) and associated documents, Hamburg, Germany

James Martin Center for Nonproliferation Studies, (2016) Combating the spread of weapons of mass destructtion with training & analysis, Monterey Institute of National Studies. http://cns.miis.edu/stories/090806_cw_dumping.htm – data dostępu – 21.02.2016 r

Knobloch T, Bełdowski J, Böttcher C, Söderström M, Rühl N-P, Sternheim J (2013) Chemical Munitions Dumped in the Baltic Sea. Report of the ad hoc Expert Group to Update and Review the Existing Information on Dumped Chemical Munitions in the Baltic Sea (HELCOM MUNI) Baltic Sea Environmental Proceedings. HELCOM

Korendovych V (2012) Enhancing abilities to counter WMD proliferation – Ukrainian Experience. Paper presented at the Countering WMD Threats in Maritime Environment, Riga

Korotenko KA (2000) Chemical weapons dumped in the BALTIC SEA: facts, tendency and prediction of level and scales of possible ecological disaster. John D. and Catherine T. MacArthur Foundation, Moscow

Medvedeva N, Polyak Y, Kankaanpää H, Zaytseva T (2009) Microbial responses to mustard gas dumped in the Baltic Sea. Mar Environ res 68:71–81

MERCW (2006) Modelling of ecological risks related to sea-dumped chemical weapons. MERCW project deliverable 2.1 – synthesis report of available data. ISBN 978-951-53-2971-4, prepared by Missaen T et al. (2006): www.mercw.org

Neffe S, Bełdowski J, Fabisiak J, Kasperek T, Popiel S (2011) Report concerning the issues included in Decisions no. 23 and 24, made at the 1st Meeting of the Expert Group concerning updating and reviewing available information on chemical weapons dumped in the Baltic Sea "HELCOM MUNI 1/2010. Chief Inspectorate of Environmental Protection, Warsaw

Nord Stream Project (2010) Environmental monitoring in Danish waters, Document nro G-PE-PER-MON-100-05070000. http://www.nord-stream.com. Accessed 15 Feb 2017

Nord Stream Project (2011) Environmental Monitoring in Danish Waters, 2011, Document Nro G-PE-PER-MON-100-05070011 http://www.nord-stream.com. Accessed 15 Feb 2017

Sanderson H, Fauser P, Thomsen M, Sorensen PB (2008) Screening level fish community risk assessment of chemical warfare agents in the Baltic Sea. J Hazard Mater 154(1–3):846–857. doi:10.1016/j.jhazmat.2007.10.117

Sanderson H, Fauser P, Thomsen M, Sorensen PB (2009) Human health risk screening due to consumption of fish contaminated with chemical warfare agents in the Baltic Sea. J Hazard Mater 162(1):416–422. doi:10.1016/j.jhazmat.2008.05.059

Sanderson H, Fauser P, Thomsen M, Vanninen P, Soderstrom M, Savin Y, Khalikov I, Hirvonen A, Niiranen S, Missaen T, Gress A, Borodin P, Medvedeva N, Polyak Y, Paka V, Zhurbas V, Feller P (2010) Environmental hazards of sea-dumped chemical weapons. Environ Sci Technol 44(12):4389–4394. doi:10.1021/Es903472a

Silva JAK, Chock T (2016) Munitions integrity and corrosion features observed during the HUMMA deep-sea munitions disposal site investigations. Deep-Sea res Pt ii 128:14–24. doi:10.1016/j.dsr2.2015.09.001

Walker PF (2012) Ocean-dumped chemical weapons: history, challenges, prospects. Materiały z International Workshop Polish Naval Academy, Gdynia, pp 16–32

Wichert U (2012) Personal communication (2012). Archival documents available on request

Chapter 2
Suitability Study of Survey Equipment Used in the MODUM Project

Miłosz Grabowski, Stefano Fioravanti, Robert Been, Federico Cernich, and Vitalijus Malejevas

Abstract Modern seafloor mapping techniques are based almost entirely on acoustics systems. With the usage of appropriate equipment, it is possible to examine seabed in terms of its morphology and also sub-bottom structures. Moreover, detection of objects placed on the bottom surface and buried in sediments is possible with proper gear as well. Above-mentioned actions were one of the crucial aims of the MODUM project (Towards The Monitoring Of Dumped Munition Threat). To achieve it, surveys with certain selected acoustic equipment were done together with data processing. The aim of this chapter is to present those systems, its possibilities, principles of working and results obtained thanks to it. Three different side scan sonar systems are described: used for seafloor surface scanning and detecting objects on it. Also, a sub-bottom profiler, which allows looking on seabed structures and detecting buried targets within, is presented. Additionally, a findings validation device, which is the remotely operated vehicle, is characterized in this chapter as well. Cooperation of all these systems gave good results in terms of detecting sunken munition in the Baltic Sea.

M. Grabowski (✉)
Institute of Oceanology, Polish Academy of Sciences, ul. Powstańców Warszawy 55, Sopot 81-712, Poland
e-mail: grabowski@iopan.gda.pl

S. Fioravanti • R. Been • F. Cernich
Science and Technology Organisation – Centre for Maritime Research and Experimentation, Viale San Bartolomeo, 400, La Spezia, SP 19126, Italy

V. Malejevas
The Environmental Protection Agency, Juozapavičiaus st. 9, Vilnius, LT 09311, Lithuania

J. Bełdowski et al. (eds.), *Towards the Monitoring of Dumped Munitions Threat (MODUM)*, NATO Science for Peace and Security Series C: Environmental Security, DOI 10.1007/978-94-024-1153-9_2

2.1 Introduction

During the recent years increasing research effort is observed in relation to the investigation of the state and spatial distribution of chemical weapon dumped after the II World War in the Baltic Sea. This task is not easy due to a small number of historical data, which could give an answer to where the dumpsites are located directly. However, a series of projects was conducted (MERCW, CHEMSEA, as examples) with an aim to locate dumping grounds and distribution of munition pieces.

As it was shown in outcomes of the CHEMSEA project, numerous potential Chemical Warfare (CW) targets are scattered on the Gotland Deep bottom, which was one of the primary dumpsites on the Baltic Sea according to historical documents (Bełdowski et al. 2016). Thus, it must be assumed that sought targets are not gathered in one place in areas of Gotland Deep, Gdańsk Deep, Bornholm Deep, and Slupsk Furrow, which were selected for the survey in the framework of MODUM project. Also, there is a strong need to perform visual validation of detected contacts with a usage of remotely operated vehicle (ROV). It is motivated by the fact, that during the CHEMSEA project, in the area of Gotland Deep 39.260 CW contacts were found with a usage of side scan sonar, but only 17.267 appeared to be probable-munition targets, which is about 48% of total CHEMSEA findings (Bełdowski et al. 2014).

Finally, a necessity of dumpsites mapping, and areas that are neighboring to them is big due to significant marine traffic, development of the maritime industry, and fishermen activities on the Baltic Sea. The presence of CW targets scattered on the sea floor of the Baltic Sea is a big threat for above-mentioned actions, which justifies undertaking of the survey.

Contemporary methods of seabed mapping are based on an application of acoustics systems. It is a consequence of the fact, that electromagnetic waves are absorbed in the water. Hence, employment of the optical systems is insufficient due to its limited ranges from 1 to 50 m in the Baltic Sea (Dera and Sagan 1990). On the other hand, acoustic impulses that are mechanical waves are characterized with significantly lower attenuation. The occurrence of this phenomenon results in much longer distances of acoustic waves propagation, which allows for examining the sea bottom even in areas of the deep water.

In pursuance of the adequate detection, classification, and cataloging the CW targets in the dumping areas, the technologically advanced acoustical equipment is involved in the investigations. One of the most popular and efficient types of equipment for those tasks are side scan sonars and sub-bottom profilers.

Typical frequencies associated with underwater acoustics are between 10 Hz and 10 MHz. The propagation of sound in the ocean at frequencies lower than 10 Hz is usually not possible without penetrating deep into the seabed, whereas frequencies range for underwater applications is rarely higher than 1 MHz due to the rapid absorption within the water column. Most systems used today for seabed mapping make use of a single acoustic frequency (Anderson et al. 2008) because different frequencies interact with the seabed or objects in different ways, which requires more sophisticated sensor to capture the desired information. For example, high frequency sonar can measure accurately the water seabed depths, whereas sub-bottom layers are better observed by lower frequencies. This is due to the decreasing sediment sound absorption with decreasing frequency.

Side scan sonars are a specific type of acoustics systems that allows creating two-dimensional images of the seabed. It is possible due to the specific location of transducer arrays on the towfish, with an angle of 40°–60° (Blondel 2009). Generated acoustic impulses are traveling through the aqueous medium, reaching the seafloor and undergo processes of reflection and refraction from it. Next, reflected impulses are recorded as a change of voltage on the sonar transducers and base on its image of the bottom is created. Due to the specific shape of transmitted impulses, it is possible to obtain echoes from a big surface in relatively short time. The shape of ensonified area is similar to ellipse, expanded in the direction from the transducer (Fig. 2.1). Under the towfish, a strip of not ensonified bottom is formed due to the geometry of the system. Hence, it is desirable to conduct the survey with overlaying the measuring profiles. The side scan sonar survey can be conducted with traditional towed towfish or with system mounted on the hull of underwater vehicles. In this chapter three types of sonars will be described: two towed behind research vessel and one mouthed on the hull of autonomous underwater vehicle (AUV). Additionally, sonars fixed on the remotely underwater vehicle (ROV) will be characterized in this chapter.

Classical Sub-bottom profilers (SBPs) are single frequency sonars that aim to explore the first layers of sediments below the seafloor over a thickness commonly reaching several tens of meters. It has been for many years a fundamental tool for oceanography and offshore engineering due to the ability of this system to determine physical properties of the seafloor and to identify geological layers below the seafloor (Mc Quillin et al. 1984). Sediment structure is directly observed by measuring the elapsed time of the received reflections of the acoustic energy when it encounters boundaries between layers of different properties.

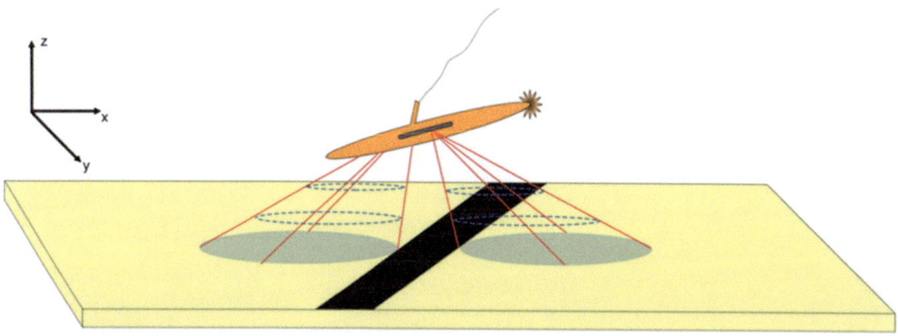

Fig. 2.1 Working scheme of towed side scan sonar with marked ensonified fields and nadir directly below the system

2.2 Autonomous Underwater Vehicle (AUV) IVER2

The IVER-2 Autonomous Underwater Vehicle (AUV), manufactured by Ocean Server Technology Incorporation was operating in the framework of NATO SPS Monitoring of Dumped Munitions (MODUM) project in the areas of Gdańsk Deep, Bornholm Deep, Gotland Deep and Little Belt. It is an autonomous robot, which is able to conduct predesigned missions of seafloor scanning with high accuracy. The AUV is equipped with side scan sonar, towed magnetometer, and environment sensors (Fig. 2.2). During the MODUM project IVER performed 36 missions in areas of interest and was able to identify 742 potential targets.

2.2.1 Sonar Performance

The AUV was integrated with high-resolution, dual-frequency side scan sonar – UUV-3500, produced by Klein Marine Systems, Inc. Sonar frequencies are 455 kHz and 900 kHz, but during missions, only higher frequencies were in use to obtain photo-like images. UUV-3500 SSS generates wideband FM chirp pulses (1, 2, 4 and 8 ms) with swath range 350 m @ 455 kHz and 150 m @ 900 kHz. The wideband FM Chirp provides crispy higher-resolution and quality images of targets. But, on the other side only smaller areas are examined and sea operations are more ship time-consuming in compare to traditional towed side scan sonars. The AUV was also equipped with towed magnetometer to detect magnetic anomalies during surveys, simultaneously with hull-mounted side scan sonar.

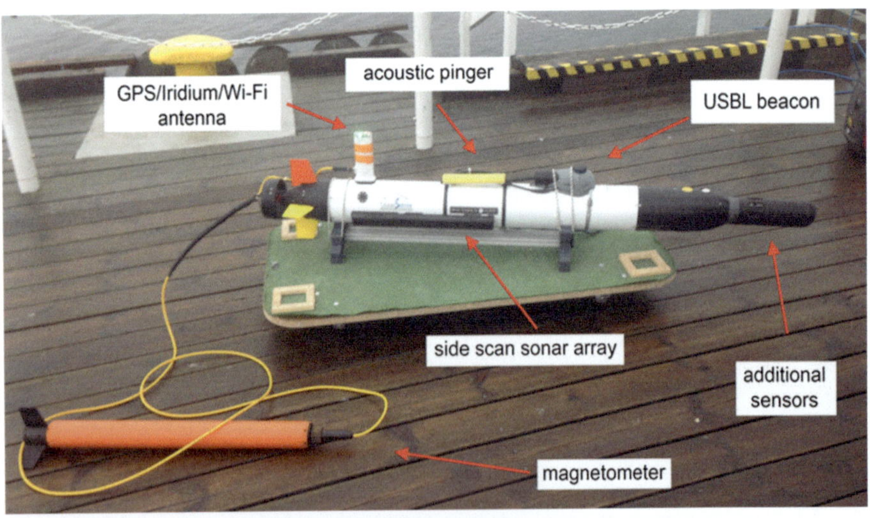

Fig. 2.2 Autonomous Underwater Vehicle IVER2 with towed magnetometer

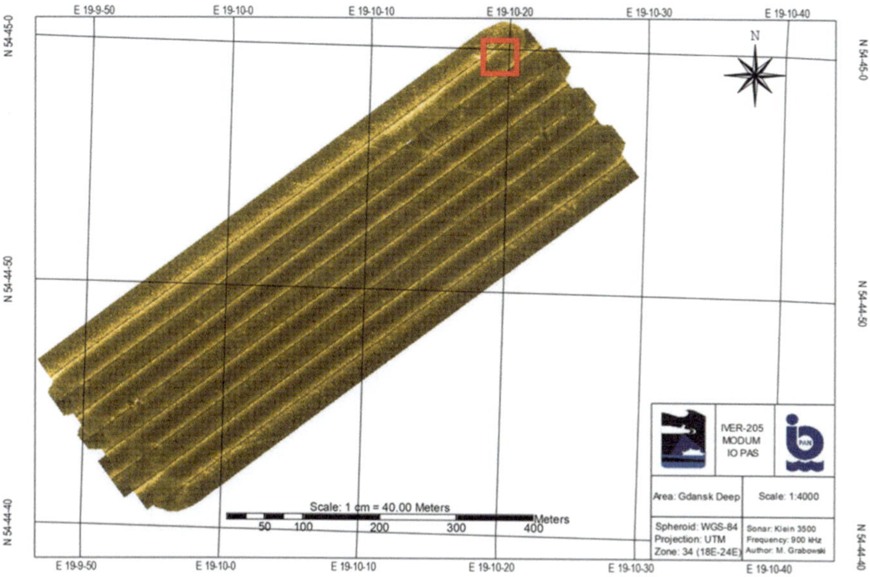

Fig. 2.3 An example mosaic from one of the missions. Target is marked with the *red rectangle*. Resolution: 0.1 meters per pixel

After every mission, data analyses were performed to localize targets. In addition, for every IVER's mission, high-resolution sonar mosaics were prepared (Fig. 2.3). Example target is shown in Fig. 2.4. Magnetometer data (Fig. 2.5) were correlated with sonar data, which was a base for planning Remotely Operating Vehicle (ROV) missions.

2.2.2 AUV Operation

AUV hull-mounted sensors can provide following information about the water column: salinity/conductivity, temperature, pH, oxygen saturation, chlorophyll concentration, and turbidity. Another advantage of the IVER2 AUV is launching/recovering system with special cocoon designed by Ocean Server Technology Inc. It is very useful especially when the AUV is operating from the vessels with a rather high deck (Figs. 2.6 and 2.7).

2.2.3 Precision of Navigation

The AUV is fully integrated with mission planning software, which allows to designing missions on the digital maps (Fig. 2.8). It is also equipped with GPS receiver and Iridium satellite system and Doppler Velocity Log (DVL). The DVL is

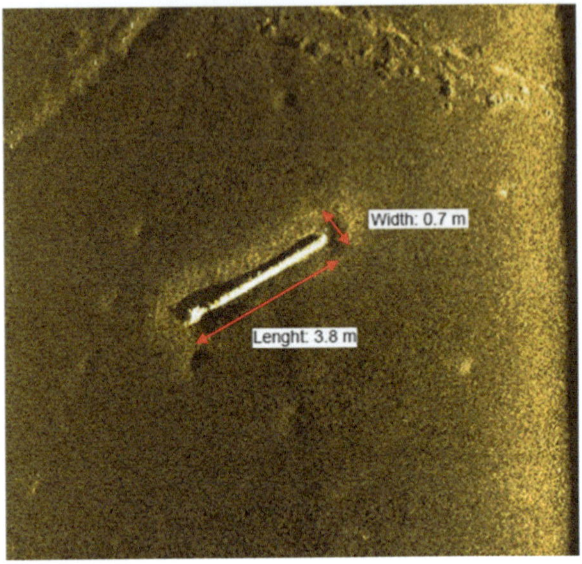

Fig. 2.4 Zoom to the red rectangle from the Fig. 2.3 – an example target detected with IVER2 AUV

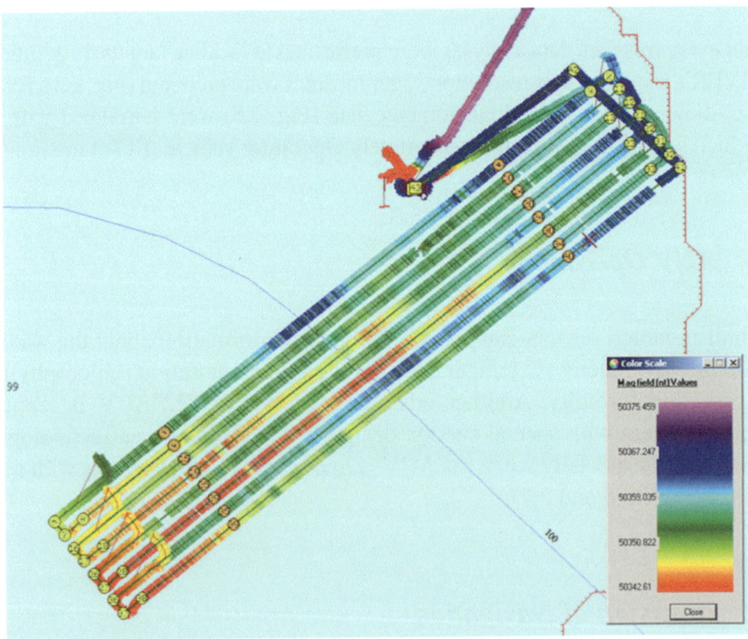

Fig. 2.5 An example of magnetic field measurements performed with magnetometer towed behind the AUV. Magnetic field variations are expressed in nanoteslas [nT]

Fig. 2.6 IVER2 AUV in cocoon ready to launch (Photo by Jacek Bełdowski)

Fig. 2.7 Recovering of the AUV (Photo by Marta Kacprzak)

responsible for positioning while AUV is underwater. When AUV cannot calculate the position, it floats up to the surface to retrieve GPS signal and then it continues a survey. The above-mentioned solution provides very precise navigation without layback and undulating issues in compare to traditional towed side scan sonars. Additionally, third-party Ultra Short Base Line system (USBL) provides online visualization of the AUV underwater position.

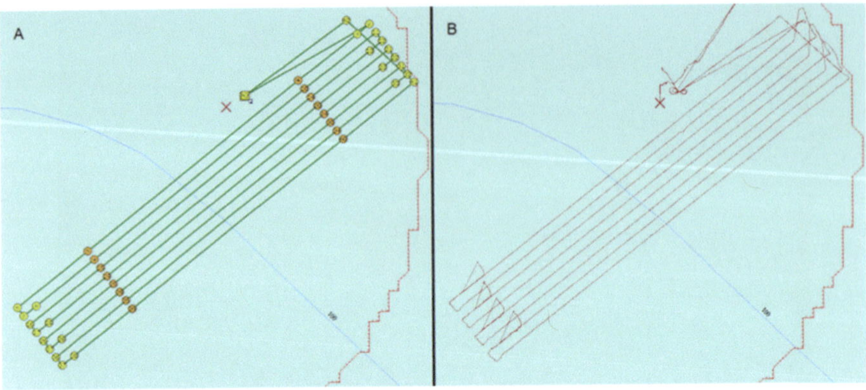

Fig. 2.8 Planned mission (**a**) and conducted mission (**b**)

Autonomous Underwater Vehicle IVER2 is a powerful, advanced tool for studying a seabed and a water column. It can provide high-quality data in relatively short time at the significant area of survey. It is compact designed, easy to transport, low-price and user-friendly device. Hull-mounted side scan sonar, together with towed magnetometer are perfect for precise seabed mapping and searching of sunken objects such as Chemical Warfare Agents (CWA).

2.3 Edge Tech DF – 1000 Side Scan Sonar (SSS)

Another hydroacoustic device used in MODUM project is a towed side scan sonar, DF – 1000 model, manufactured by Edge Tech. It is mainly applied to perform a preliminary bottom scan of the dumping area, in order to detect big obstacles, which can be a threat for the AUV during its missions. The basic set consists of towfish with integrated sonar (Fig. 2.9), a deck unit with CODA data acquisition system, based on Linux operating system (Fig. 2.10a.), and additional VHF radio with built-in GPS (Fig. 2.10b).

2.3.1 Sonar Performance

Edge Tech DF – 1000 system features simultaneous 4-channel data acquisition in two frequencies: 2–100 kHz port and starboard and/or 2–500 kHz port and starboard as well. Pulse length is 0.1 ms @ 100 kHz and 0.01 ms @ 500 kHz. Vertical beam width for 100 kHz frequency is 1.2° and for 500 kHz – 0.5°. Sampling rate during a survey is 24 kHz per channel. With upper-mentioned features it is possible to obtain sonar data in two resolutions: standard resolution – 100 ± 10 kHz and

Fig. 2.9 Towfish integrated with a side scan sonar

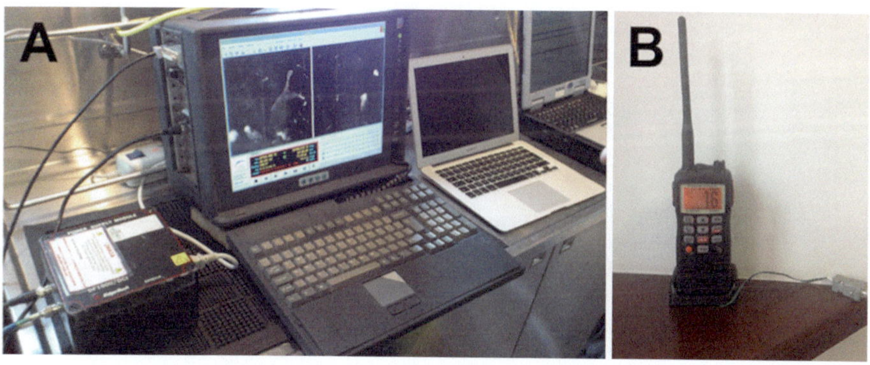

Fig. 2.10 (**a**) – CODA deck unit, (**b**) – VHF radio with built-in GPS

high-resolution 400 ± 20 kHz. The towed sonar sending harmonic pulses usually gives lower-resolution images.

After and during the survey, targets that could be a potential threat for the AUV (wrecks, ghosts nets near the bottom) are localized. Its positions are specially taken into consideration when AUV missions are planned. An example target is shown in Fig. 2.11.

Additionally, for every set of sonar data, mosaics of the sea bottom were prepared with CODA GeoSurvey and/or SonarWiz software (Fig. 2.12).

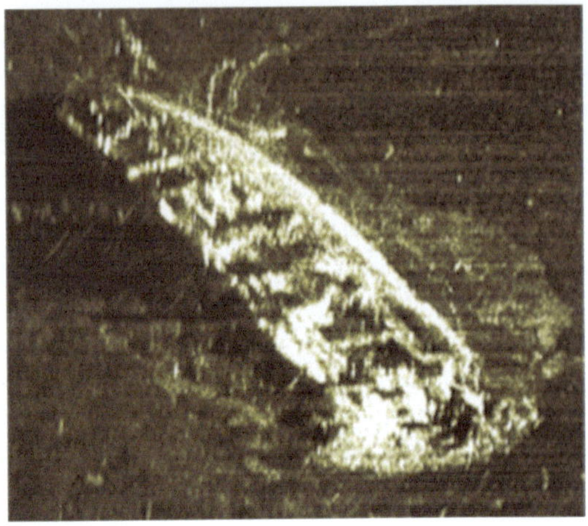

Fig. 2.11 Wreck on the seafloor of the Bornholm Deep. The side scan sonar was towed about 5.5 m above the bottom

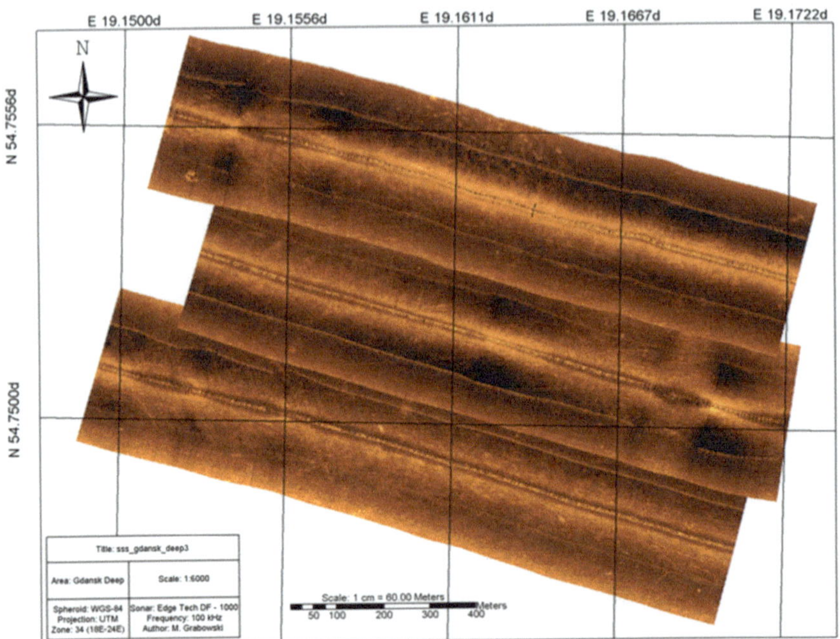

Fig. 2.12 An example of a mosaic created from data collected with Edge Tech DF – 1000 sss in the Gdańsk Deep area

2.3.2 Edge Tech DF – 1000 Operation

The DF – 1000 SSS can operate up to 1000 m depths. It is light and can be used in every vessel equipped with a suitable winch. Although the manufacturer declared the maximum tow speed about 12 knots, to obtain the good images quality it is highly recommended to perform a survey with the speed up to 4 knots. However, in the case of the used setup we observe some problems with correct positioning of the towfish. It is undulating what results in gaps in recorded data. In compare to the AUV, with a similar amount of ship-time a towed side scan have delivered significantly more images of the seabed and providing rapid visualization of a bottom. However, the system lacks the accuracy in navigation.

2.3.3 *Precision of Navigation*

The greatest weakness of the used DF – 1000 system is the precision of the navigation. It comes both with proper layback presets, and undulating. Another issue is that the SSS is operating next to the port side of the s/y Oceania. The ship is not equipped with a dynamic positioning system that is why usage of SSS significantly depends on the sea state. Even application of the additional GPS during the survey does not guarantee proper tracking of the towfish. Positioning errors even up to 100 m were noted.

Employing the Edge Tech DF – 1000 side scan sonar not equipped with underwater positioning system results in not very precise localization of bottom laying targets. Images quality is fairly correct, however, the pseudo-harmonic pulses have disadvantages comparing to broadband chirp sounding signals used in the more modern systems and even at higher frequencies resolution is not suitable for recognition and classification purposes. Also, the bottom tracking problem appears occasionally. Moreover, the performing of an exact survey at high winds from a relatively small ship is a hard task due to above-named issues. Nevertheless, it is recommended to use it, especially in areas where sunken objects occur highly above the bottom level to prevent damage of the AUV. Overall, the DF – 1000 SSS is a decent tool for preliminary surveys, but at the best inadequate for the precise searching and monitoring of the sea-dumped chemical weapon munition targets.

2.4 Klein 3900 Side Scan Sonar

Dual - frequency side scan sonar system KLEIN 3900 is high - resolution digital sonar for use is search and recovery missions, manufactured by Klein Marine Systems, Inc. During NATO SPS project “Towards the Monitoring of Dumped Munitions Threat” side scan sonar system was used to identify object over seabed in Gotland chemical munitions dumpsite, in Lithuanian EEZ (Fig. 2.13).

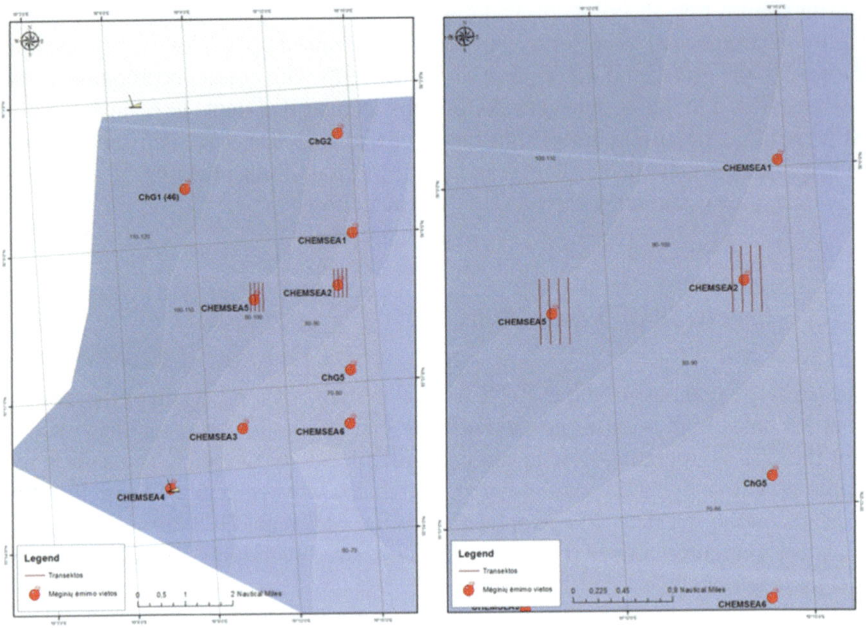

Fig. 2.13 Map of Gotland chemical munitions dumpsite, in Lithuanian EEZ were side scan sonar missions were accomplished (Compiled by Vitalijus Malejevas)

2.4.1 Sonar Performance and Operation

The system 3900 is single beam side-looking sonar intended for high-resolution survey use. The Series 3900 Sonar System equipment consists of single beam sonar instrumented towfish, a Transceiver and Processing Unit (TPU). The model is a selectable dual - frequency system with 445 kHz and 900 kHz. Using lowest frequency at 445 kHz, maximum swath can be reached till 150 m, while highest frequency at 900 kHz, maximum swath can be reached till 50 m. Sonar can reach maximum 200 m depth.

Several side scan sonar missions were accomplished for basic seabed screening (Figs. 2.14 and 2.15). Moreover, bathymetric data was collected by single beam echosounder. For that reason such system were used only for general view, just to avoid big obstacles. After screening AUV equipment were used to get pictures of better resolution.

Firstly, for investigation lines were planned for bottom coverage in the certain area. Areas were selected according to data, with the biggest possibility of chemical warfare objects presence.

During the mission, all data were recorded to the deck unit. Post processing of collected data was conducted and suspected targets were analyzed (Fig. 2.16).

Fig. 2.14 Launching of the Klein 3900 side scan sonar (Photo by Algridas Stankevičius)

Fig. 2.15 Side scan sonar prepared for launching (Photo by Algirdas Stankevičius)

Dual – frequency side scan sonar system KLEIN 3900 is appropriate equipment for search and identification of objects. Especially this type of sonar is suitable for shallow waters.

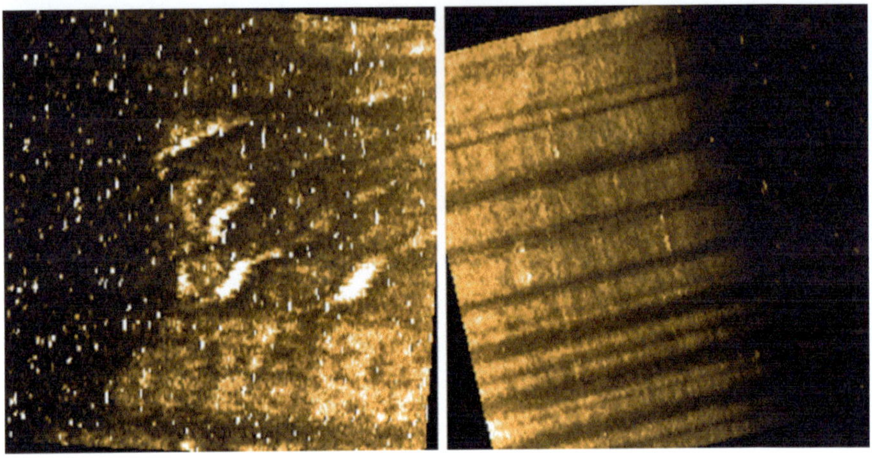

Fig. 2.16 An example of pictures from Klein 3900 side scan sonar

2.5 Falcon Seaeye Remotely Operated Vehicle (ROV)

During the MODUM project, Remotely Operated Vehicle (ROV) Falcon Seaeye, manufactured by Saab Seaeye Limited was used in areas, where sonar surveys were performed, and potential CWA targets for future investigation were selected. Application of the ROV in the project was connected with two tasks: the first one was to conduct a visual inspection of detected objects, which were chosen based on AUV's missions, and the second one was to collect sediment samples in close proximity of above-mentioned targets.

2.5.1 ROV Performance

After every AUV mission, a number of targets were selected as potential CWA objects. There is a big need, to confirm it with the visual inspection to get 100% confidence. Unfortunately, it is not efficient to send diver for this task, due to large depths, for example in Gdańsk Deep, Gotland Deep, and Bornholm Deep.

This type of actions is dangerous, time consuming and expensive. A perfect tool for this activity is work class ROV, such as Falcon Seaeye.

The Falcon Seaeye ROV is light and portable vehicle – 60 kg. It can be operated from relatively small vessels with a crane, but it requires a large power input: from 100 to 270 V AC @ 2.8 kW. A maximum operating depth of the ROV is 300 m with an umbilical length of 450 m. The Seaeye Falcon is equipped with 4 vectored and 1 vertical thrusters, which can generate 50 kgf of forward trust without additional payload. It provides a cruising speed of up to 3 knots under the water. With a vertical thrust of 13 kgf, it can carry the payload equal to 14 kg. Additionally, Falcon is equipped with a high-resolution color camera, mounted together with Teledyne BlueView sonar on 180° tilt platform and two 6400 Lumens LED lights with variable intensity (Fig. 2.17). Precise navigation is provided by Tritech Sea Super SeaPrince DST sonar.

The aforesaid hardware is essential when it comes to performing underwater jobs, in areas where visibility is limited, due to lack of sunlight and occurrence of the marine snow (small, organic particles floating above the sea bottom). These conditions are prevalent in Gdańsk Deep, Gotland Deep and in Bornholm Deep, where sought objects are set at the depth from 80 to 120 m. Hence, applying only a video camera is not enough, due to limited optical visibility, up to 8 m (Fig. 2.18). A solution for this issue is the BlueView sonar. It works with two frequencies:

Fig. 2.17 Front of the Falcon Seaeye ROV

Fig. 2.18 HD camera image – a visual inspection of the "torpedo" target, found in the area of Gdańsk Deep, at the 97.7 m depth. Marine snow is visible

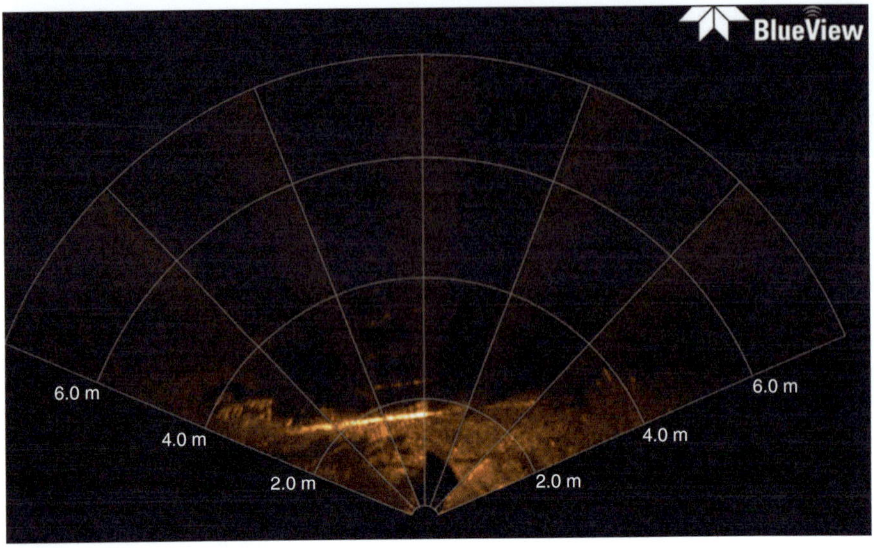

Fig. 2.19 The "torpedo" target. The image recorded by BlueView sonar. About 2 m above the target

900 kHz and 2250 kHz, and provides acoustical "visibility" in maximum ranges of 100 m and 10 m, respectively. The field of view of this device is 130° horizontally and 20° vertically, with a number of beams equal to 768. Another advantage of the BlueView sonar is its mounting system. The transducers, together with the HD camera are fixed on tilted platform. This solution provides easy inspection of the targets from different angles (Fig. 2.19).

2.5.2 Precision of Navigation

Two independent devices provide precise navigation of the ROV. The first is an Ultra Short Base Line (USLB) MicronNav USBL System manufactured by Tritech. With the transponder mounted on the hull of the ROV and transducer placed below the sea surface, at depth about 10 m, the system, which is paired with the GPS is able to fix ROV's position under the water via acoustic pings. The vehicle operator can observe the position, and depth of the vehicle relative to the vessel and on this basis navigation is conducted. The USBL system works with frequency band from 20 to 28 kHz and provides tracking range up to 700 m horizontally and 200 m vertically in the Baltic Sea. An operating beamwidth is equal to 180°.

Another device responsible for navigation is Tritech Super SeaPrince DST sonar. It generates omnidirectional, chirp pulses with frequencies from 500 kHz to 900 kHz. Maximum range is set to be 100 m, with beamwidth 38° vertically and 2.3° horizontally. Super SeaPrince sonar is a good tool to detect big objects such as

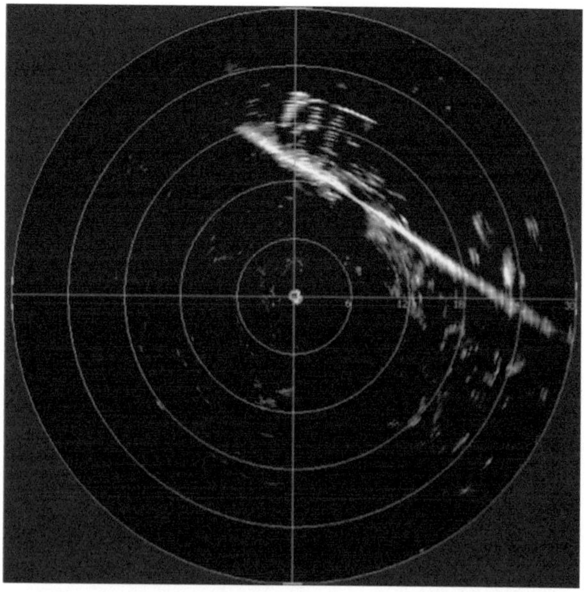

Fig. 2.20 Sonogram recorded by Super SeaPrince DST sonar with visible shipwreck about 12 m from the ROV

wrecks and other obstacles, as well as smaller objects (Fig. 2.20). It provides a live preview of ROV surrounding area.

MicronNav USBL system and Super SeaPrince DST sonar can be operated via SeaNet software simultaneously, with the background of a bathymetrical map or sonar mosaics to obtain the best precision of navigation.

Additionally, Falcon Seaeye ROV is equipped with auto heading and depth features, which are very useful during diving and resurfacing.

Remotely Operated Vehicle Falcon Seaeye with all additional equipment described in this chapter is a great tool for verification of targets detected by the AUV. It is easy to use, even for a not very experienced operator and significantly reduces ship-time. In relatively short time, it is possible to examine many targets in selected dump-side (Fig. 2.21).

An additional advantage is a possibility to collect samples from hazardous areas. Also, thanks to its performance, the ROV can be a great tool for various types of missions, such as, for example, recovery of small sunken objects.

2.6 Sub-bottom Profiler Edge Tech SB216S

2.6.1 Equipment Description

The function of a sediment profiler is to record echoes from interfaces between sedimentary layers that correspond to differences in acoustic impedance. The movement of the support platform will allow reconstruction of a vertical cross-section of

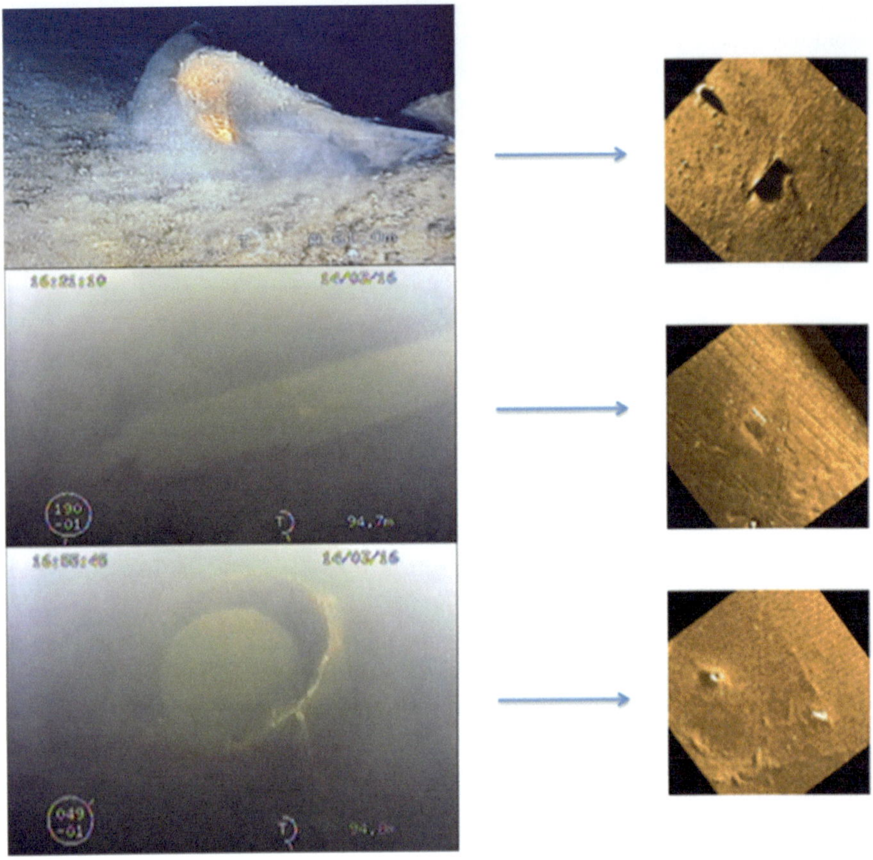

Fig. 2.21 An example targets detected by the AUV (*right*) and their verification by the ROV (*left*)

the sedimentary environment obtained as an image of boundaries between layers. Good horizontal resolution requires a directivity pattern with a very narrow opening angle. The directivity pattern is the transducer directional sensitivity of transmission and/or receiving as illustrated in Fig. 2.22. The directivity pattern of an antenna depends on the transducer geometry, and frequency.

For the same transducer geometry, higher frequencies give narrow opening angles and lower frequencies gives wider opening angles.

The transducer dimension design is based on the desired beam pattern. The beam pattern is a dimensionless and a relative parameter of the transducer. It is a function of the operational frequency, aperture angle, and size and shape characteristics of the vibrating surface.

The mathematical expression '*sin* function' for the normalized directivity pattern that gives the transducer sensitivity of the plane circular piston transducer is:

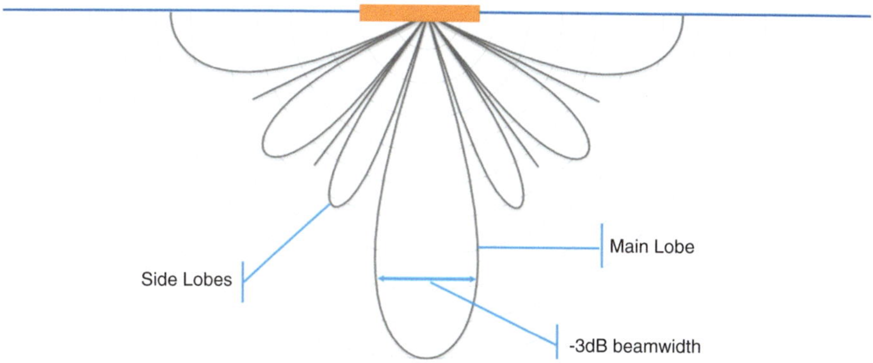

Fig. 2.22 Schematic image of transducer beam pattern scaled in dB

$$D(\theta)=\left[\frac{2J(ka\sin\theta)}{ka\sin\theta}\right]^2$$

where J is the Bessel function of first order, k the wave number, a is the radius of the transducer, and θ is the aperture angle.

The main objective from a SBP is to obtain the deepest penetration depth with highest vertical resolution also known as range resolution. Vertical resolution is the ability of sonar system to distinguish between two or more objects on same bearing but at different ranges (Mulhearn 2000). In principle, vertical resolution depends mainly on the transmitted pulse duration of the CW. The range resolution can easily be estimated by:

$$\Delta r = \frac{ct}{2}$$

where: c = sound velocity and t = pulse duration.

This means that in order to obtain high range resolution (i.e. short Δr) very short pulses are needed. However, in order to obtain deeper penetration, the transmitted signal has to have enough energy to such that the pulse can be detected from the noise. Since the power is limited due to cavitation, the only option is emit a long pulse. For this reasons many sub-bottom profilers use chirp signals to obtain high resolution. A chirp is a frequency modulated signal, where the pulse is emitted with a modulated frequency. The frequency modulation is processed at the receiver to focus the pulse to a much shorter value and hence obtain the desired resolution.

The acoustic wave interaction with the seabed depends partly on the impedance contrast between two layers. Impedance is a medium characteristic equal to the product of the density and propagating sound speed. Large impedance contrast between water and rocky seabed with a considerable smooth surface means that the seabed surface behaves as an almost perfect reflector. On the other hand, at softer sediments, the acoustic impedance mismatch is much less which means that larger

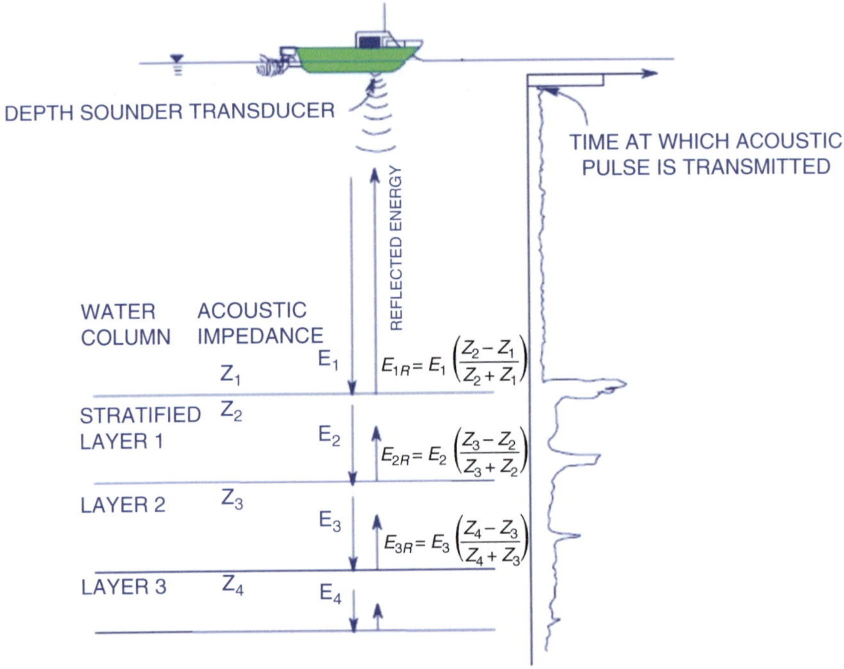

Fig. 2.23 Acoustic impedance changes in different sediment densities

energy will be able to penetrate this boundary. Each time the signal encounters a different material, a portion of the energy is reflected and recorded by the system (Fig. 2.23). The percentage of the acoustic energy reflected at each layer surface is a function of the relative densities, sound speeds and angle of incidence at the two layers.

The reflection coefficient depends on the impedance contrast between two mediums. For the high frequency signal, a portion of the signal energy will be reflected at the first layer. The remaining energy will be highly absorbed, and the wave transmitted into the layer progressively becomes unable to reach the substratum. On the contrary, the low frequency signal is subjected to lower sediment attenuations and the remaining energy can easily penetrate into deeper layers until it is totally absorbed, or meets other sediment layer with high impedance contrast such as clay-rock interface. The percentage of acoustic energy reflected at each interface surface is a function of the relative density and sound speed of the two layers known by the impedance contrast. An equation for the acoustic reflectivity of an underwater surface is given in Fig. 2.23. This equation is valid only for the simplified case in which the change in material composition from one layer to another occurs in a short vertical length compared to the wavelength of the incident signal.

The system adopted is an Edge Tech SB216S Profiling System, a high resolution wideband frequency modulated (FM) sub-bottom profiler that uses Edge Tech's proprietary Full Spectrum chirp technology to generate cross-sectional images of the seabed and collect digital normal incidence reflection data over many frequency ranges (Fig. 2.24 and Table 2.1). The system transmits an FM pulse (also called a "chirp pulse") that is linearly swept over a full spectrum frequency range.

In all sonar systems, higher frequency content is invariably associated with an increase in resolution and a decrease in penetration. Chirp technology reduces the trade-off between signal range and image resolution.

The sound pulse is the sonar system's probe. The system's frequency, or bandwidth, determines the resolution of this probe. Naturally, a finer, high-frequency, broad-band probe is more discriminating.

Fig. 2.24 Edge Tech SB216S towfish, deck unit, and deck winch

Table 2.1 Edge Tech SB216S characteristics

	SB-216S CMRE
Frequency range	2–20 kHz
Pulse type	Frequency Modulated
Pulses (user selected)	2–16 kHz, 2–12 kHz, 2–10 kHz, 2–20 kHz
Vertical Resolution (depends on pulse selected)	6–10 cm
Penetration (typical)	6 meters
In coarse calcareous sand in clay	80 meters
Beam width (depends on center frequency)	17°–24°
Size (centimeters)	105 (L) × 67 (W) × 40 (H)
Weight	76 kg
Cable requirements	3 shielded twisted pairs (5 used)
Maximum operating depth	300 meters
Tow speed	3–4 knots optimal, 7 knots maximum safe operation

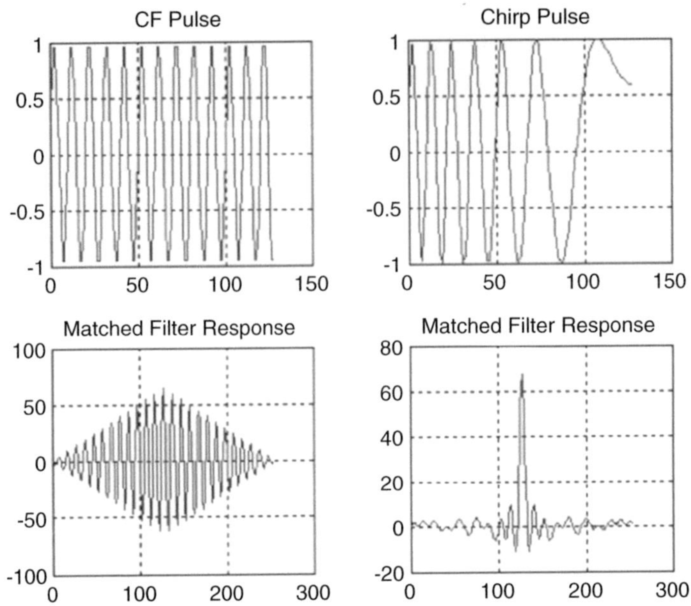

Fig. 2.25 Chirp Technology: comparison between the ultimate resolution of a rectangular constant frequency pulse and a chirp pulse of the same duration

As explained before, for sub-bottom sonar, sound energy transmitted to the seafloor is reflected off the boundaries between layers of different densities. The first boundary is between the water and the seafloor itself. As layers of clay, sand and various other sediments succeed each other, they create interfaces that reflect sound. It is the energy reflected from these boundaries that the system uses to build the image. Likewise, for side scan sonar, backscatter from the seafloor produces an image of the seafloor contours.

Edge Tech's Full Spectrum chirp technology has several distinct advantages over conventional sub-bottom profiling systems. Chirp pulse compression is a signal processing technique commonly used by radar, sonar and echography to increase the range resolution as well as the signal to noise ratio (Fig. 2.25).

The use of separate acoustic projectors and receivers enable:

- Simultaneous transmission and reception of acoustic signals
- High repeatability of the transmitted signals to enable sediment classification,
- High signal-to-noise ratio (SNR) for improved acoustic imagery
- High resolution for measurement of fine sediment layering
- Additional processing gain for energy efficiency
- Gaussian shaped amplitude spectrum of the outgoing pulse to preserve resolution with sediment penetration
- Reduction of side lobes for minimal destructive signal scattering caused by the sediment when profiling near the bottom.

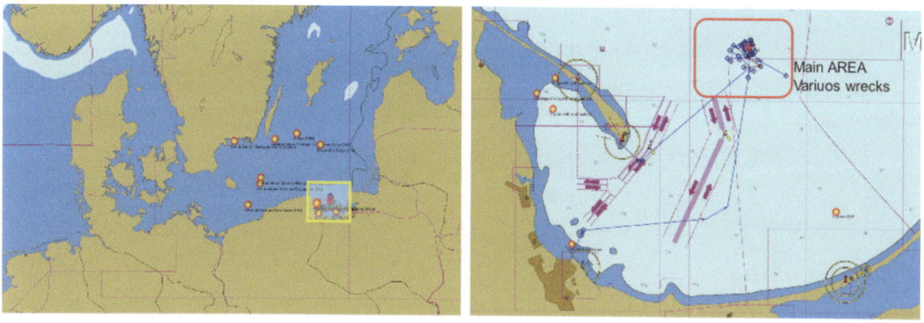

Fig. 2.26 Area of interest

2.6.2 Experimental Campaign

The plan for the cruise was mainly focused on the following activities:

1. AUV/Side scan imaging of the area to detect possible obstacles to navigation
2. Sub-bottom profiling missions for sediment characterization
3. AUV/Side scan/Magnetometer missions to verify possible proud and buried contacts
4. Sediment sampling on selected areas of interest
5. Sediment chemical analysis for the detection of pollutants
6. Passive sampler installation in selected sites
7. Passive sampler recovery

The area was selected by the Institute of Oceanography based on historical data and survey they did in the past with ROV and chemical passive samplers (Fig. 2.26). Data collected in the past has detailed the presence of pollutants in the area. Water depth was pretty constant varying between 97 and 102 m, with some documented wrecks on the seafloor.

The Gdańsk Deep is covered with mud and clayey mud; the thickness of the uppermost, unconsolidated layer reaches 1 m. The sedimentation rate in the area is estimated to be 1.8 mm/year, meaning the objects in question could be covered by 11 cm of sediments (Bełdowski and Pempkowiak 2007). Bottom water is usually anoxic, with periodic flushes of North Sea water, oxygenating the area during the medium and major inflows.

2.6.3 Data Collection

During the trial the ship and the sea state have allowed the collection of a high quality geophysical acoustic dataset. In order to better characterize the various layers the intersections among the various tracks have been acquired and processed (Fig. 2.27). The dataset can provide a really accurate geophysical characterization of the area.

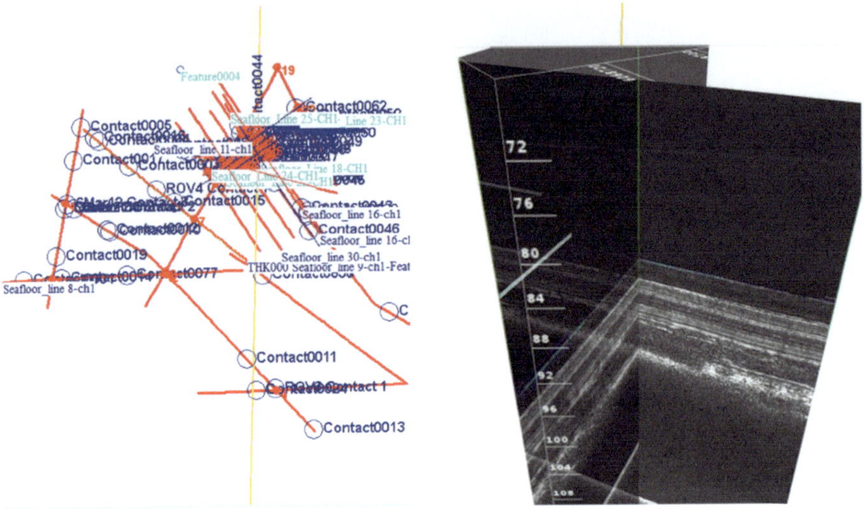

Fig. 2.27 Runs, intersections, anomalies and 3D plot of geological stratification

In summary we have collected 30 transects on the area of interest that allows:

- Characterization of layers
- Search for anomalies in the soft sediment layer for possible large buried targets
- Classify bottom characteristics for prediction of target burial

2.6.4 Data Analysis: Sub-bottom Data

After the first runs we have clearly identified that the whole area is characterized by a thick layer of soft mud that is acoustically semi-transparent. As a consequence heavy targets will be buried into the sediments. In addition in the zone there is a strong presence of gas in the sediments that generate many false targets (Fig. 2.28 and 2.29).

In the muddy area conventional side scan sonar and ROV mission will have very low probability of detection. However, a small area is characterized by the rise of the harder sediment layer. In this area object burial is less probable, and therefore, side scan sonars and ROV are effective tools for searching ammunitions on the seafloor (Fig. 2.30).

2.6.5 Sub-bottom Anomalies

We have identified several anomalies in the SBP data. The examples are listed in Table 2.2, together a sample image, with their location and depth properties.

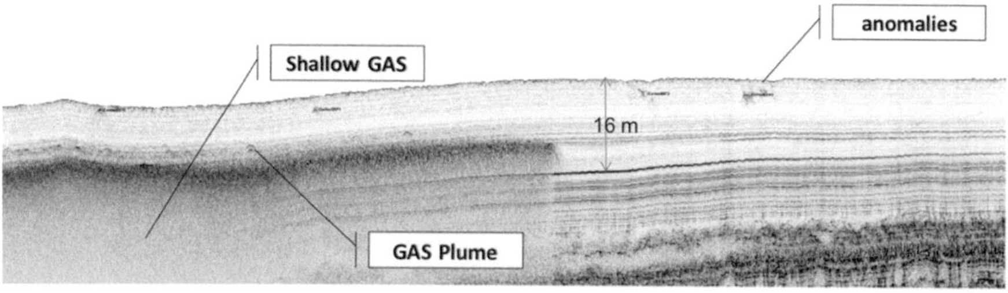

Fig. 2.28 Gassy sediments

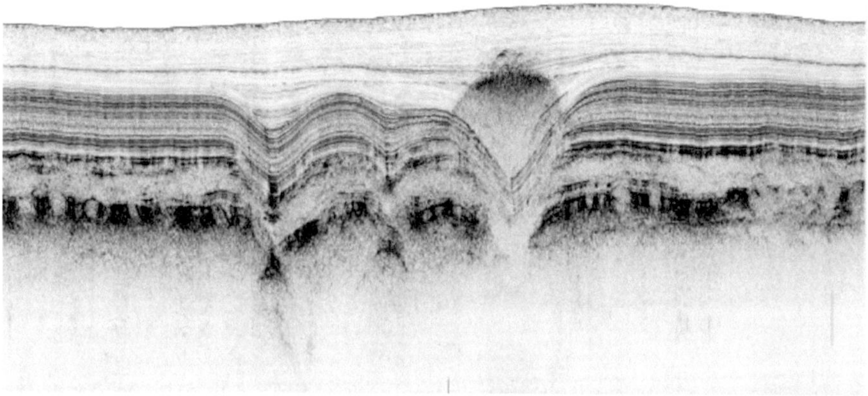

Fig. 2.29 Gas plumes and hard layer variations

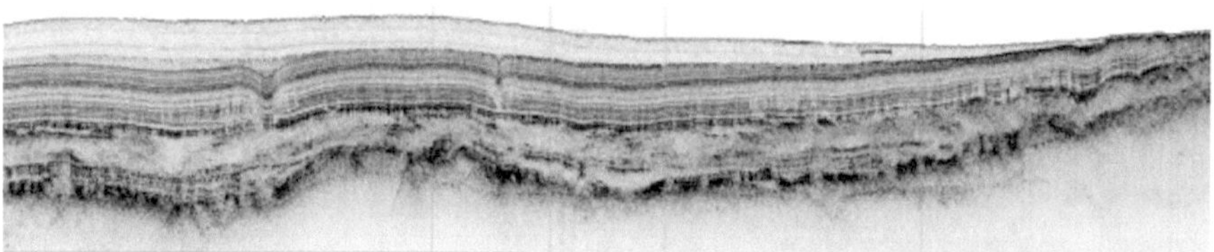

Fig. 2.30 Sediment rise

Due to the nature of the sediment, most of them are probably gas bubbles in the sediments but this can be verified by further acoustic data post processing (i.e., signal phase analysis), that, however, was outside the scope of the project.

By using SonarWiz software we have provided projection of the sonar data, both sub-bottom and side scan, on a map (Fig. 2.31).

Table 2.2 Examples of contact database entries

Target Image	Target Info
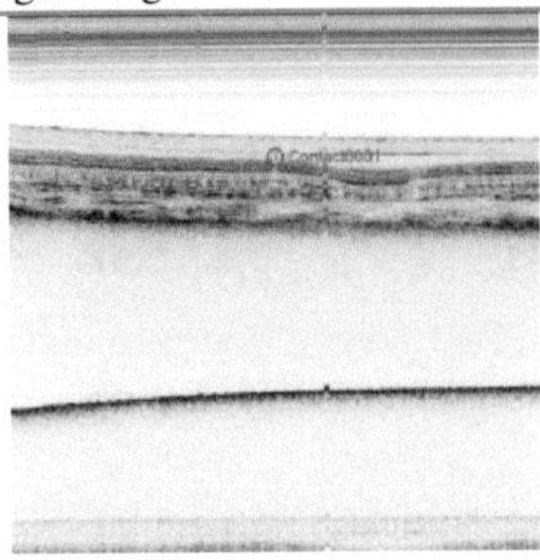	Contact0001
	Sonar time at target: 9/17/2015 7:54:51 AM
	Click position
	54.7333414816 19.1108845081 (WGS84)
	0.0000000000 0.0000000000 (NAD27LL)
	54.7333414816 19.1108845081 (LocalLL)
	(X) 378363.28 (Y) 6066756.07 (Projected Coordinates)
	Map projection: UTM84-34 N
	Ping number: 2395
	Range to target: 29.13 m
	Fish Height: 25.13 m
	Heading: 0.000°
	Line Name: Line 8
	Water depth: 0.00 m
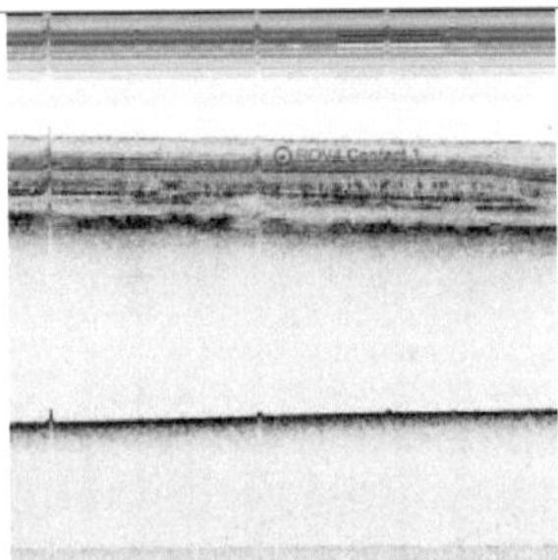	ROV4 contact 1
	Sonar time at target: 9/17/2015 9:39:04 AM
	Click position
	54.7449080267 19.1452966208 (WGS84)
	0.0000000000 0.0000000000 (NAD27LL)
	54.7449080267 19.1452966208 (LocalLL)
	(X) 380612.78 (Y) 6067983.79 (Projected Coordinates)
	Map projection: UTM84-34 N
	Ping number: 8524
	Range to target: 28.39 m
	Fish Height: 25.13 m
	Heading: 0.000°
	Line name: Line 9
	Water depth: 0.00 m
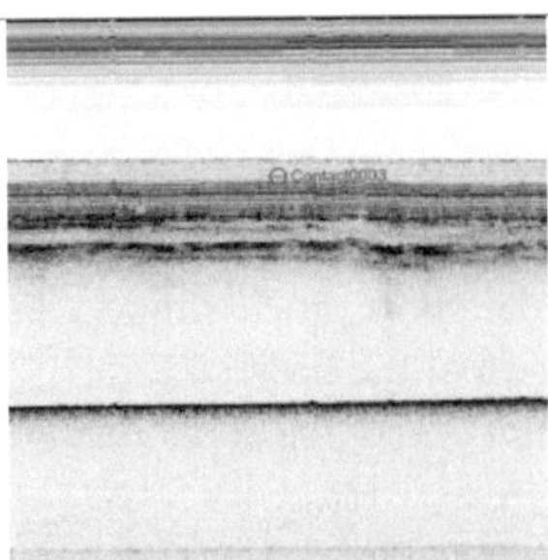	Contact0003
	Sonar time at target: 9/17/2015 9:45:00 AM
	Click position
	54.7478097554 19.1386705768 (WGS84)
	0.0000000000 0.0000000000 (NAD27LL)
	54.7478097554 19.1386705768 (LocalLL)
	(X) 380194.88 (Y) 6068317.92 (Projected Coordinates)
	Map projection: UTM84-34 N
	Ping number: 10,762
	Range to target: 31.44 m
	Fish Height: 27.92 m
	Heading: 0.000°
	Line name: Line 9
	Water depth: 0.00 m

(continued)

Table 2.2 (continued)

Target Image	Target Info
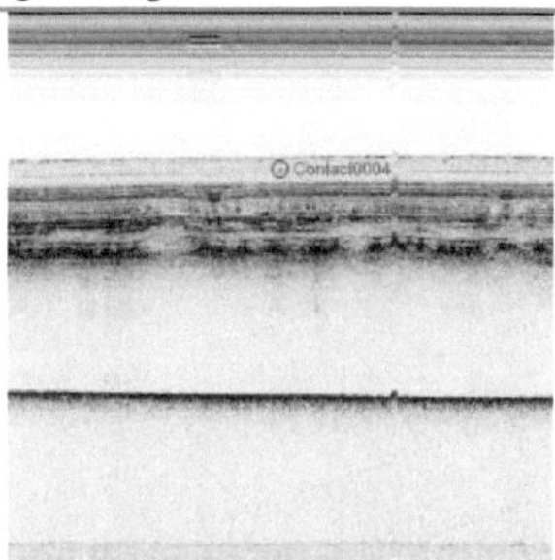	Contact0004
	Sonar time at target: 9/17/2015 9:51:18 AM
	Click position
	54.7510716300 19.1315895362 (WGS84)
	0.0000000000 0.0000000000 (NAD27LL)
	54.7510716300 19.1315895362 (LocalLL)
	(X) 379748.82 (Y) 6068692.93 (Projected Coordinates)
	Map projection: UTM84-34 N
	Ping number: 13,148
	Range to target: 31.02 m
	Fish Height: 28.72 m
	Heading: 0.000°
	Line name: Line 9
	Water depth: 0.00 m
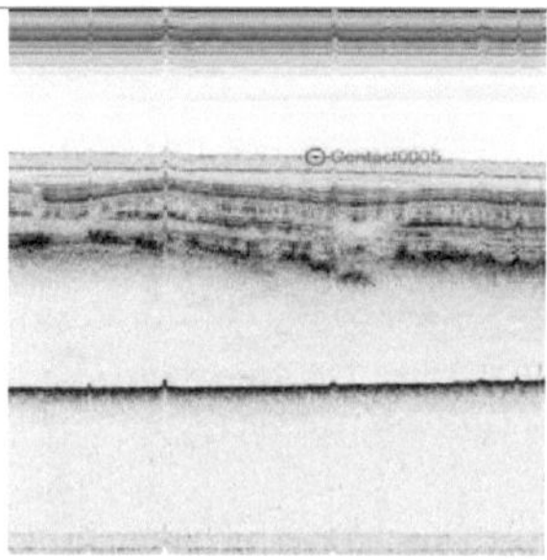	Contact0005
	Sonar time at target: 9/17/2015 9:54:55 AM
	Click position
	54.7529098817 19.1274834351 (WGS84)
	0.0000000000 0.0000000000 (NAD27LL)
	54.7529098817 19.1274834351 (LocalLL)
	(X) 379490.05 (Y) 6068904.49 (Projected Coordinates)
	Map projection: UTM84-34 N
	Ping number: 14,512
	Range to target: 29.44 m
	Fish Height: 29.17 m
	Heading: 0.000°
	Line name: Line 9
	Water depth: 0.00 m

2.7 Summary and Conclusions

Employing of all above-mentioned equipment gave good results in searching and detecting chemical warfare targets as well as a conventional munition. During the MODUM project, 36 detail survey and identification missions were conducted with a usage of the IVER2 AUV, in the areas of Bornholm Deep, Gotland Deep, Gdańsk Deep and Little Belt (Fig. 2.32). Total coverage area of those activities was 8.4 square kilometers and 742 potential bottom-lying targets were selected, based on collected data. The average density of the targets per square kilometer was 143 [1/km^2].

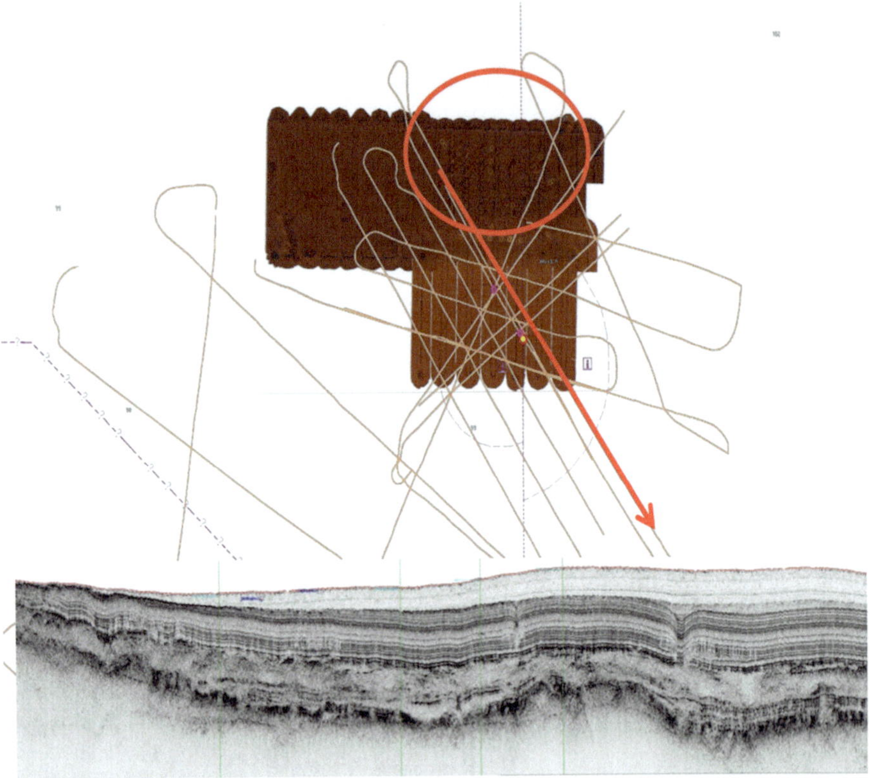

Fig. 2.31 Side scan, sub-bottom data and contacts mapped on the same image

Centre for Maritime Research & Experimentation (CMRE) participation in the Baltic sea trial with direct understanding on conducting test on dumped munition sites was a very valuable experience thanks to the multidisciplinary information collected. The possibility to merge information coming from acoustic imaging devices (i.e., side scan sonars), acoustic geophysical devices (SBPs), and magnetometers together with chemical analysis is an important asset.

More in details, sub-bottom and sediment data analysis are crucial for planning the AUV and ROV missions, and subsequent chemical analysis of the seafloor. In these particular areas, the seafloor is characterized by a thick soft mud layer and therefore object released on the seafloor will be probably buried.

Seabed mapping in terms of searching and identifying of CWA objects is not an easy task. It requires good quality equipment, experts for data acquisition and its processing, lots of ship-time and good weather conditions. However, the collaboration of equipment described in this chapter gives significant results. Gathered outcomes allowed for the implementation of further project objectives.

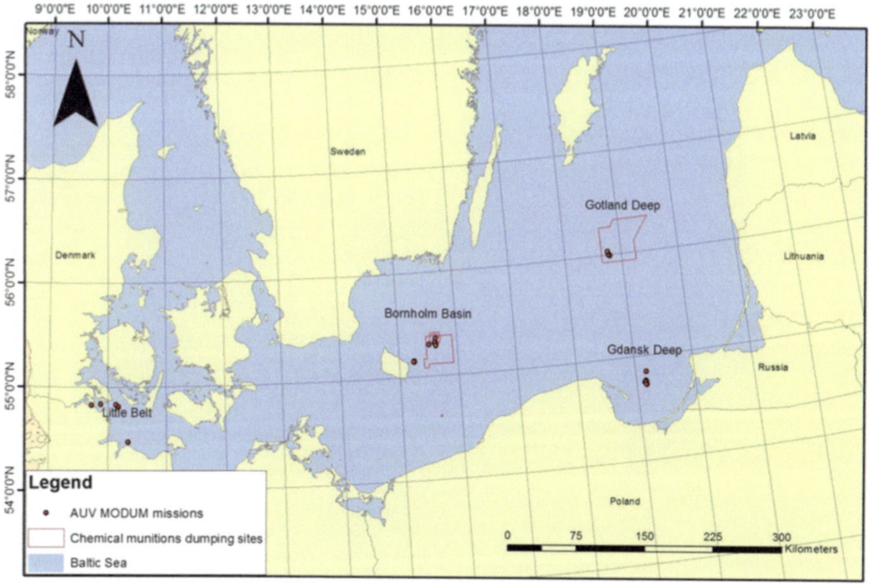

Fig. 2.32 General map of all AUV missions on the Baltic Sea (Compiled by Jan Majcher)

References

Anderson JT, Van Holliday D, Kloser R, Reid DG, Simard Y (2008) Acoustic seabed classification: current practice and future directions. ICES J Mar Sci 65:1004–1011

Bełdowski J, Pempkowiak J (2007) Marcury transformation in marine coastal sediment as derived form mercury concentration and speciation changes along source/sink transport pathway (southern Baltic). Estuar Coastal Shelf Sci 72:370–378

Bełdowski J, Fabisiak J, Popiel S, Östin A, Olsson U, Vanninen P, Lastumaki A, Lang T, Fricke N, Brenner M, Berglind R, Baršiene J, Klusek Z, Pączek B, Söderström M, Lehtonen K, Szubska M, Malejevas V, Koskela H, Michalak J, Turja R, Bickmeyer U, Broeg K, Olejnik A, Fidler J (2014) CHEMSEA Findings. Gdańsk, Institute of Oceanology Polish Academy of Sciences, p 31. 9788393660919

Bełdowski J, Klusek Z, Szubska M, Turja R, Bulczak IA, Rak D, Brenner M, Lang T, Kotwicki L, Grzelak K, Jakacki J, Fricke N, Östin A, Olsson U, Fabisiak J, Garnaga G, Rattfelt Nyholm J, Majewski P, Broeg K, Söderström M, Vanninen P, Popiel S, Nawałka J, Lehtonen K, Berglind R, Schmidt B (2016) Chemical Munition Search & Assessment – an evaluation of the dumped munitions problem in the Baltic Sea. Deep-Sea Res II Top Stud Oceanogr 128:85–95

Blondel P (2009) The handbook of Sidescan sonar. Springer, Berlin/Heidelberg/New York/Chichester, p 45. 9783540426417

Dera J, Sagan S (1990) A study of the Baltic water optical transparency. Oceanologia 28:77–102

Mc Quillin R, Bacon M, Barclay W (1984) An introduction to seismic interpretation: reflection Seismics in petroleum exploration. Gulf Publishing Company, USA. 9780872017733

Mulhearn PJ (2000) Modelling acoustic Backscatter from Near – Normal Incidence Echosounders – Sensitivity Analysis of the Jackson Model. Technical note (Defence Science and Technology Organisation (Australia)), DSTO-TN-0304

Chapter 3
Results of Acoustic Research in the CM Deploying Areas

Zygmunt Klusek and Miłosz Grabowski

Abstract Short overview of problems connected with acoustic detection and recognition of chemical munition deployed in the Baltic Sea after the II WWW is demonstrated.

Results of 36 high-frequency acoustic scanning missions carried out with the IVER-2 AUV over a chemical munition dumpsite in the Bornholm Basin, Gdansk Deep, Little Belt and south-western part of Gotland Deep are presented. The main goal of the investigations was to image structure of the dumpsite and to positioning targets for purposes of taking probes of environment for further chemical analysis.

The data support of the earlier findings wide variety of types of objects at the sea bottom – both historical items as wracks or sea mines and contemporary litter.

It is generally observed a low contrast between acoustic shadows and backscattered signals, which are likely due to the sediments properties in the Baltic Deeps.

In the Bornholm Basin and Gdansk Deep the distribution of the buried objects seems to be rather concentrated around surrounded by diffusive spacing of objects due to fishery activity in the areas. The results of complementary studies on the acoustic properties of the bottom demonstrate the dependency of probability of findings with the type of sediments.

Finally some aspects of probability to detect of is given.

3.1 Introduction

Underwater acoustics is considered as the most practical remote sensing method in fishery, physical oceanography, marine geology and mapping the sea bottom. It is well acknowledged that acoustic signal scattered at the sea bottom comprises information about water – sediment interface as its roughness, sediment type, grain-size

---

Z. Klusek (✉) • M. Grabowski
Institute of Oceanology, Polish Academy of Sciences,
Powstańców W-wy street 55, 81-712 Sopot, Poland
e-mail: klusek@iopan.gda.pl

J. Bełdowski et al. (eds.), *Towards the Monitoring of Dumped Munitions Threat (MODUM)*, NATO Science for Peace and Security Series C: Environmental Security, DOI 10.1007/978-94-024-1153-9_3

spectra, porosity, presence of gas bubbles, material density, etc. Depending on frequency of sounding acoustic signals – in the case of frequencies above tens of kilohertz sound is usually scattered only at the surficial layer but at lower frequencies could penetrate into the sediments and the basement material.

Scientists now regard acoustical active sounding as a cost-effective method to image seabed roughness, sediment types and small-scale details of the benthic habitat and recognise objects on the sea bottom or buried in sediments.

The backscattering of sound depends on a number of physical characteristics of scanning signals as the frequency of the system, technical configurations of exploited systems, or geometry of acoustic beams and incidence angle. Above technical characteristics of geological parameters of sea bottom as porosity of sediments roughness induced by their physical or chemical characteristics define the acoustic scattering on the seabed surface.

Fully understanding the interaction of sound with the seabed is sometimes discouraging, mostly due to variety, both due to the physical structure of the sediments and especially in shelf seas as result of the benthic community.

However, sidescan sonars provides crispy images (sonographs) from which we would recognise not only objects resting on the bottom surface but also use as a tool for diagnosis and understanding the seabed microtopography and the nature of sediments (as example Blondel 2009). These devices are used also in specifications of bottom biocenosis like seagrass meadows, or appreciation of the anthropogenic impact on the sea bottom.

In the chapter we present different aspects of the acoustic recognition of many years before deployed targets under specific conditions of the Baltic Sea bottom.

In the second section we shortly discuss qualitatively theories of sound scattering at the sea bottom regarding three information: intensity of reflection, scattering of sound and frequency dependences reflected their significance in recognition of targets deployed on the sea bottom.

The theory of sound backscattering at specific kind of targets is included in Sect. 3.3. Short setup characteristics and their suitability tests used in detection of presence of chemical munitions in the Baltic Sea are given in the next Sect. 3.4. Further presentation of methods of signal and image processing aimed at improving detection and classification of targets is given. Statistics of surveys, acoustic detection and classification of targets and results of ground truthing by optical observations comprise Sect. 3.5. Some characteristics of sediments obtained with acoustic methods during the project with examples is given in Sect. 3.6 and critical aspects of probability of the detection and recognition of targets at the Gotland Deep as the case study are discussed. The summary terminates this chapter.

3.2 Theory of Sound Scattering at the Sea Bottom

From the 60's of the last century, intensive investigations were performed to determine a correspondence between properties of backscattered at the sea bottom sound signals and characteristics of sediments.

The first functional relationships were developed from very limited field data and frequency bands and were mainly empirical.

As a result, a variety of inverse acoustic methods have been put forward and adopted in the literature to obtain geoacoustic parameters of sediments from results of measurements of the backscattered acoustic signals.

The main and the straightforward parameter described reflection on the is a complex reflection coefficient defined by

$$V(\theta)=\frac{P_r}{P_i} \tag{3.1}$$

where: P_r and P_i are the complex amplitudes of the incident and reflected waves, measured or more commonly recalculated to distance of 1 m above the sediment–water interface scattering area.

Usually reflection coefficient in logarithmic scale is preferred and the Reflection Level -RL is computed -

$$RL = 20\,logV = 10\log\frac{P_r^2}{P_i^2} \tag{3.2}$$

where P_r^2 is the squared pressure amplitude of the reflected wave averaged over some number of transmissions.

In the first order of approximation, the basic theoretical equation which represents reflection and transmission of acoustic waves respectively at and through the plain boundary of two homogeneous media, would be given in the form:

$$|V(\theta)|=\frac{(m\cdot cos\theta - M_2)^2+M_1^2}{(m\cdot cos\theta + M_2)^2+M_1^2} \tag{3.3}$$

where |V| is the modulus of the reflection coefficient.

The phase of the reflection coefficient is given as

$$\phi = arctan\left[2\cdot\frac{M_1 cos\theta}{m^2 cos^2\theta}-M_1^2-M_2^2\right] \tag{3.4}$$

where $n = n_0(1 + i\alpha)$ complex index of refraction, the real part of refraction is $n = c_2/c_1$, m the bulk density contrast between the two media, $m = \rho_2/\rho_1$ and ρ_2 and ρ_1 are the densities of medium 2 and 1 respectively and c_2 and c_1 the sound speeds of these media and α - sound attenuation in the lower half-space depends on frequency [nep/m], $\alpha > 0$.

$$A=\sin^2\theta - n_0^2\left(1-\alpha^2\right); B = 2n_0^2\alpha;$$

$$M_{1,2} = \sqrt{1/2}\sqrt{\sqrt{A^2 + B^2} \pm A};$$

On the basis Eq. (3.3) we could predict how much sound is reflected into the water column and how much penetrates into sediments. Additionally, on the basis of the same equation we could predict a critical grazing angles above which, a substantial fraction of the sound energy incident on the seafloor is transmitted through the water/sediment interface. Consequently, we would estimate possibility to detect objects bury in sediments.

In the case of the vertical sounding, Eq. 3.3 is simplified to the form:

$$|V| = \frac{\rho_1 c_1 - \rho_2 c_2}{\rho_1 c_1 + \rho_2 c_2} \tag{3.5}$$

The formula (3.5) is valuable as the simple measure of acoustic hardness of sediments.

However, for the real water-sediment interface it has been proven that this relation is not to be rigorous, as for example in the case of porous seabed.

Sound propagation in porous materials can be described in the frame of Biot's models (see, e.g. Williams et al. 2001). Although, application of the theory to the seabed is complex, and for prediction of the propagation and scattering properties of sediments several geo-parameters are required as the input to the model (Jackson et al. 1986). To avoid complexity, the sediment frequently is characterized as fluid-like media, and therefore, wave interaction with the seabed is discussed in terms of a fluid-fluid assumption.

Usability of such theory to predict detection of covered objects, is constrained by scattering from the rough interface, sediment layering structures and from volume inhomogeneities of density, elasticity and sound speed within the sediment presented by shallow gas presence or cobbles.

Scattering at the volume inhomogeneities would led to false alarms as well as to distortion of real targets images.

In shallow water, bottom organisms would alter locally the physical properties of the sediments and create seabed structures affecting the acoustic reflections from the seabed.

Even in the face of such complexity, over only a few decades much progress has been made in predicting acoustical seabed scattering from the basic physics of the scattering processes.

The diversity of the seabed and the underlying strata has stimulated acousticians to formulate various theoretical and numerical models to describe bottom scattering.

In general, wave scattering problems in statistically rough surfaces are challenging classical problems in theoretical physics and frequently are solved using approximations and numerical modelling.

Basic concept is presented below using simple relevant mathematical scattering model.

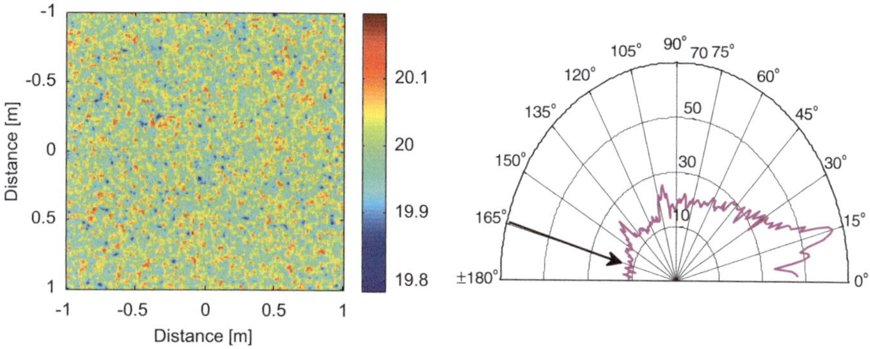

Fig. 3.1 Representative angular dependence of a high frequency signal scattering on a rough interface water-sediments. On the left the image of the water-bottom interface with amplitude spectra of irregularities according to 2D the Goff-Jordan spectrum. On the right results of numerical modelling (radius is in dB scale)

In the Fig. 3.1 result of numerical modelling demonstrates the effects of surface roughness on the plane-wave scattering and presents angular distribution of scattered acoustic energy for the water-sediments boundary with the 2D Goff–Jordan spectrum for bottom roughness. The 2D Goff–Jordan spectrum for bottom roughness is defined by the formula -

$$w(k) = A\frac{k_0^3}{k^2 + k_0^2} \tag{3.6}$$

where k_0 – space wave number characteristic for diverse types of sediments roughness and here A – amplitude factor, k- wave number of acoustic wave.

Presented here data are outcome of a class of "point scatterer" model that based on the Huygens Principle concept, saying that each point on the rough surface acts as a secondary radiator of incident waves. The model can be easily implemented on a computer and would be used in interpretation of results for different type of sediments. The results are presented for the incident angle 70° in polar coordinates where the radius is in logarithmic scale. The example is representative for wave scattering at moderate rough surface – main lobe of scattered sound is around mirror reflection angle. We see that only minor portion of energy is scattered back to the source.

3.3 Sound Backscattering at Munition and Methods of Detection

Due to the complexity of the scattering phenomenon, theories, even very elaborated and complicated, have limited effectiveness in prediction precisely many characteristics of scattering at irregular targets and would be used only as a starting point in the interpretation of results.

Probability of successful detection and recognition of targets mainly depends on the physical/acoustic cross section of the target and the wavelength of acoustic waves. In the case of comparable wavelength and characteristic size of a target the success rate is independent of an object surface shape.

Above the size and object physical properties, the probability of detection of any target in side scan sonar images depends also on seabed properties. Generally, it is higher for an uncovered flat acoustically-hard sediments and lower for a soft, irregular or covered with plants.

The existences of ripples, boulders or trawl marks diminish possibility to distinguish targets and/or shadow shapes in a sonar image from bottom irregularities. During scanning in all areas under investigations echoes from apexes and crossing of trawlmarks gives the highest number of false targets.

Due to low frequency of pinging or to high speed of the fish towing images of targets sometimes cover only a few pixels or interfere with other non-target seafloor irregularities or natural objects within a pixel cluster making impossible to properly classify.

For bigger objects of buried in sediments, characteristic texture in sonar images are suitable tool in target detection for the reason that they contain more independent parameters, such as entropy, contrast, variability or similarity. Texture of the target images would be also useful in the detection process at the only small number of pixels level (Fig. 3.2).

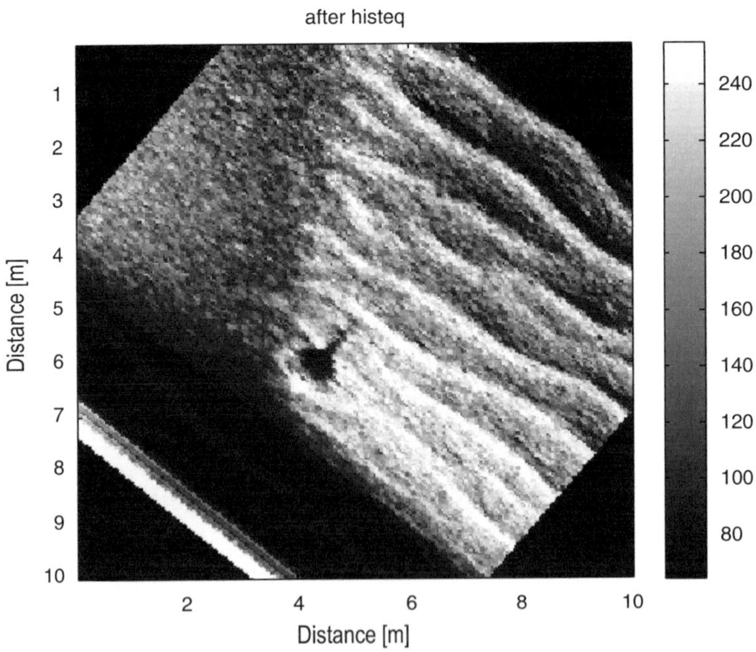

Fig. 3.2 Image of a mine and its shadow on the hard bottom. The data were collected during test of IVER-2/Klein 3500 configuration at the NATO Centre La Spezia, Italy

Semi-buried or fully buried objects after many years of organic matter deposition which commonly occurs in the Baltic Sea Deeps results in distorting contours and changed of the image of object and diminishing shadow proportions decisively limiting the effectiveness of conventional high frequency sonars.

Attractive features of the low frequency sonars include the enhancement of possibilities of detection and recognition buried in sediment targets relatively to the most common high frequency sonars improving the sensitivity of systems. This feature makes the low frequency sonar system robust in case of submerged targets.

In principle spectral analysis of low frequency returning signals in audio or low frequency ultrasound range due to the excitation of structural resonances in the target allows to discriminate between mine-like objects and boulders. These advantageous attributes have created interest in this methods and attempts to recognize different models of sea mines using multifarious signal analysis were published in series of papers (Tesei et al. 2000, 2002).

However, currently, only a few researchers and research laboratories are working on employment this method into practice.

3.4 Setup Characteristics

Acoustical devices used in the active targets detection as sonars and echosounders work by the principle of transmitting and receiving the sound waves that are reflected, diffracted or scattered off from the sea bottom and objects lying on the surface of the sediment-water interface or buried under the surface.

Side scan sonars consist of two or more transducer arrays commonly mounted on each side of a sonar towed body ('fish') or mounted on the board of an Autonomous Underwater Vehicle (AUV).

We could differentiate them on the basis of employed frequencies as they work in broadband frequency range from tens of kilohertz up to almost 1 MHz. First constructed sonars and echo sounders developed in 60s of the twenty century used frequencies in tens of kilohertz band. But in the last decades the evolution of frequency range towards of low and very high frequencies is observed.

At frequency around 100 kHz maximum ranges of observation would reach 200 to 300 m per side, forming total swaths widths of up to 400–600 m, with a typical maximum resolution of 0.15/0.2 m. At 1000 kHz ranges would be reduced to 75–100 m per side, but both the range and along track resolutions are improved. As the sound attenuation increases with salinity in brackish water of the Baltic Sea with salinity 6–7 PSU, real range in very high frequency significantly improves if comparing to oceanic water.

Using so extensive frequency range of sounding signals we are able to get available different information about the seabed's properties and character of detected targets.

Regarding complexity of instruments used in target detection and recognition we recognise evolution from simple one-beam echo-sounders looking vertically at the sea bottom, through sidescan sonars (SSS) towards multibeam systems (Multi-Beam Echo-Sounders MBES, as example). In our opinion, a separate category represents broadband with complex signals utilities and parametric sonars.

Due to shape of a transducer and adopted methods of beamforming a conventional side-scan sonar has a very narrow beam along track and a very wide beam across track. Horizontal narrow beam help to receive good resolution along track, and allows to get more detailed targets silhouettes. Though, in the case of too narrow band we observe higher signal variability. Vertical wide beam results in extended of across track range of bottom imaging. In the consequence portion of the beam is looking up to the sea surface. Therefore, in the case of shallow water scanning we observe regularly echoes straight from the sea surface as well as from bubble clouds generated by breaking waves. Interferences between signals scattered from seabed and sea surface may deteriorate sonar images masking the seabed targets (as example Badiey et al. 2000).

Typical horizontal beam widths are from 0.2° to 2° and vertical beam width usually tens of degrees (40–60°). The beam angles of the transducers are typically fixed and maximum of directivity function (main sonar lobe) is usually inclined with grazing angle less as vertical beam width.

Around centre of sonograms a narrow strip in sonograms on each side of the 'nadir' are blank. Moreover a range resolution near nadir is significantly diminishing.

The characteristic transverse resolution and range resolution of the side-scan sonar systems as the geometry of across-track and along-track footprints respectively are presented in literature (Fleming 1976, Blondel and Murton 1997).

As a matter of good practice to achieve full 100% coverage of the seabed, swaths should be overlapped by at least 50%. However, in everyday practice the reliable information relating to objects protruded or rest on the seabed swaths are when area is overlapped by 20–30%.

Survey speed, sonar altitude above the seabed and the swath range determine the quality of the sonograms.

Diagram in Fig. 3.3, illustrates some principles of the SSS technique:

1. vertical orientation of the main beam with respect to the towfish;
2. increasing of spreading losses with distance in vertical view;
3. change in seabed incidence angle of the beam across the track;
4. schematic of an scattering at fish or sea surface bearing the disturbances most commonly observed.

The spatial sampling constraint is that the distance covered by an AUV or ship between following pings should be less than one-half the length of the horizontal aperture.

Another opposing condition put on ping repetition rate is that echoes from the most distant bottom area should return before the termination of received echoes from the subsequent ping (range is one-half the distance sound travels between pings). Both conditions are limiting factors affected the efficiency of covering search area.

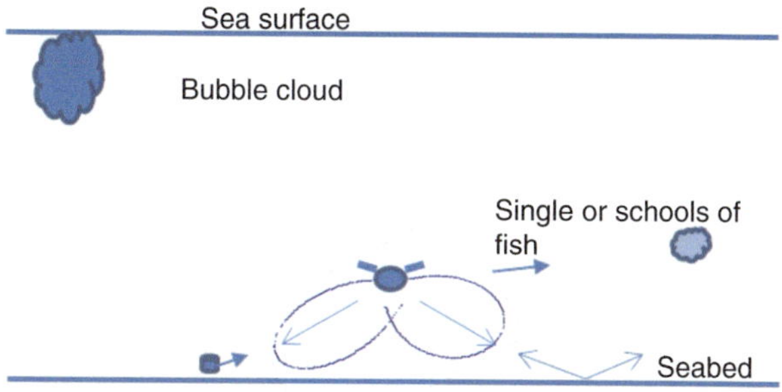

Fig. 3.3 Principle of the SSS technique and sources of disturbances in recorded sonographs

Depending on the aim of the survey we select or pinging less frequently and moving more slowly with broader span of bottom image, or move more rapidly, pinging with higher rate and work to shorter swath.

The high resolution sonographs help to better understand sedimentary and hydrodynamic processes as internals waves or bottom currents acting upon the shape and material of seabed.

In the most recent decades we observe introduction into practice more advanced interferometric sidescan sonar systems based on the idea of Synthetic Aperture Radars with extraordinary high resolution at frequencies around 100 KHz, and also low frequency sonars with frequencies working around 10 KHz.

Due to lower attenuation of frequency sound in sediments, usually decreasing linearly with lessening frequency, using low-frequency techniques allows to acquire the reconstruction of shape of buried objects in the first meters of depth.

Utility of low-frequency sonars working in broadband audible frequency range of 2–15 kHz with broadband Ricket sounding signals was proposed and further successfully tested.

The main predominance this kind of technique over conventional high-frequency sonars are considerable better penetration of signals into the sediment at these frequencies and that ensonification of sediments could be performed above some critical grazing angle.

Analysis of different spectral components of broad frequency echosignals of this sonar type help to exploit spectral features of returned echoes from buried targets, allows to recognise the munition shell resonances, structural response, etc.

It seems that low-frequency sonars are only useful solution in detection and recognition of objects which sank into semiliquid sediment or were buried due to material advection or rain of organic material from upper water layers.

Recently another possibilities of buried object recognition (as marine cable) at low single frequencies were proposed using parametric sonar, however with different success.

Some sonars use the broadband technologies (chirp) to gain high quality imagery at a range exceeding range of single frequency sonars at the same frequency.

During operation in the CHEMSEA and MODUM project a set of side scan sonars were involved. A workhorse of the CHEMSEA scanning was the Klein 3000

SSS towed from R/V "Baltica", the ship provided by the Swedish Maritime Agency (SMA). The centre frequency of the sonar is 455 kHz that gives 20 cm along-track resolution. As the main goal of the CHEMSEA project performed by R/V "Baltica" was to cover the vast area about 2000 sq. km. The Klein 3000 was usually towed at 7 knots with a 300 m swath at an altitude of approximately 15 m.

Another intensively used sonar, although at more limited extension, was Edge Tech DF1000 sonar (supplied with two scanning frequencies 105/390 kHz) towed from S/Y "Oceania", with typical ship speed 3–3.5 knots. The Edge Tech was used for classification of objects in the south-western part of the Gotland Deep, in unofficial area of dumped munition in the Gdansk Deep (PEEZ) and in Bornholm Deep. Due to presence of wracks and fishing ghost nets abandoned by fishing vessels after catching and tearing on wrecks, this sonar was regularly used during reconnaissance transects before planning missions with the Iver2 AUV in the Bornholm Deep.

On-board of the AUV Iver2 is mounted the Klein 3500 SSS. The fish is equipped with many sensors including an altimeter sensor that allows it to follow the bottom and "fly" at a constant altitude. Moreover, measurements of such parameters as heading, pitch, and roll are recorded in the fish memory for further improving image.

The AUV has on-board inertial navigation systems, which together with GPS fix recorded, on the start of the each diving help to pinpoint positions of targets in sonograms.

The employed AUV is compact and relatively light and easy to use by operators. But, small AUVs when typically move at speeds 1.5–2.0 m/s^{-1}, are influenced by bottom currents which in the case of strong current would significantly affect stability of tracks and subsequently data quality.

During the IVER-2 missions quite frequently we observe the sonar instability manifested in effects of rolling and heading oscillations, which occurred at the beginning of a leg or when a current was significant.

In the Bornholm basin registered bottom currents would reach values 0.35 m/s, (Bulczak et al. 2016) substantially influencing a planned track. Example of performance of the IVER-2 during one of missions in the Bornholm Deep is shoved in Fig. 3.4. In subsequent panels are showed pitch, roll, the AUV's heading and height above the seabed. On the basis of the data is expected that the observed during missions the vehicle pitch, roll and heading motion would be also a source of additional image distortion.

As a rule of thumb, the towfish work at a height above the seabed of 10–20% of the horizontal range.

Before the start of each track the AUV checks its position with the GPS system. In this way the dive position is actuated at the start. At the end of the track the AUV checks in position again to measure and correct drift while surveying.

AUVs are less suited for scanning in fishing areas due to possible collision with local fishing activity or ghost fishing net entanglement.

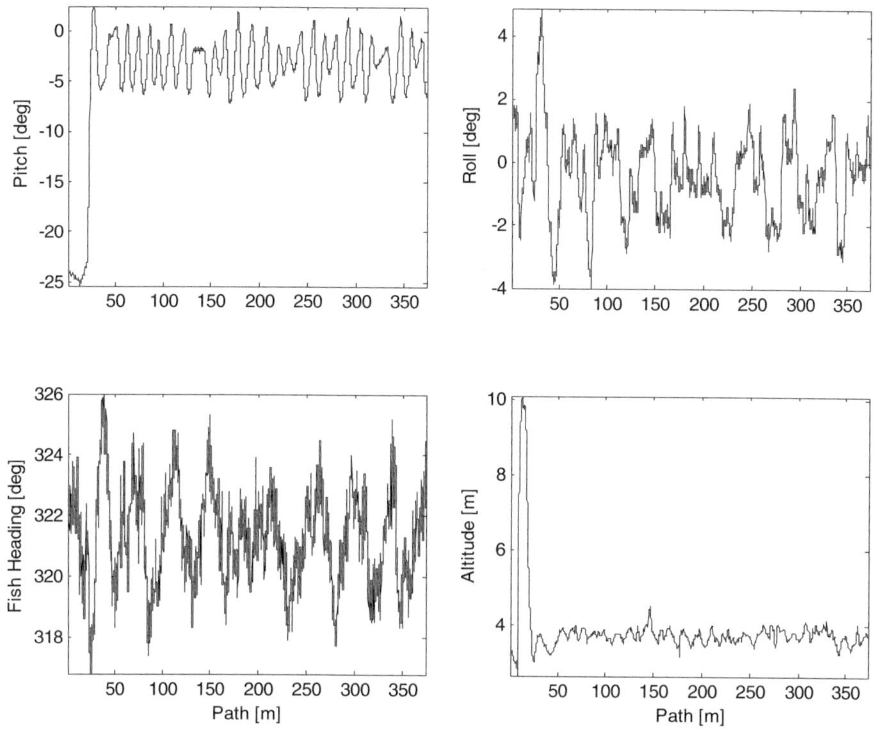

Fig. 3.4 Illustration of performance of the IVER-2 during a mission in the Bornholm Basin. In subsequent panels are pitch, roll, the AUV's heading and height above the seabed

3.5 Signal and Image Postprocessing

The geometry of sounding with sidescan sonars entails slant-range distortion. It means that sample footprint increases with the increasing incident angle of observation (distance from the nadir), and the distortion in position of the imagery is nonlinear function of time (sampling).

Therefore, at the first step of a time series correction we recalculate the horizontal resolution of the raw image into the natural units (distances on the seabed) using an 1D interpolation procedures, taking into account that that in all cases the water column portion of an echo time series is firstly removed.

Returned signals are affected by geometrical spreading of a beam which in the theory would be compensated by a time varied gain (TVG) usually incorporated into the sidescan sonar. Above the diminishing of returning signals due to geometrical spreading an echo signal is convolution of two functions: angular dependency of backscattering from the sea bottom determined by sediments type and beam pattern of transducer. The all factors would be compensated by the digital time varied gain at the processing stage.

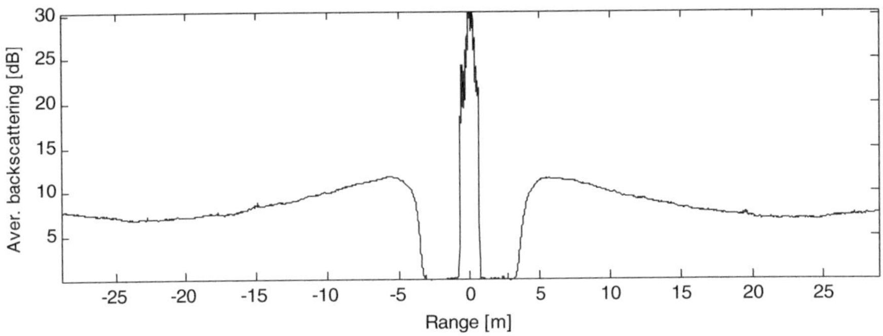

Fig. 3.5 Averaged over hundreds of pings characteristic time series of echosignals in area with relatively smooth bottom (the Bornholm Deep) received after removing local speckles from munitions and other objects presented in the transect

Averaged over hundreds of pings characteristic time series of echosignals from the AUV in area with relatively smooth bottom (from the Bornholm Deep) received after removing local speckles from munitions and other objects is presented in Fig. 3.5.

The process of classification of targets includes segmentation of sonographs into discrete areas exhibiting a particular acoustic signature. This signature is thought of as being characterized by the amplitudes of the backscatter in the imagery and the interrelations between pixels or texture within regions of the image.

The segmentation is carried out manually by visual analysis on the data after the first two steps of corrections. There is advantages to use only corrected images as they are typically of a higher resolution than the mosaic imagery.

Due to acoustic impedances contrast between water and sediments an metallic object show up as bright spots in sonograph with adjacent shadows that face perpendicular away from the sonar track. The shadows are best visible on the hard well backscattering relatively flat bottom as gravel or sand.

Example of image of well-defined mine like object on the hard bottom obtained by the AUV sonar in the NATO training ground, La Spezia, Italy.

Features of various shapes and sizes can be detected by the shadows, however we should keep in the mind that the size of the shadow varies as a function of beam grazing angle (distance from the sonar) and feature dimensions.

Target detection or their feature identification involves a broad range of techniques, including detection of local maxima in images (speckles) or more sophisticated methods designed to identify characteristic features of typical shapes of mine-like objects or their shadows (Fig. 3.6).

Target detection or feature extraction involves a broad range of techniques, including recognition of local maxima in images (speckles), recognition of typical shadow shapes and more complex methods designed to recognize discrete features by an object shape or in the case of more extended objects the texture.

Targets of interest sometimes comprise only a few pixels in the image or are mixed with other nontarget seabed cover connected within a pixel cluster.

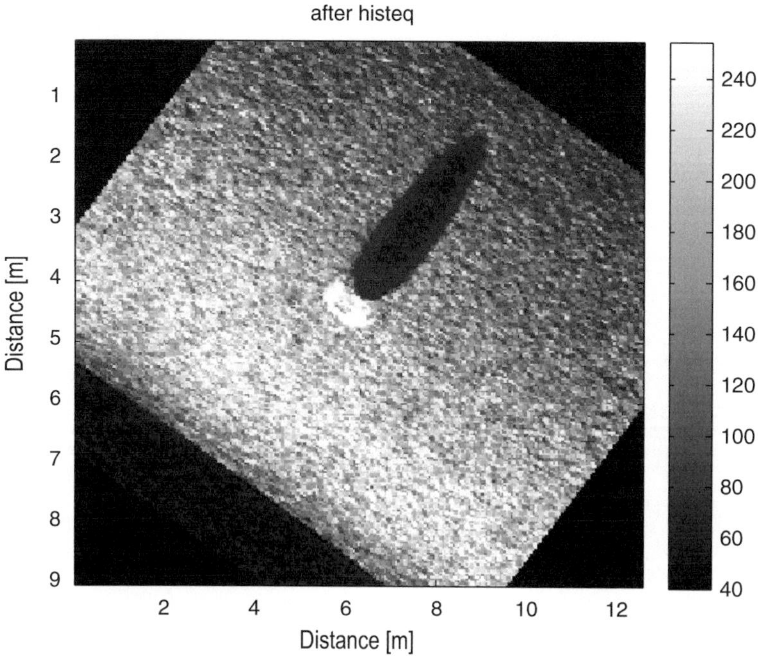

Fig. 3.6 Example of a sonograph of well-defined target deployed on the hard bottom. Picture achieved with the sidescan sonar Klein 3500 mounted on the IVER2 AUV

Texture in images are useful in target detection because they contain a large contiguous set of properties, such as contrast, variability, similarity and discriminability, which can be used also to detect targets at smaller number of pixels level.

Backscatter images are basis textural analysis through object based image analysis. Image texture is quantified by the spatial variation of grey level values.

A common approach uses the idea put forward by Haralick et al. (1973) where texture measures are derived by comparing the values within a window (usually eight contiguous pixels).

Where it was possible shape features were extracted from the shadow and highlight as: area; elongation; solidity; eccentricity; ratio of highlight to shadow area; ratio of highlight to shadow height; minimum distance between highlight and shadow; and horizontal alignment of shadow and highlight. Also, the curvature of the contour at small, medium, and large scale were acquired.

From a SSS imagery the snippets were processed and then conducted a classification stage.

The details of the more accurate classification processes on the basis of abovementioned features are out of the scope of this paper.

Example of sonograph with targets munition like objects found in the Bornholm deployment area are showed in Figs. 3.7 and 3.8. Here we present seven particular variables used to explore the relationship between the backscatter from the objects

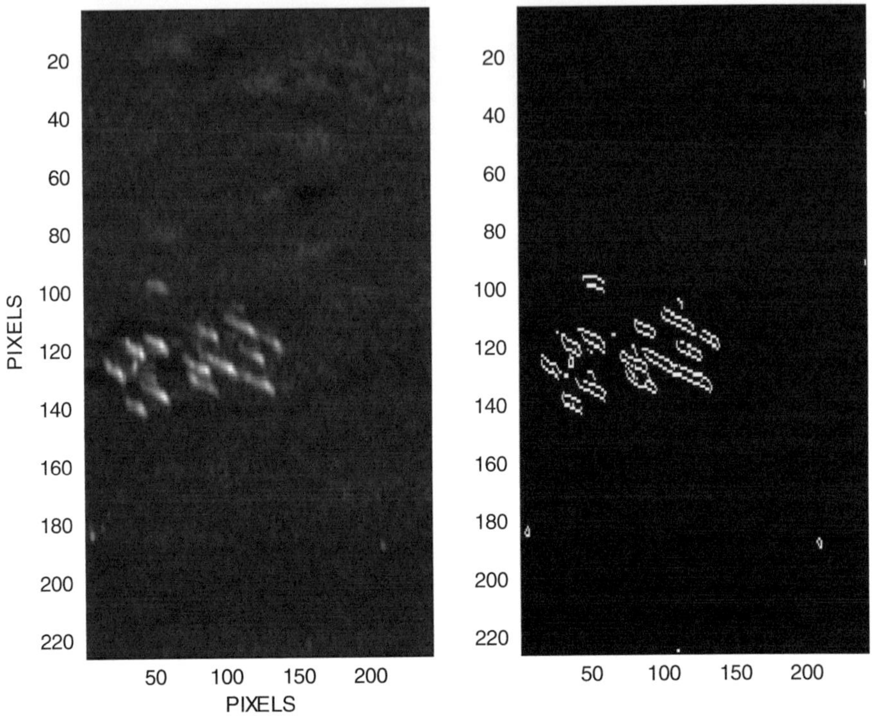

Fig. 3.7 Sonograph of preprocessed and segmented image of a cluster of gas containers found in the Bornholm CM deployment area (on the left) and their contours. One pixel along x-axis (across track) is equal 0.011 m, and along y-axis 0.065 m

and the seabed. Four of them are computed on the basis of the co-occurrence matrix from the image i.e. contrast, homogeneity, correlation and energy (in upper two panels). The next three are: local standard deviation of image (STD), local entropy of grayscale image and local range of image.

It was found that the most significant parameters useful for targets detection are the contrast, STD, entropy and local range computed on the grayscale data 0–225.

For greater clarity, the axes in the Figs are not in dimension units (in metres) but as pixel's numbers. One pixel along x-axis (across track) is equal 0.011 m, and along y-axis 0.065 m.

3.6 Statistics of Surveys Aimed at Acoustic Detection and Classification of Targets

The team of IO PAS in co-operation with the Polish Navy Operators conducted a visual and acoustic survey of the main deployments in the Southern and Central Baltic Sea i.e. Gotland Deep, Bornholm Basin, Gdansk Deep and in shallow waters

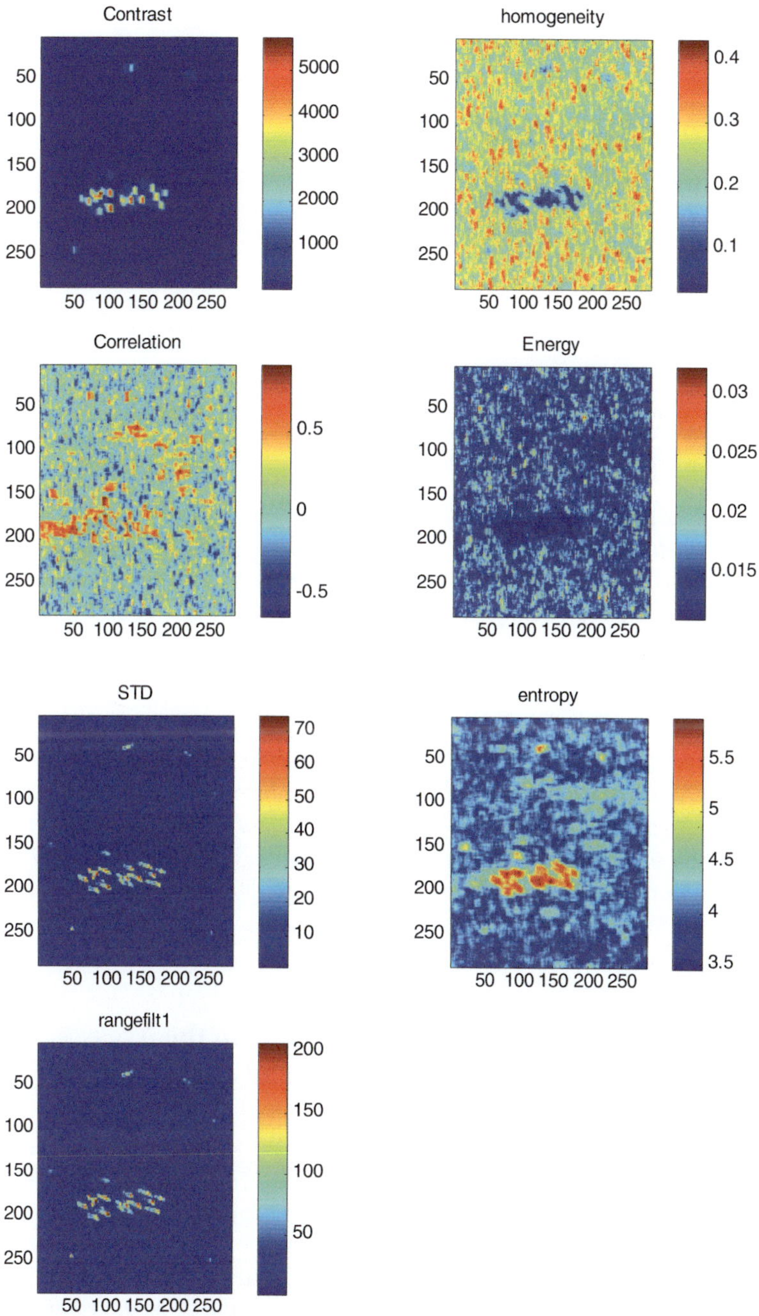

Fig. 3.8 Variables used in the interpretation of relationships between the backscatter from objects and the seabed. Four of them are computed on the basis of the co-occurrence matrix from the image i.e. contrast, homogeneity, correlation and energy (in *upper* two panels). The next three are locals: standard deviation of image (STD), entropy of grayscale image and range of image (*below*)

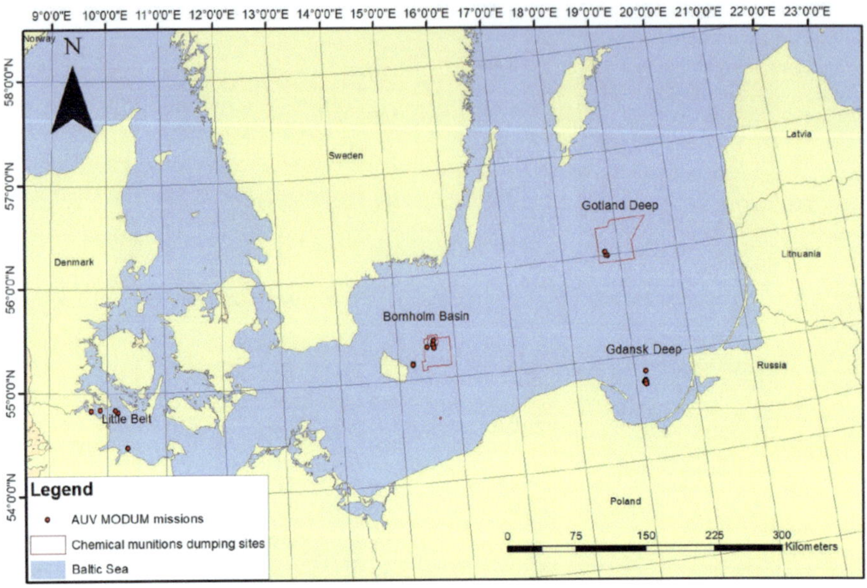

Fig. 3.9 Map of chemical and conventional munition deployment areas with performed missions of the AUV

of the Little Belt (Map in Fig. 3.9). The survey was carried out from S/Y Oceania in spring and summer seasons of 2014/15 and 2016. Pre-determined missions of the IVER-2 AUV were designed on the basis of earlier investigations with less detailed equipment. To avoid obstacles, and reducing the risk of losing the setup in advance of the AUV's surveys a preliminary inspection by the Edge Tech DF 1000 towfish hauling in the tow cable was completed. Altogether 36 successful missions of the IVER-2 were launched (Table 3.1. and Fig. 3.10).

3.7 Acoustic Classifications of Sediments in the Area and Critical Aspects of Detection and Classification

A capability to recognise characteristics of seabed such as roughness acoustic reflectivity/scattering strength and mechanical shear strength is useful as a means of identifying those areas where object detection and characterisation are likely to be difficult; for example inside of soft sediments.

For some decades it has been accepted that character of the seabed surface, e.g. surface roughness, sediment type, grain size spectra, could be predicted on the basis of the different descriptors of acoustic echoes. When sound penetrates into sediments, the echoes also include information about the sediments layering.

Table 3.1. Statistics of detected and classified objects in different areas of the Baltic Sea with the AUV sonar

Deployment area	Data/Mission number/Total number of targets	1st category	2nd category	3-rd category
Bornholm basin	March 2016 /1/14	3	7	4
	March 2016 /2/17	3	7	7
	March 2016 /3/42	12	17	13
	March 2016 /4/87	21	30	36
	March 2016 /5/17	3	7	7
	March 2015/1/6	1	3	2
	March 2015/2/12	6	3	3
	March 2015/3/45	11	19	15
	March 2015/4/20	8	11	1
	March 2015/5/3		2	1
	March 2015/6/10	3	4	3
	March 2015/7/17	3	9	5
	March 2015/8/26	6	7	10
Gdansk deep	August 2014 /1/14	4	5	5
	August 2014/2/20	9	7	4
	September 2015/1/17	7	5	5
	September 2015/2/40	15	10	15
	September 2015/3/53	25	17	11
	September 2015/4/47	17	19	11
	September 2015/5/39	12	18	8
	September 2015/6/64	33	22	9
	September 2015/7/9	3	2	4
	March 2016/1/10	4	4	2
Gotland deep	August 2014/1/10	6	4	0
	August 2014/2/11	2	3	6
	August 2014/3/3	0	1	2
	August 2014/4/30	6	7	
	August 2014/5/10	5	1	4
	August 2014/6/9	3	3	3
	August 2014/7/3	0	2	1
Little belt	August 2015/1/8	4	3	1
	August 2015/1b/9	7	1	1
	August 2015/1c/8	4	3	1
	August 2015/2/3		1	2
	August 2015/3a/3b/5	2	3	3

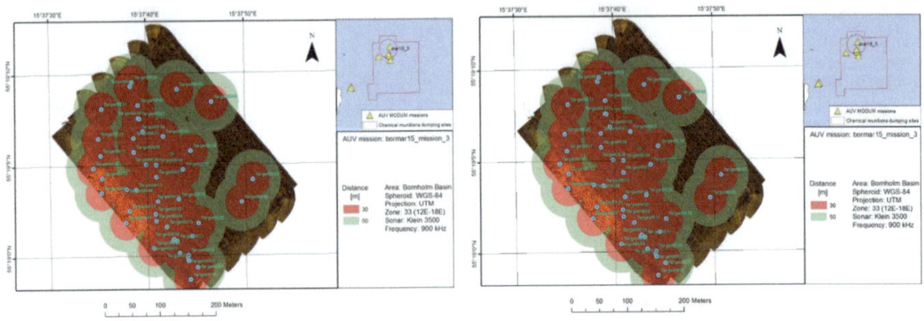

Fig. 3.10 Overview of the of space distribution of detected targets in the area of primary deployment of the CW in the Bornholm dumpsite obtained with the IVER-2

The spatial distribution of different properties of backscattering with vertical sounding at 30 kHz in the Polish economic zone of the Baltic Sea was performed in 90's (Klusek et al. 1994). It was stated that in the two deeps – the Gdansk Deep and the Bornholm Basin acoustic signals have penetrated deeply into the soft silty and clayey bottom. The characteristic duration of the echo signal from the bottom in the both Deeps is 40 or even 50 times longer in comparison with that of the soundings signals. It was also found that in the both Deeps there are areas with shallow gas presence in marine sediments of biogenic origin with characteristic pockmarcks.

In areas of non-gaseous fine sediments (e.g., the mud or fine sand) are in the uppermost bottom layer, the low frequency acoustic signal sounds deeper which leads to a longer time of response. It is reflected in the form of returned time series as the positively skewness (towards the deeper layer) of an echo. In the case of hard sediments placed in the uppermost layers, echo's maximum is at the beginning of time series and deeper layers are less visible in echograms.

As a measure of sediments transparency time of return between the first local maximum on the water-sediment boundary and absolute maximum in the echo and echo envelope skewness were selected as predictors of sediment layering (Klusek et al. 1994). This approach is presented here for 12 kHz single-beam ODOM echosounder.

Time of return between the first local maximum on the water-sediment boundary and absolute maximum in the echo within the study subarea range from zero to almost 25 ms and as is depicted in Fig. 3.11. Each point represents mean value for 500 consecutive pings for each of the subarea. In the upper part of the figure the transects are shown, down the bottom depth and colours as a measure of the sediments transparency are depicted (Fig. 3.12).

In the Gotland Basin munition deployment area, which is presented as the case study in shallower region hardy sediments prevail. Consequently, the most frequently detected mine-like targets are lying on the bottom surface. As the results of the local bottom currents the targets are frequently accompanied with a cavity in the ground. In the deeper part of the area, most likely the hard sediment is covered with soft sediments.

Probability of the detection and recognition of targets depends not only on the type of sediments but also on the distance from the sonar track. It was observed that

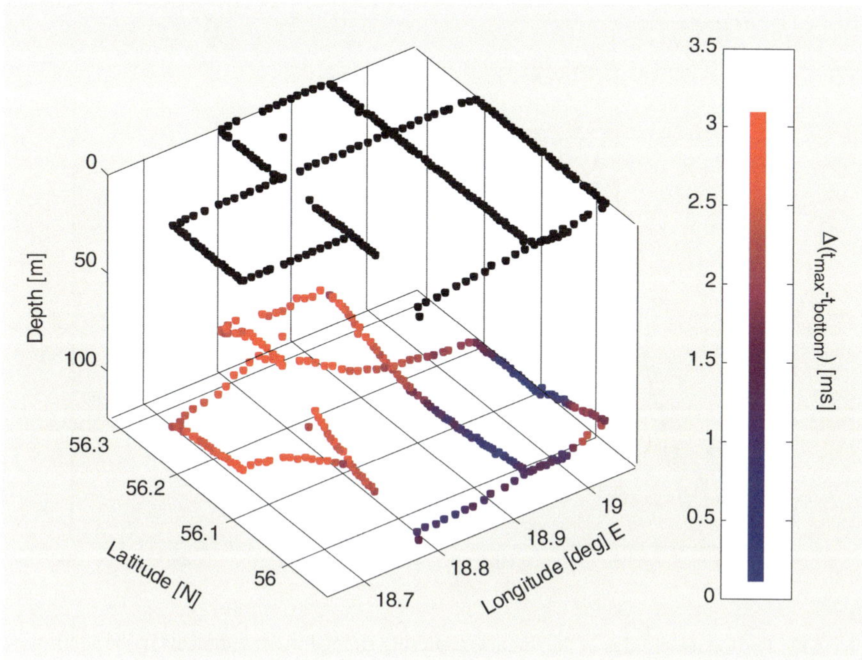

Fig. 3.11 Track of soundings at 12 kHz is with black points. One of the bottom echo parameters is given - time of return between the first local maximum on the water-sediment boundary and absolute maximum in the echo within the study subarea

the density of detected targets is not evenly spaced in relation to sonar's track. The highest surface density of findings is at some distance from the track and single targets are located along lines parallel to the track (as example as presented in Fig. 3.6 in Missiaen et al. 2010). A common problem encountered when surveying is that at low grazing angles small targets partly covered by sediments or found in trawl marks can be masked and their detection by the sonar operator or data analyst would be impeded. On the other side in the strip near nadir target's images are deformed and shadows are small or do not occur and small objects are not detected.

Critical aspects of probability of the detection and recognition of targets as the function of distance from a track was analysed at the compiled data base of objects detected in the Gotland Deep during the CHEMSEA project from R/V "Baltica" and S/Y "Oceania".

Histograms of number of detected targets as the function of the distance from the SSS transects confirms that in the slice of the seafloor almost directly under the towfish, we observe lower probability of small target detection. Similarly at the furthest distance from the towfish the also detection is declining.

This fact requires us to perform dense survey line spacing which ensure that the nadir zone is always viewed off to the side of an adjacent survey line pass. However in the case of the large-scale inventory, as during the CHEMSEA, we can only recal-

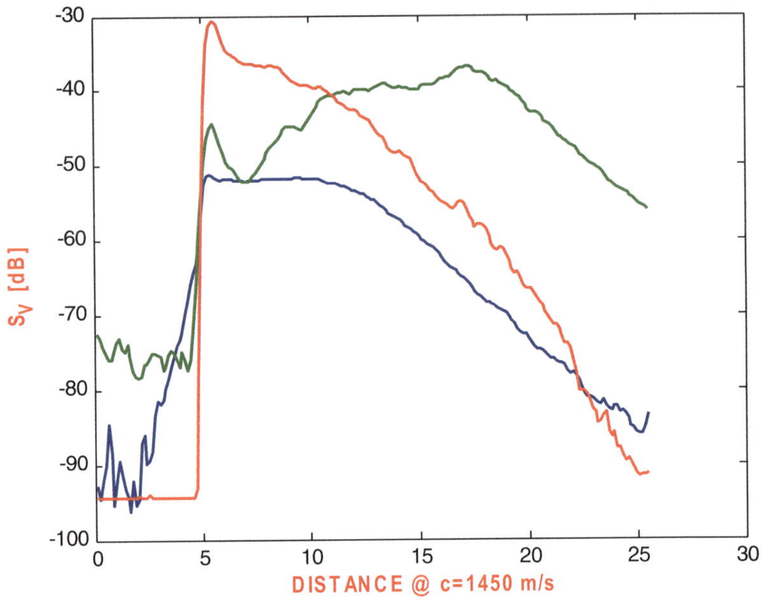

Fig. 3.12 Distinctive profiles of the volume scattering strength from sediments in the south-west of the Gotland Deep dumping area documented at 12 kHz. The red curve is for the most acoustically hardy sediments uncovered with mud, the blue one for the hard bottom covered with semiliquid acoustically transparent sediments. The green is for bottom with the greatest transparency

culated possible number of classified targets by the factor 9/6 the number found on the basis of images analysis. Similar rates are expected for other deployment areas.

Histogram of recognized as the first class targets with distance from the sonar track in selected area from the Gotland Deep is presented in Fig. 3.13. The highest number of detected by the Swedish Maritime Agency operators targets is situated at the distances around 100 m with diminishing number towards the nadir.

3.8 Summary

Detection and positioning of chemical munition at four grounds was performed with high resolution sidescan sonars launched at the IVER-2 AUV. It had allowed to take water and sediment probes for further analysis.

However, in the process the inventory of chemical munition in the Baltic deeps the success rate was diminished due to many factors, as:

- outside of area of primary deployment due to large number of old deployed and contemporary litter material on the sea bottom we face difficulties in classification of the IIWW objects;
- different munition types;

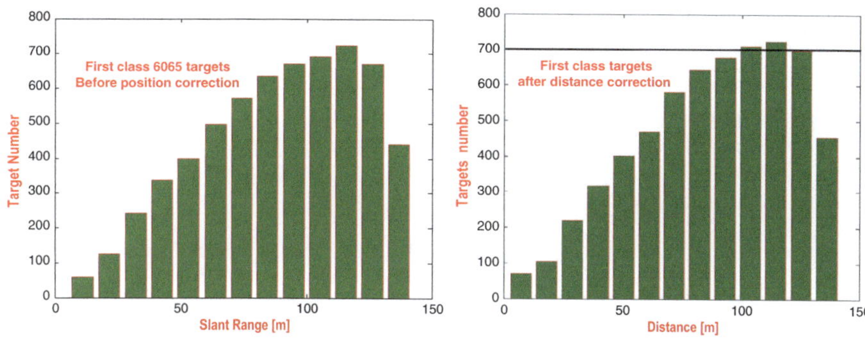

Fig. 3.13 Number of detected objects in relation to distance from the track of a sidescan sonar

- 60 years after deploying – shell eroded,
- some objects are partly covered by sediments;
- in some part of the area crossing trawling scars give false echoes.

The data being gathered in the frame of the CHEMSEA and MODUM projects collected in the database, with primary purposes cataloguing detected objects provide useful source of information for others future investigations.

Some expected applications of the acoustic images database in the future are –

environmental planning data base to help taking environmentally sound decisions by marine administrations to diminish anthropogenic impact;
basic academic marine research related applications as ecology (environmental zoning), sediment patchiness, or geomorphology due to higher resolution and bottom coverage by acoustic survey;
high data quality are assumed to meet the requirements as the base for better modeling of bottom currents, circulation processes in the area, prediction of diffusion of pollution from corroded munition or sediment's transport;
identification of small-scale morphological structures on the seafloor - such as ripples, stones, boulders gives valuable information for habitat mapping purposes;
data base shows different objects as sea mines, improves positioning of historical/archaeological findings as airplanes or wrecks;
due to observed high density of trawl marks should consequently improve the monitoring for fishing vessels operating within the CM-areas.

In the case of the Baltic deeps where high rates of sedimentations of organic matters are observed covering of items of munition by sediments is quite common. In this situation broadband, low frequency sonars should be promising method of detection and classification.

References

Badiey M, Mu Y, Simmen JA, Forsythe SE (2000) Signal variability in shallow-water sound channels. IEEE J Ocean Eng 25(4):492–500

Blondel P, Murton BJ (1997) Handbook of seafloor sonar imagery. Publisher John Wiley & Sons, Chichester, p 314

Blondel P (2009) The handbook of Sidescan sonar. Springer et Praxis Publishing, Berlin et Chichester, p 316

Bulczak AI, Rak D, Schmidt B, Beldowski J (2016) Observations of near-bottom currents in Bornholm Basin, Slupsk Furrow and Gdansk Deep. Deep-Sea Res II Top Stud Oceanogr 128:96–113

Fleming BW (1976) Side Scan Sonar: A Practical Guide. Int Hydrogr Rev 1 III(1):65–92

Haralick RM, Shanmugam K, Dinstein I (1973) Textural features for image classification. IEEE Trans on Systems, Man and Cybernetics SMC-3:610–621

Jackson DR, Winebrener DP, Ishimaru A (1986) Application of the composite roughness model to high-frequency bottom backscattering. J Acoust Soc Am 79:1410–1422

Klusek Z, Tegowski J, Szczucka J, Śliwinski A (1994) Characteristic properties of bottom backscattering in the southern Baltic Sea at ultrasound frequencies. Oceanologia 36(1):81–102

Missiaen T, Söderström M, Popescu I, Vanninen P (2010) Evaluation of a chemical munition dumpsite in the Baltic Sea based on geophysical and chemical investigations. Sci Total Environ 408:3536–3553

Tesei AWLJF, Maguer A, Løvik A (2000) Target parameter estimation using resonance scattering analysis applied to air-filled, cylindrical shells in water. J Acoust Soc Am 108:2891–2900

Tesei A, Maguer A, Fox WLJ, Lim R, Schmidt H (2002) Measurements and modelling of acoustic scattering from partially and completely buried spherical shells. J Acoust Soc Am 112:1817–1830

Williams KL, Grochocinski JM, Jackson DR (2001) Interface scattering by poroelastic seafloors: first-order theory. J Acoust Soc Am 110:2956–2963

Chapter 4
Chemical Analysis of Dumped Chemical Warfare Agents During the MODUM Project

Martin Söderström, Anders Östin, Johanna Qvarnström, Roger Magnusson, Jenny Rattfelt-Nyholm, Merike Vaher, Piia Jõul, Heidi Lees, Mihkel Kaljurand, Marta Szubska, Paula Vanninen, and Jacek Bełdowski

Abstract MODUM project continued the work on monitoring of the chemical weapons (CW) dumped in the Baltic Sea started in previous projects. As a new aspect, on board analysis methods – headspace gas chromatography-mass spectrometry (GC–MS) and capillary electrophoresis (CE) – were developed and tested in laboratory conditions and during cruises. The GC–MS method could be successfully applied on board to verify that collected sediment samples contained degradation products for sulfur mustard, one of the major chemical warfare agents dumped in Baltic Sea. This method could in future project be used during cruises to redirect sample collection in order to make most of the available ship time. Other part of the analysis task during MODUM project was the work done at the reach back laboratories. These analyses were done to both verify the results obtain on board and to fully identify the chemicals related to the sea-dumped CW agents. Reach back analysis of CW-related chemicals were done on sediment samples collected around a wreck in Bornholm Deep (same samples as analyzed on board) and on monitoring samples collected in Bornholm, Gotland and Gdańsk Deeps. The samples from Bornholm and Gotland Deeps are in line with previous findings. Samples from

M. Söderström (✉) • P. Vanninen
Finnish Institute for Verification of the Chemical Weapons Convention, University of Helsinki, Helsinki, Finland
e-mail: martin.soderstrom@helsinki.fi; paula.vanninen@helsinki.fi

A. Östin • J. Qvarnström • R. Magnusson • J. Rattfelt-Nyholm
Swedish Defence Research Agency, Umeå, Sweden
e-mail: anders.ostin@foi.se; johanna.qvarnstrom@foi.se; roger.magnusson@foi.se; jenny.nyholm@foi.se

M. Vaher • P. Jõul • H. Lees • M. Kaljurand
Tallinn University of Technology, Tallin, Estonia
e-mail: merike@chemnet.ee; piia.joul@gmail.com; heidi_lees@hotmail.com; mihkel@chemnet.ee

M. Szubska • J. Bełdowski
Institute of Oceanology, Polish Academy of Sciences, Sopot, Poland
e-mail: szubi@iopan.gda.pl; hyron@iopan.gda.pl

J. Bełdowski et al. (eds.), *Towards the Monitoring of Dumped Munitions Threat (MODUM)*, NATO Science for Peace and Security Series C: Environmental Security, DOI 10.1007/978-94-024-1153-9_4

Gdańsk Deep are in line with previous findings that this area has been used as a dump site. Additionally, α-chloroacetophenone (CN) was found in the area for the first time. In addition to the analysis of CW-related chemicals, a new method was developed for measurement for arsenic concentrations in sediment samples. A method was also developed for arsenic speciation, which could help in estimation of the source of arsenic in the sediments.

Abbreviations

AED	atomic emission detector
AMDIS	Automatic Mass spectral Deconvolution and Identification Software
APCI	atmospheric pressure chemical ionisation
Asb	arsenobetaine
BGE	background electrolyte
BPA	butylphosphonic acid
C^4D	capacitively coupled contactless conductivity detector
CBRN	chemical, biological, radiological, nuclear
CE	capillary electrophoresis
CHEMSEA	Chemical Munitions Search and Assessment, an EU-funded project
CW	chemical warfare
CWA	chemical warfare agent
CWC	Chemical Weapons Convention
DMA	dimethylarsine
EDEA	ethyldiethanolamine
EEZ	extended economic zone
EMPA	ethyl methylphosphonate
ESI	electrospray ionization
FOI	Swedish Defence Research Agency, Umeå, Sweden
GC	gas chromatography
GC–MS	gas chromatography-mass spectrometry
GC–MS/MS	gas chromatography-tandem mass spectrometry
HD	sulphur mustard
HRMS	high resolution mass spectrometry
ICP-MS	inductively coupled plasma mass spectrometry
IOPAS	Institute of Oceanology of the Polish Academy of Sciences
L	Lewisite
LC–HRMS	liquid chromatography-high mass spectrometry
LOD	limit of detection
MDEA	methyldiethanolamine
MMA	monomethylarsine
MODUM	Towards the Monitoring of Dumped Munitions Threat, a NATO-funded project
MPA	methylphosphonic acid

MS	mass spectrometry
MUT	Military University of Technology, Warsaw, Poland
PMPA	propyl methylphosphonate
PPA	propylphosphonic acid
ppb	part-per-billion (e.g. μg/kg)
PrSH	propane-1-thiol
SIM	selected ion monitoring
SRM	selected reaction monitoring
TDG	thiodiglycol
TDGO	thiodiglycol sulfoxide
TDGOO	thiodiglycol sulfone
TEA	triethanolamine
TTÜ	Tallinn University of Technology, Estonia
UHPLC	ultra-high performance liquid chromatography
VERIFIN	Finnish Institute for Verification of the Chemical Weapons Convention, University of Helsinki, Finland
WWI	First World War
WWII	Second World War
XRF	X-ray fluorescence spectrometry

4.1 Introduction on Sea-Dumped CWAs in the Baltic Sea

Chemical warfare (CW) is strongly associated with the First World War (WWI) during which the technological development allowed the production of large quantities of chemicals required for weaponization. These weapons were used in attempts to break or defend static defense lines of the trench warfare (Coleman 2005). After WWI, extensive development of tactics and stockpiling of chemical weapons took place by several states. Therefore, at the beginning of the Second World War (WWII), there were large quantities of weapons stored either in containers or directly in munitions. However, due to the modernization of tactics, e.g. German Blitzkrieg, with fast moving units resulted in making the chemical weapons less useful. One consequence of this was that at the end of the war there were large unused stockpiles of chemical munition in most countries taking part in the war (Coleman 2005; Hammes 2004).

The authorities of the occupation zones in Germany reported that they had found ca. 300,000 t CW materials (BSH 1993). According to the Potsdam agreement (1945) Germany should be disarmed and those responsible for each occupations zone shipped the munitions to remaining fronts or destroyed them. Several methods of destruction were tested and about 85% of the seized chemical weapons were dumped at sea (Arison 2013). The allies controlling western Germany – US and British authorities – dumped the CW materials in Skagerrak while Soviet authorities controlling eastern parts of Germany dumped the materials in the Baltic Sea (Knobloch et al. 2013).

4.1.1 Dumping Sites, Munitions and Chemicals

The chemical warfare agents (CWA) dumped in the Baltic Sea included sulfur mustard (HD), Clark I (possibly also Clark II), Adamsite, α-chloroacetophenone (CN), phosgene, tabun (GA), phosgene and, possibly, nitrogen mustard. Dumping of Lewisite I is mentioned in some reports, but this has not been confirmed. Lewisite I would typically contain also Lewisite II and III. Some known and verified mixtures contained technical grade Clark I, which is called arsine oil. It is known that Germans preferred to add arsine oil in sulphur mustard to produce winter-grade agent, while Russia and USA preferred Lewisite as the additive (Arison 2013). The structures of dumped chemicals are shown in Fig. 4.1. It should be remembered that also conventional munition containing large amounts of explosives were dumped. Additionally, chemical munitions contain also explosives for dissemination the chemical agent. However, explosives are not discussed here.

The dumped CW are not covered by the Chemical Weapons Convention (CWC; entry into force April 29, 1997) since this convention does not include materials that were buried before 1977 or dumped at sea before 1985 (OPCW 1993). Dumping of hazardous waste, including chemical and conventional munitions, has been a method of disposal until the London Convention of 1972 entered into force in 1975 (IMO 1972). Further limitations to sea dumping of waste were set London Protocol of 1996, which entered into force in 2006 (IMO 1996).

The CWA will typically degrade in contact with water by hydrolysis. Additionally, sulfur and arsenic-containing chemicals will produce oxidation products in environment. For arsenic-containing chemicals hydrolysis occurs fast in aqueous conditions.

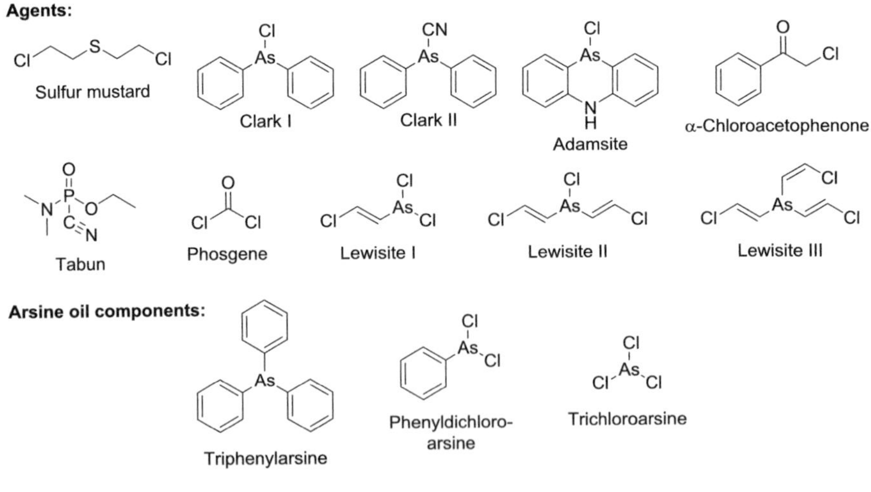

Fig. 4.1 Structures of the chemical warfare agents dumped in the Baltic Sea

Sulfur Mustard The sulfur mustard is an agent that causes chemically induced blister on the skin and affects also eyes and lungs. Mustard causes casualties requiring medical treatment rather than working as a directly lethal agent. In its pure form mustard agents is a colourless liquid, with melting point of 14 °C. For technical mixtures, the melting point is lower, but still above 0 °C. Often, in order to improve the physical properties sulphur mustard was produced in mixtures e.g. with arsine oil (Germany) or Lewisite (United States and Russia).

In typical environmental samples, sulfur mustard is expected to form – through hydrolysis – thiodiglycol, (Munro et al. 1999) which can then oxidize. In sediment samples, thiodiglycol and its oxidation product, thiodiglycol sulfoxide, are seldom found. Instead mustard heel-related compounds are observed. Mustard heel consists of water soluble degradation-polymerisation products – first observed in stored mustard munitions – and cyclic degradation products, *e.g.* 1,4-dithiane and 1,4-oxathiane (Rohrbaugh and Yang 1997; Wagner et al. 1999). These cyclic chemicals have frequently been used as markers for sulfur mustard leaking objects (Tørnes et al. 2006). Degradation of mustard into cyclic products is shown in Fig. 4.2.

In dumped munitions, similar processes cause the hydrophobic sulphur mustards to polymerize on its surface in contact with water when the container deteriorates. The result of this is that sulfur mustard encapsulates itself. These polymer-covered lumps are occasionally caught by fishermen during trawling. Active mustard contained in the lumps may cause serious hazard to people coming into contact with the lumps. Analysis of the lumps have identified active mustard after 50 years on the sea floor. Additionally, by-products from the synthesis, e.g. sesqui mustard, and components of the arsine oil have been identified (Mazurek et al. 2001). This type of information can be helpful for investigations related to the origin of the munitions (Hanaoka et al. 2006).

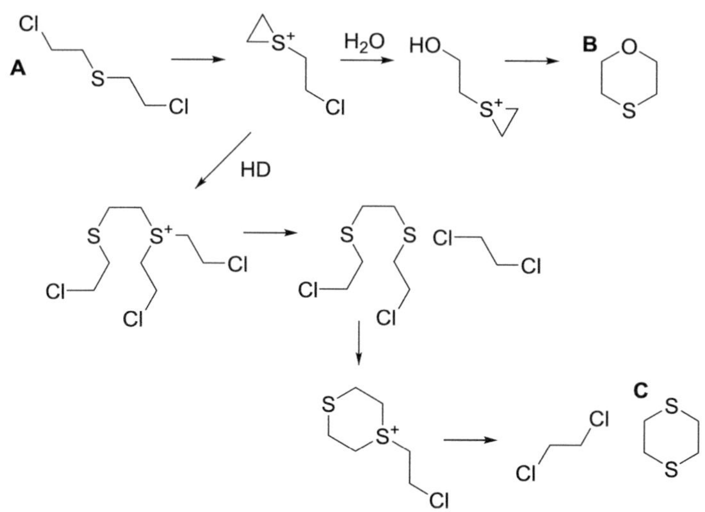

Fig. 4.2 Degradation of sulfur mustard (**A**) in marine sediment or munitions forming cyclic degradation products such as 1,4-oxathiane (**B**) and 1,4-dithiane (**C**)

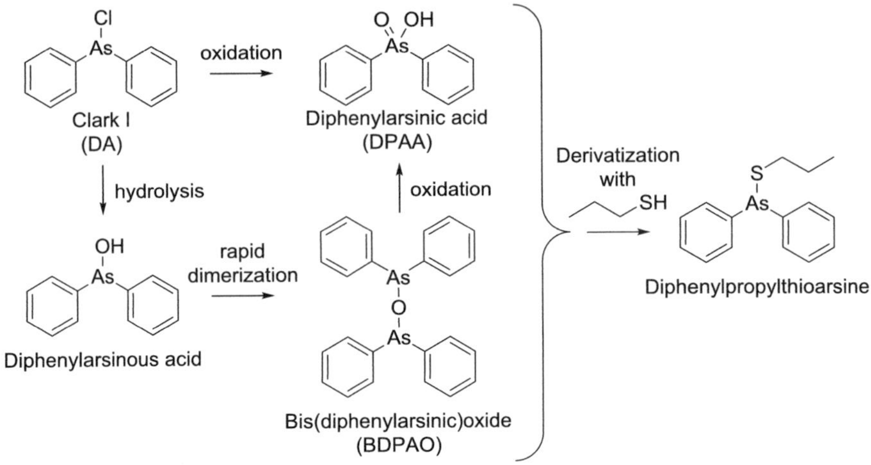

Fig. 4.3 Degradation and derivatization reactions for Clark I

Arsenic-Containing Chemicals Arsenic-containing chemicals have historically been classified as riot control chemicals. However, currently they are considered to be too toxic for any riot control purposes. German munitions could have been with filled Adamsite or Clark as the main agent. Additionally, winter-grade sulfur mustard munitions contained technical Clark I (arsine oil).

Chemicals with arsenic-chlorine bond hydrolyze in aqueous conditions quickly. For some chemicals, e.g. Clark I, the hydrolysis product can form a dimer. An arsenic-containing chemical and its hydrolysis product (as well as the dimer) can oxidize to form one common oxidation product. Oxidation using hydrogen peroxide can be as a sample preparation procedure to convert the various degradation products into the oxidized form. All these chemicals – including the chlorine-containing agent – can also be converted to propane-1-thiol (PrSH) derivative in sample preparation. Reactions for Clark I are shown in Fig. 4.3.

Based on observations, Clark I and Adamsite are seldom found as lumps. Based on analysis of sediment samples, their degradation products are spread widely on the sea floor.

Target Chemicals Based on known degradation reactions and experiences from previous studies a list of CWA-related target chemicals in Baltic Sea has been compiled. This list is presented in Table 4.1.

4.1.2 *Arsenic Concentration in Baltic Sediments*

Additional threat from the arsenical CWA is a possible elevation of arsenic levels in the environment close to chemical munition dumpsites. Although arsenic load to the Baltic Sea mainly results from natural sources (bedrock weathering, volcanic and

Table 4.1 Summary of target chemicals

Chemical	Structure	Analysis method (sample fraction)
Sulfur mustard (H)	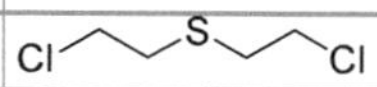	GC–EI/MS/MS (intact)
Thiodiglycol (TDG)	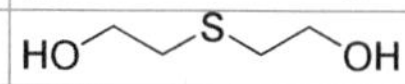	LC–API/MS/MS (intact + oxidized)
Thiodiglycol sulfoxide	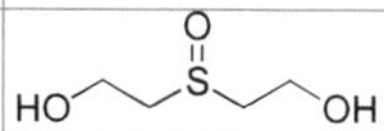	LC–API/MS/MS (intact)
1,4-Dithiane	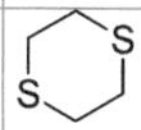	GC–EI/MS/MS (intact)
1,4-Oxathiane	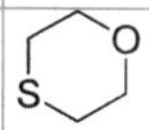	GC–EI/MS/MS (intact)
1,4,5-Oxadithiepane	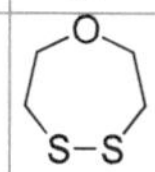	GC–EI/MS/MS (intact)
1,2,5-Trithiepane	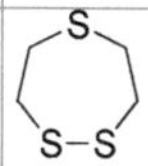	GC–EI/MS/MS (intact)
Adamsite (DM)	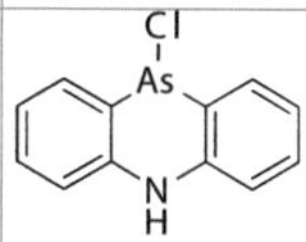	LC–API/MS/MS (intact)
5,10-Dihydrophenarsazin-10-ol 10-oxide	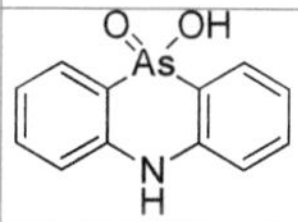	LC–API/MS/MS (oxidized)
Clark I (DA)	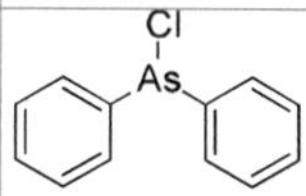	*Not analyzed*
Clark II (DC)	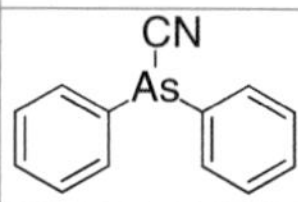	*Not analyzed*
Diphenylarsinic acid (DPAA)	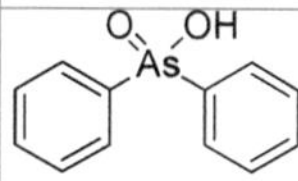	LC–API/MS/MS (oxidized)
Diphenylpropylthioarsine	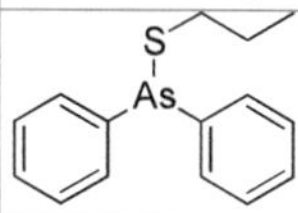	GC–EI/MS/MS (PrSH)
Triphenylarsine (TPA)	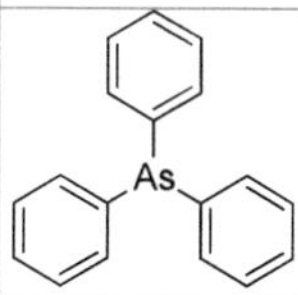	GC–EI/MS/MS (intact) LC–API/MS/MS (oxidized)
Triphenylarsine oxide	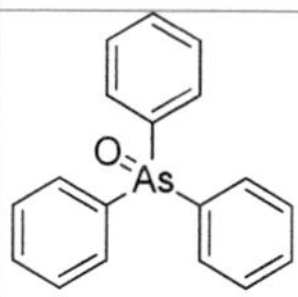	LC–API/MS/MS (oxidized)

(continued)

Table 4.1 (continued)

Chemical	Structure	Analysis method (sample fraction)
Phenyldichloroarsine (PDCA)	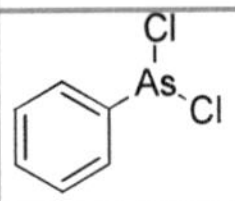	*Not analyzed*
Phenylarsonic acid (PAA)	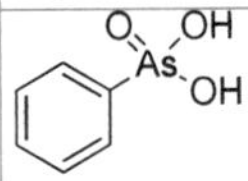	LC–API/MS/MS (oxidized)
Dipropyl phenylarsonodithioite	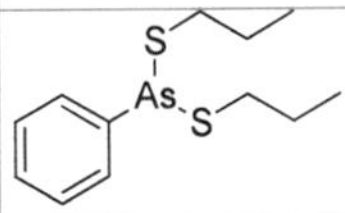	GC–EI/MS/MS (PrSH)
α-Chloroacetophenone (CN)	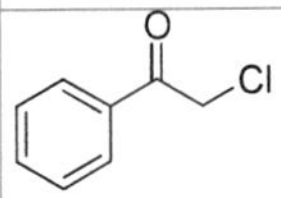	GC–EI/MS/MS (intact)
Lewisite I (L1)	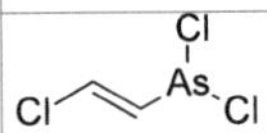	*Not analyzed*
Dipropyl (2-chlorovinyl)-arsonodithioite	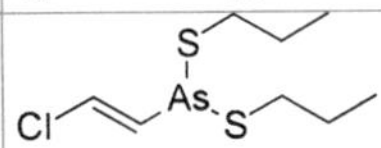	GC–EI/MS/MS (PrSH)
Lewisite II (L2)	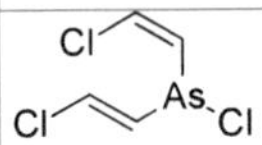	
Bis(2-chlorovinyl)arsinic acid	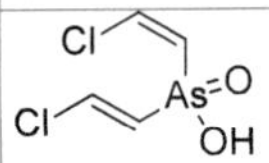	LC–API/MS/MS (oxidized)
Bis(2-chlorovinyl) propylthioarsine	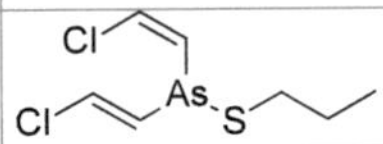	GC–EI/MS/MS (PrSH)
Tabun	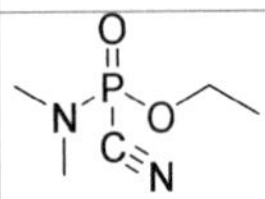	GC–EI/MS/MS (intact)

The agents are marked with **bold**

geothermal processes) and on-land anthropogenic activities (coal burning, smelting, mining, pesticide usage etc.), leakage from large amount of arsenic-containing CWA can be considered as a localized source of arsenic to the sediments.

Concentrations of total arsenic in Baltic Sea bottom sediments show high spatial variability and depend from number of factors, including: type of arsenic source, distance to the source, type and properties of the sediment, environmental conditions (e.g. redox conditions, pH). Different authors report background concentrations of arsenic in Baltic Sea sediments ranging from 5 to 20 µg/g.

In areas of the Kattegat, Belts Sea, Baltic Proper, Gulf of Riga and the Gulf of Finland (with prevailing aleurite-clay surface sediments) the mean arsenic concentration is 34 µg/g. The highest arsenic concentrations in the Baltic Sea are observed within the Bothnian Sea and range from 167 to 216 µg/g (Uścinowicz 2011). The main cause of such elevated values in this region is the Swedish heavy industry. Recently the quantity of arsenic released from this source into the Bothnian Bay have decreased due to emission restrictions, however the influence of past anthropogenic activity remains quite evident within the bottom sediments of the Baltic Sea (Borg and Jonsson 1996). In the eastern part of the Gotland Basin and in the Gdańsk Basin elevated values of arsenic occur locally, reaching 70 μg/g (Emelyanov 2007). In the southern Baltic Sea arsenic occurs in concentrations ranging from less than 5 μg/g up to 29 μg g in sediments. The concentration depends on the grain size. In sands the concentrations are usually below 5 μg/g, and in muddy sands they reach values between 5 and 15 μg/g. In aleurite clays the average value of arsenic concentrations is 12 μg/g, with higher concentrations in the surface clay layers of 15–29 μg/g. Maximum concentrations of arsenic measured in the Bornholm Basin in an unpolluted area reached 29 μg/g in fine, muddy sediments (Uścinowicz 2011).

Previous studies (e.g. MERCW, CHEMSEA) show, that there are significant differences in arsenic concentrations in sediments between areas under the influence of dumped CWA and distant from the dumpsites (Bełdowski et al. 2016a, b). Southern part of the Bornholm Basin is one of the most arsenic contaminated areas of the Baltic Sea (Uścinowicz 2011, MERCW 2006). One of the hot spots characterized by elevated concentrations of arsenic is situated in close vicinity to the chemical weapon dumpsite in the Bornholm Deep. There also appears to be a strong influence of fishing harbors based on normalized arsenic/iron concentrations (Uścinowicz 2011). In the Bornholm Deep area measured the highest arsenic concentration (210 μg/g) in fine grained pelitic mud sample (Sanderson et al. 2008). Emelyanov et al. (2010) found up to 210 and 277 μg/g arsenic in fine-grained sands to pelitic clayey mud and in aleuro-pelitic mud sediments, respectively, from chemical weapon dumping sites. In the most recent project CHEMSEA mean concentrations of arsenic in Bornholm Deep dumping area was established on 17 μg/g, which is not above the geochemical background value, but is still significantly higher than in surrounding reference areas (c.a. 5–6 μg/g) (Bełdowski et al. 2016a, b). In sediments from Gotland Deep, within the Lithuanian Exclusive Economic Zone (EEZ) also elevated values of arsenic concentrations were found near the chemical munitions dumping sites (average 10 μg/g) when compared to surrounding locations (average 2 μg/g). However these concentrations are relatively low compared to those reported in other Baltic and North Sea investigations (Knobloch et al. 2013, Garnaga et al. 2006). Bełdowski et al. (2016a, b) also reported elevated arsenic concentrations in sediments near the chemical munitions dumping sites up to 24 μg/g (average 13 μg/g) when compared to surrounding reference locations (average 6 μg/g).

As a result of microbial processes occurring on the sea bottom, arsenic can be introduced back to the biogeochemical cycle and accumulate in organism. Arsenic is assumed to be toxic to plants, animals and human and is considered as a potential carcinogen. It disrupts enzymatic processes in cells, causes cell walls breakdown, inhibits mitochondria functions, affects proteins formation by its high affinity to thiol groups, inhibits phosphate insertion to DNA, affecting transmission of genetic information (Harkabusová et al. 2009, Bissen and Frimmel 2003, Nicholas et al.

2003, Niedzielski et al. 2000). Chronic exposure to elevated arsenic concentrations can cause disturbance in nervous system and heart diseases. It is also assumed that bladder and lung cancer may be caused by chronic arsenic poisoning (Flora 2015). As fish and seafood are the main sources of arsenic in human diet it is very important to investigate and monitor the marine environment for arsenic levels (Flora 2015).

4.2 Chemical Analysis On-Site by Gas Chromatography – Mass Spectrometry

Sampling for environmental analysis is performed in the vicinity of interesting objects in order to study potential leakage. However, the targeting of the sampling equipment in 100 m deep water is most often difficult and a direct hit on an object which may contain high levels of toxic material cannot not be excluded. An example of such object is shown in Fig. 4.4. Furthermore, the condition of deteriorating dumped objects often makes it difficult to determine if they are of relevance or not. Especially, thickened sulfur mustard will form solidified lumps that are difficult to distinguish from other objects on the sea floor. The sampling missions in marine environment are costly due to ship and personnel time. Therefore, it is of interest to have capability during on ship analysis to detect sulfur mustard-related chemicals. This way the sampling team would know already at sea how to safely handle and pack the samples. Additionally, they can redirect the sampling efforts, if neccessary.

Fig. 4.4 Image of a dumped object from MODUM project

On ship screening of samples can be done with detector equipment developed for Hazmat-teams/CBRN-units in order to rapidly indicate predicted toxic compounds. However, these detectors are developed for detection of CWA in military context and must be applied with caution when used for dumped munitions. For example, the flame photometric detector AP2C (Proengin, Saint Cyr l'Ecole, France) indicates vapour containing the element sulphur which can be a detection of both sulfur mustard and hydrogen sulphide from anaerobic degradation of organic matter in the sediment. Ion mobility mass spectrometry is also a common technique for the detection of chemical warfare agents. During sediment sampling, this type of detectors will warn if a direct hit on active material is collected. However, these detectors are most often not fit for purpose since sediment contains mostly hydrolysed products at ppb-levels.

On the other hand, the development of ruggedized laboratory instrument provides the sensitivity and selectivity required for on ship analysis for identification and semi quantification of hot spots of sulfur mustard degradation products (Fish et al. 2010). The use of such instruments can help sampling teams to focus on relevant dumped objects and wrecks. Obtained analysis results may later be validated by sending the samples to reach back laboratory to be analysed with hyphenated mass spectrometric techniques *e.g.* GC–MS/MS and LC–HRMS after extraction of sediment.

4.2.1 On Ship Analysis with Deployable Headspace GC–MS

Analysis of sulfur mustard and its degradation products from sediments collected at WWII dumpsites in Skagerrak and Baltic Sea has been demonstrated with headspace trap gas chromatographic-mass spectrometric (GC–MS) analysis on ship and in reach back analysis (Fig. 4.5) (Magnusson et al. 2016, Røen et al. 2010). One advantage of performing this analysis on ship is that it will provide information onsite that is crucial for effective use of expensive ship time at sea during sampling campaigns.

The ruggedized GC–MS system used during the MODUM project was originally developed to be used by first responder handling hazardous material. The system includes software (Automatic Mass spectral Deconvolution and Identification Software (AMDIS, National Institute of Standards and Technology, Gaithersburg, MD, USA) that can automatically work through obtained data-files and provide the analysis results. Thus, this will provide a rather unexperienced user with rapid data interpretation and results. This software was originally produced as a tool for identification of the CWC but can be adapted for any GC–MS analysable compounds. For an experienced user the ruggedized gas chromatograph mass spectrometer will provide gas chromatographic (GC) properties with rather broad peaks (due to technological limitations) and normal quadrupole full scan mass spectrometry (MS) data. Thus, the on ship analysis may be performed at two levels dependening on the skill of the operator. It is preferred that the sampling team handle the analytical

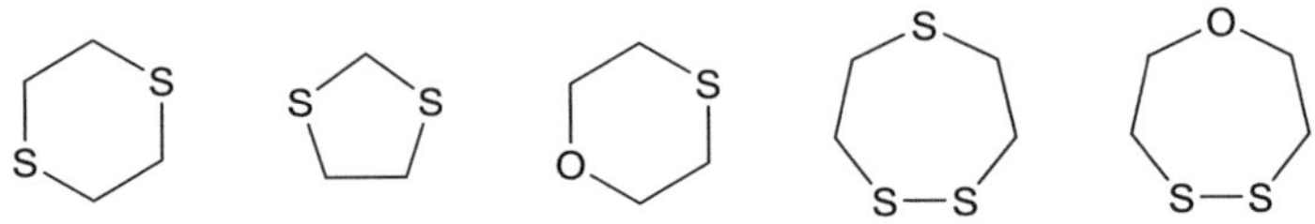

Fig. 4.5 Sulfur mustard metabolites 1,4-dithiane, 1,3-dithiolane and 1,4 thioxane, 1,2,5-trithiepane and 1,4,5 oxadithiepane identified by Røen et al. (2010) and Magnusson et al. (2016)

equipment themselves in order to minimize the involved crew on board. In this case the team will use automatized data convolution for the identification of target chemicals based on linear retention index and good spectra match factors (>85%) from the analysis. During the MODUM project the analytes in Fig. 4.5 were identified using full scan analysis. The limit of detection using automatic AMDIS identification was 15–80 ppb (dependent on analyte). Sending the in-field collected sample files for further data processing by experienced chemist will decrease the limits of detection to 2–10 ppb and semi-quantification can be performed based on in-field added internal standard. Results from an investigation of wrecks at Bornholm deep is shown in Fig. 4.6. As compared to reach back analysis performance, the on-site analysis may fail to identify all individual compounds in each sample. However, the analysis will still provide identification of sulfur mustard leakage in the sample which is fit for purpose in the field.

The headspace trap GC–MS method for on-ship analysis required extensive method development compared to the pre-set first responder methodology (Magnusson et al. 2016). The principle of the method is described in Fig. 4.7. To detect sulfur mustard markers at ppb levels in sediment all steps in the analysis had to be optimized. In the field it is desired to use minimum of sample manipulation. The only sample preparation used during the MODUM cruise was centrifugation to remove the pore water from the collected sediment. Thereafter, headspace GC–MS analysis was performed on the sediment fraction.

4.3 Analysis of Chemical Warfare Agents Degradation Products by Portable CE Instrument

4.3.1 Brief Overview of Capillary Electrophoresis

Capillary electrophoresis (CE) is an analytical technique that separates ions based on their electrophoretic mobility with the use of an applied voltage. Separation takes place in a narrow bore capillary (inner diameter 10–100 um) typically made from fused silica, less frequently from borosilicate glass or Teflon. Separation of analytes by CE is based on two electrokinetic phenomena – electrophoresis and electroosmosis. The electrophoretic mobility is dependent upon the charge of the molecule, the viscosity of background electrolyte (BGE), and the atom's radius. The rate at which the particle moves is directly proportional to the applied electric field – the greater

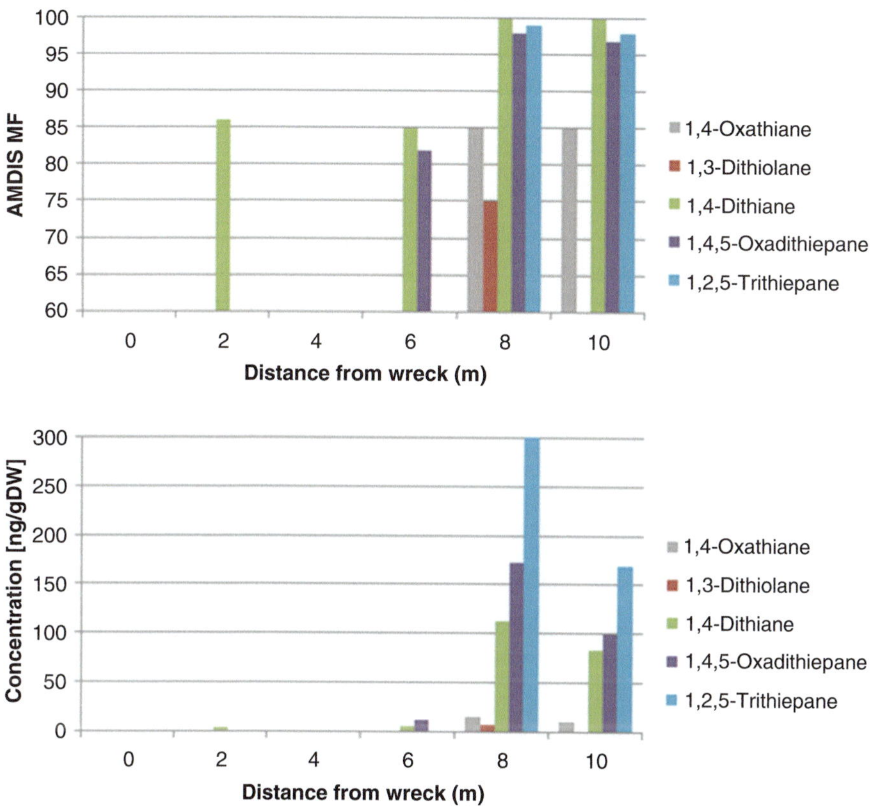

Fig. 4.6 On ship screening for sulfur mustard degradation products. (*on top*) identification of products based on match factor. For matches under 85% manual verification is recommended. (*bottom*) The same sample-data processed for semi-quantification. (Actual sampling spots can be seen in Fig. 4.11)

the field strength, the faster the mobility. Neutral species are not affected, only ions move with the electric field. If two ions are the same size, the one with greater charge will move the fastest. For ions of the same charge, the smaller particle has less friction and overall faster migration rate. CE is used most predominately because it gives faster results and provides high resolution separation. It is a useful technique because there is a large range of detection methods available (Li 1992).

4.3.2 Overview of Analysis of Chemical Warfare Agents and Its Degradation Products by Portable CE Instruments

CE has many advantages over chromatographic separation techniques in terms of potential miniaturization. To name just a few, the CE device consists only of a high-voltage supply, a capillary, small background electrolyte vessels and a detection

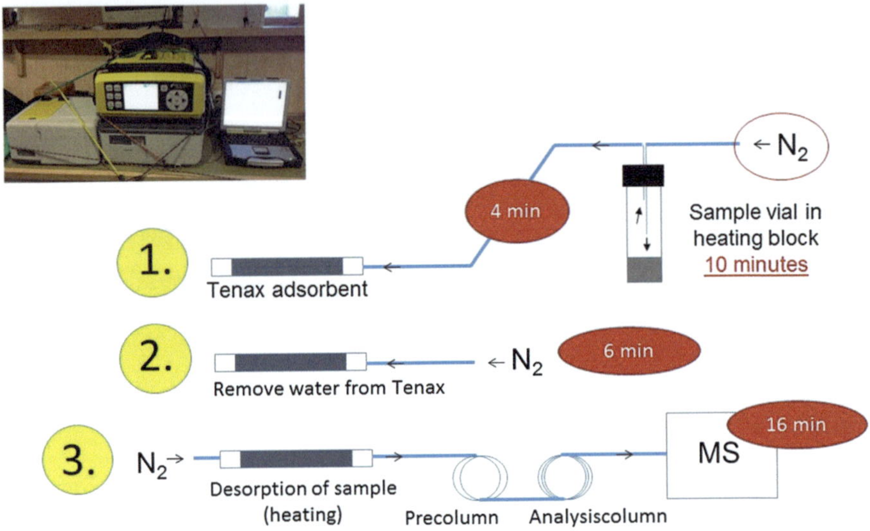

Fig. 4.7 Headspace analysis principle: (*1*) Headspace vapour from preheated sample is concentrated on an in-built tenax concentrator tube, (*2*) excess of accumulated water on the Tenax concentrator from sampling has to be removed with drying gas (N_2) optimised to minimize the losses of analytes, (*3*) desorption and GC–MS analysis

system. What is most important is that CE requires no use of complex mechanical high-pressure pumps as liquid chromatography does. Amperometric, potentiometric, and conductometric detection systems may all be used in portable capillary or microchip electrophoresis devices (Seiman et al. 2009). Conductivity detection is probably best suited to an on-site, portable CE device, because it can be miniaturized, power consumption is minimal, and it is relatively sensitive to the compounds of interest, especially nerve agent degradation products (Kubáň et al. 2011).

There are many papers in the literature where sulfur and nitrogen mustard degradation products have been analyzed by conventional CE system, while there are few papers about the separation of these compounds with portable CE instruments. Sáiz et al. (2013) developed a method with portable CE instrument and with contactless conductivity detector to determine nitrogen mustard degradation products (MDEA, EDEA, TEA) in different water samples. The method permitted the detection of the three nitrogen mustard degradation products down to 5 μM. VanderNoot et al. (2012) demonstrated an automated compact electrophoresis system suitable for on-site analysis of samples containing chemical warfare blister agents, like sulfur mustard (HD) and lewisite (L), in aqueous samples. The sensitivity via UV absorbance was modest, however, direct detection did not need any chemical derivatization and the analysis time was rapid – less than 6 min.

However, besides analysis of sulfur and nitrogen mustard degradation products, there are quite many articles in the literature about nerve agents and their degradation products separated and detected by portable CE instruments. Some of these articles are depicted here. Kubáň et al. (2011) demonstrated a portable CE system with contactless conductometric detection for rapid and accurate identification and quantification of genuine nerve agents (sarin, soman, VX) that have been deposited on various matrices (concrete, tile, soil, and vegetation). The portable CE system was used for the in situ analysis of the extracted samples. The overall duration of the process including instrument start-up, sample extraction and analysis was less than 10 min. Makarõtševa et al. (2010) developed simple procedures for the extraction of phosphonic acids from different surfaces. However, this study was mainly focused on qualitative analysis. Seiman et al. (2009) tested a fully portable CE device equipped with a capacitively coupled contactless-conductivity detector (C^4D) and a cross-injection device in laboratory conditions. The portable device could work on batteries for at least 4 h. The detection limits for different phosphonic acids were in the range of 2.5–9.7 μM.

Capillary electrophoresis microchips have also been used for the separation and detection of CWAs. The miniaturized systems, often referred as "lab-on-a-chip" devices, are attracting due to their high speed of separation, small system size, low weight and cost, minimal consumption of reagents and sample. The main CWAs separated with these kind of systems are nerve agents and its degradation products (Heleg-Shabtai et al. 2006, Pumera 2006, Wang et al. 2001, 2002, 2004), but degradation products of nitrogen mustard have also been separated (Ding and Rogers 2008).

Since CWAs and their degradation products have different chemistry and occur in a variety of matrices, many analytical methods are being accepted in inspecting laboratories, among which CE has found a niche. Fitted with various detection techniques, CE provides satisfactory limits of detection (LOD) to make it realise the analysis of real samples, especially when enforced with in-line pre-concentration (Aleksenko et al. 2011).

4.3.3 Tallinn University of Technology Research in MODUM Project

The main part of Tallinn University of Technology (TTÜ) research in MODUM project was to develop a suitable method for analysing some of the sea dumped chemical warfare agent degradation products by a portable instrument and in case of success test these analysis protocols the sites of suspected dumping.

As there are already available some analysis protocols for analysis of nitrogen mustards and nerve agents degradation products (Seiman et al. 2009, Kubáň et al. 2011), TTÜ research group applied these protocols for seawater and sediments porewater samples and the results were positive. The LOD values for five nerve agent hydrolysis products (PMPA, BPA, PPA, EMPA and MPA) were between 6.8 and 19.8 μM and for three nitrogen mustards hydrolysis products (EDEA, MDEA and TEA) were between 4.1 and 5.2 μM. All these analysis were performed by portable CE-C^4D instrument, which has been developed in TTÜ (Fig. 4.8). For on

Fig. 4.8 A portable CE-C4D instrument

board analyses the most important is the minimum size and weight of the equipment, which allows rapid and easy transportation. Therefore, a portable CE-C^4D instrument made by TTÜ is perfect, because its dimensions are 30 × 30 × 15 cm and weighs only 3.5 kg.

Since a large number of the dumped munitions in the Baltic Sea contain sulfur mustard and in aqueous environment it hydrolysis to TDG, TDGO and TDGOO, TTÜ research group developed the CE protocols for the separation of the hydrolysis and oxidation products of sulfur mustard (TDG, TDGO and TDGOO). The developed method has the potential to be transformed into a portable CE format. The baseline separation of three derivatives was achieved in less than 8 min by applying a background electrolyte composed of a 30 mM borate buffer at pH 8.5. The LODs of the TDG and its oxidation products were in the range of 98–154 ng/mL. The developed method was successfully applied for the analysis of TDG and oxidation products in seawater and pore water, utilizing the carbon aerogel-based adsorbents for sample purification and concentration (Jõul et al. 2015).

Taking into account the fact that portable methods have many benefits TTÜ research group developed a portable analysis method for the simultaneous determination of the hydrolysis products of nitrogen mustards (EDEA, MDEA, TEA) and the hydrolysis and oxydation product of sulfur mustard (TDG, TDGO, TDGOO) in seawater and pore water samples using a CE-C^4D instrument (Fig. 4.8). A solid-phase extraction, utilizing the carbon aerogel-based adsorbents for sample purification and concentration, and derivatization with phthalic anhydride was used to minimize the values of limit of detection (Kubáň et al. 2011). The separation for all six derivatives was achieved in less than 15 min, but the sample preparation is still time-consuming. For electrophoretic baseline separation of all six derivatives a BGE composed of 20 mM Tris, 0.1 M Glycine, 10% acetonitrile was used. The LOD for the analytes was 1.0–2.0 μM and the limit of quantitation was 3.3–6.7 μM.

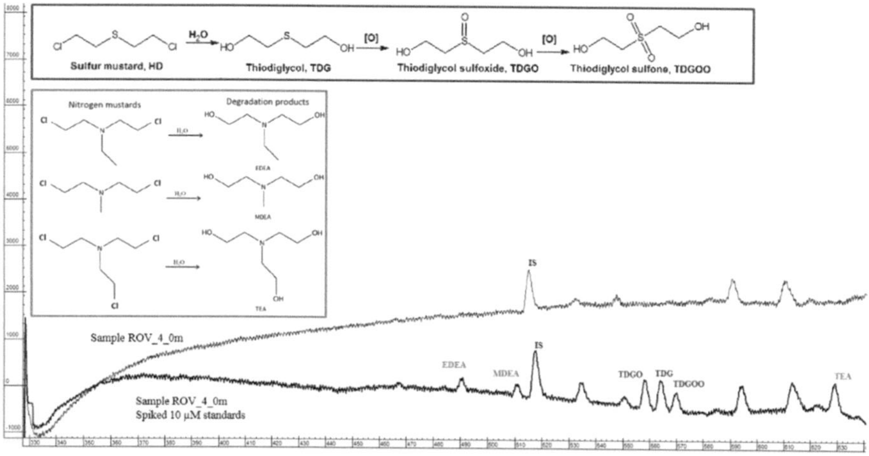

Fig. 4.9 Representative electropherograms obtained from the MODUM Bornholm cruise in March 2016 and a sulfur and nitrogen mustards degradation pathways

As planned the portable CE instrument and analysis method was tested during the MODUM cruise aboard the research vessel R/V Oceania (Institute of Oceanology of the Polish Academy of Science, IOPAS, Poland) in the Bornholm Basin between the 12th and 16th of March 2016.

All aboard analysed pore water samples (5) from the Bornholm Basin showed no traces of EDEA, MDEA, TEA or TDG, TDGO, TDGOO (Fig. 4.9). The samples were reanalysed in a laboratory using the portable CE-C^4D instrument and a conventional stationary CE–UV instrument (Kubáň et al. 2011) and it confirmed the results.

The portable CE-C^4D instrument worked well during the on board analysis and therefore it is suitable for the detection of EDEA, MDEA, TEA and TDG, TDGO, TDGOO. It has also confirmed that the method for the determination of the degradation products of nerve agents can be used for aboard analysis as well.

To conclude, a CE is not the most common analysis technique for determination of CWA degradation products, but it seems to be a very promising due its simplicity and robustness.

4.4 Reach-Back Laboratory

Beside the sulfur mustard, the World War II dumped chemical munitions contain arsenic based agents and sulfur mustard additives. To German WWII mustard was arsine oil (phenyl arsines) added while Russia and USA preferred Lewisite as an additive to sulfur mustard (BSH 1993). These additives improves the immediate impact of the sulfur mustard and lowered the freezing point of the agent mixture. Today, these additives will carry the information of the origin of the dumped sulfur mustard and most reports from the Baltic Sea demonstrates, as expected, the presence of arsine oil. The vomiting agent Adamsite and the teargases Clark I and II have also been dumped into the Baltic Sea.

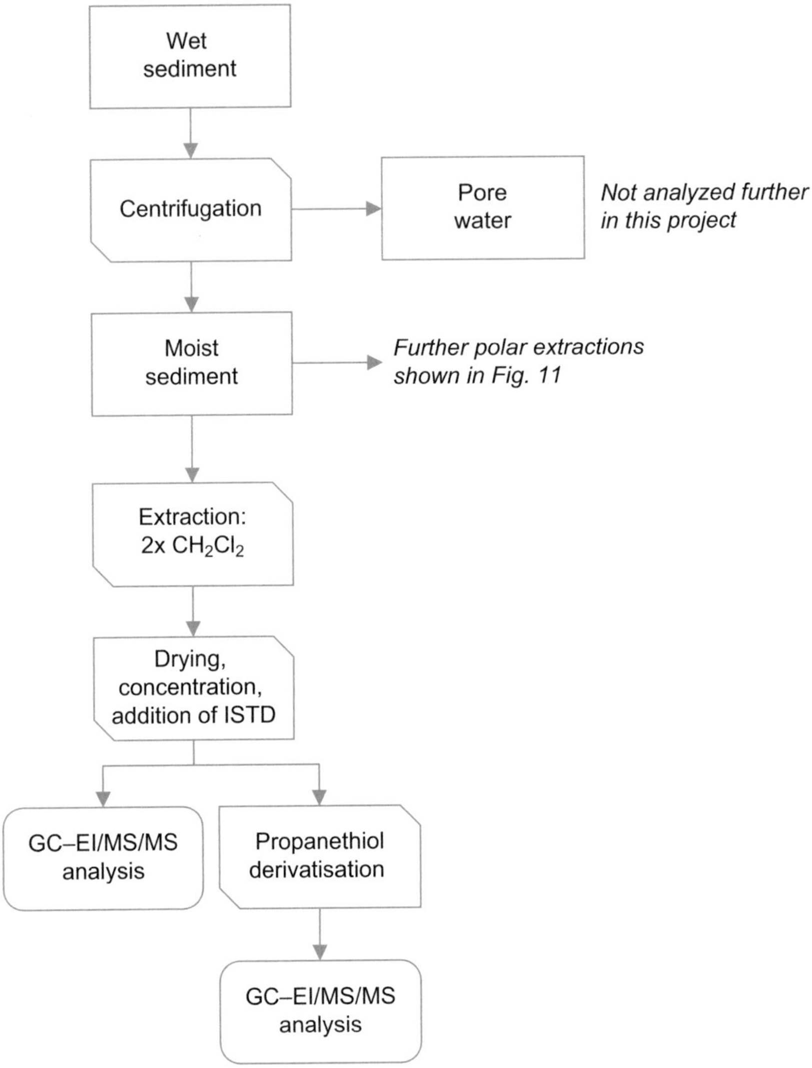

Fig. 4.10 Sample preparation of non-polar extractions for GC–EI/MS/MS analysis of intact and propanethiol derivatised target chemicals. This procedure was used by both FOI and VERIFIN

Extensive analysis of traces of dumped agents in environmental samples has been done during the MERCW, CHEMSEA and MODUM-project investigations of the dumping sites of the Baltic Sea. The preferred matrix for laboratory analysis was the sea sediment after removal of the pore water (*moist sediment*). The sample preparation of the sediment was based on two different types of extractions: non-polar and polar. The former was done using dichloromethane as solvent and was aimed for analysis by GC–MS-based analysis. Both FOI and VERIFIN used the same approach shown in Fig. 4.10. The latter extraction was done using acetonitrile – in

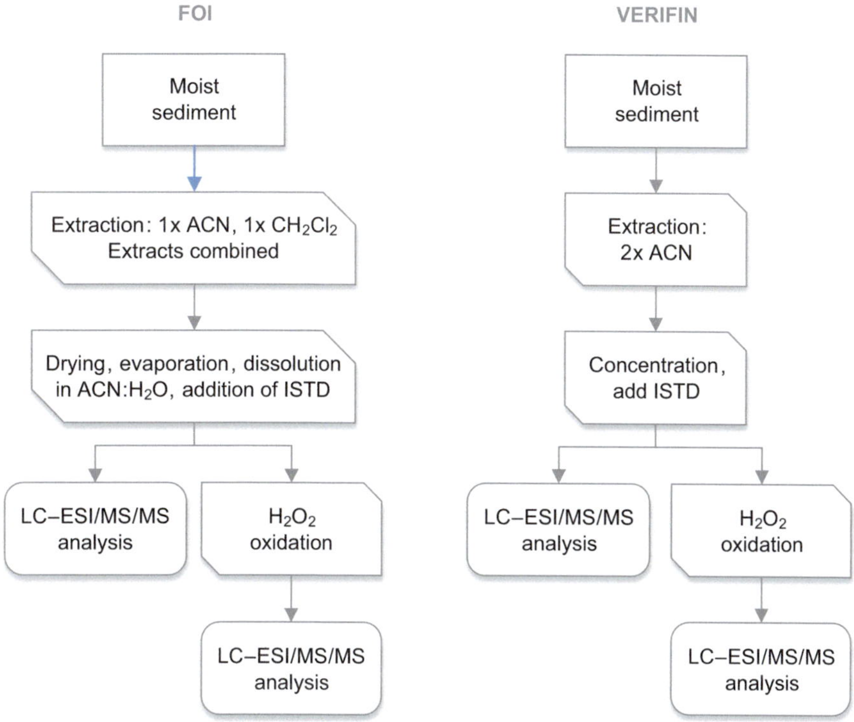

Fig. 4.11 Sample preparation of polar extractions for LC–ESI/MS/MS analysis of intact and oxidized derivatised target chemicals

FOI followed by dichloromethane extraction – mainly for LC–MS-based analysis. The polar extraction procedures are shown in Fig. 4.11.

Pore water will in principle reflect the composition of the constituents of the sediment sample but at a lower concentration. The chemicals present in the pore water has been considered to be bioavailable as they may be diffusing from the sediment to the water layer above the sediment (*near bottom water*). The chemicals could be transported further of sea animals, e.g. cod could be exposed to these chemicals when swimming in this water layer. Even though it is possible to use the pore water after a simple filtration to screen for hot spot samples with LC-MS, the concentrations may be too low for some instruments.

Separation techniques such as GC and LC combined with mass spectrometry is the state of the art tool for trace analysis in complex matrixes. The analysis of degrading war gases in sediments will require capability to detect and measure part-per-billion (ppb) levels of contaminants and even though single sector instruments have the required sensitivity, the selectivity of the instruments will be problematic. Therefore, tandem mass spectrometry is preferred since it will provide both the sensitivity and selectivity required for the screening of potential contaminates. With this choice of technology, screening of a wide range of predicted analytes can be performed. Applied to dumped chemical munitions approximately 30 different related analytes have been selected for screening during investigation of the dumpsites in the Baltic Sea (Östin 2013, Söderström 2014, Bełdowski et al. 2016a). It is important to note is that there is no one general single technique capable of this analysis but a combination of GC and LC with tandem mass spectrometry will be required.

The different arsenic-based agents are in the reach back laboratory analysed after thiol derivatisation with GC–EI/MS/MS or alternatively with LC–ESI/MS/MS after oxidation of As^{III} with hydrogen peroxide to As^{IV}. Preferably, the analyses of Adamsite should be performed with LC–MS due to its problematic derivatisation and GC–MS analysis properties. During the project CHEMSEA the LOQ in targeted LC–ESI/MS/MS analysis were determined to 1–9 µg/kg *dw* for the oxidized arsenic based compounds of interest (Östin 2013).

4.4.1 Analysis of Samples near the Wreck in Bornholm Deep

As discussed above, during the sampling campaign in the Bornholm Deep in MODUM project a two-step process was used: screening of mustard-related chemicals was performed on-board immediately after sampling and later at the reach back laboratory (FOI). The sampling was carried out in the vicinity of a wreck in the Bornholm Deep. See Fig. 4.12 for the sonar images and sampling locations.

Data for the reach back laboratory analysis is presented in Fig. 4.13. This data is the merged results from LC and GC–tandem mass spectrometry analysis. The sampling points 3–8 correspond to the screening with the on-ship analysis in Fig. 4.12 (0–10 m from wreck).

The results of the ruggedized GC–MS instrument correspond well with the reach back laboratory results.

4.4.2 Analysis of Monitoring Samples Taken in Baltic Dumpsites

In addition to the sampling campaign concentrating to the area surround the wreck in the Bornholm Deep, a more general sampling campaign was carried out in Bornholm, Gotland and Gdańsk Deeps. The sampling locations were selected based on data from previous projects, e.g. CHEMSEA. These analysis were carried at VERIFIN.

The results of these samples are shown in Table 4.2. Some of the result concentrations – marked with asterisks (*) – are below the lowest calibration standards (typically in 1 ppb range), but the identifications fulfill identification criteria.

As seen in previous projects, the sediment contamination varies in both concentration and composition. Some samples – e.g. first Bornholm Deep sample (see Table 4.2) – contain chemicals related to one single type of munition; here winter-

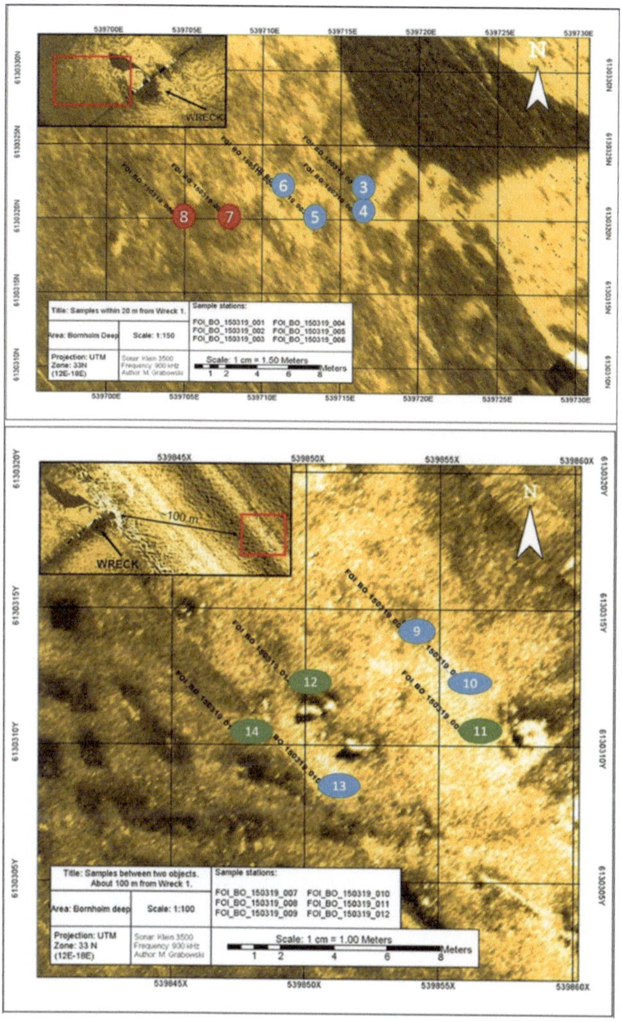

Fig. 4.12 Sampling positions near the wreck in Bornholm Deep. Colours of the sampling points correspond to the findings shown in Fig. 4.13

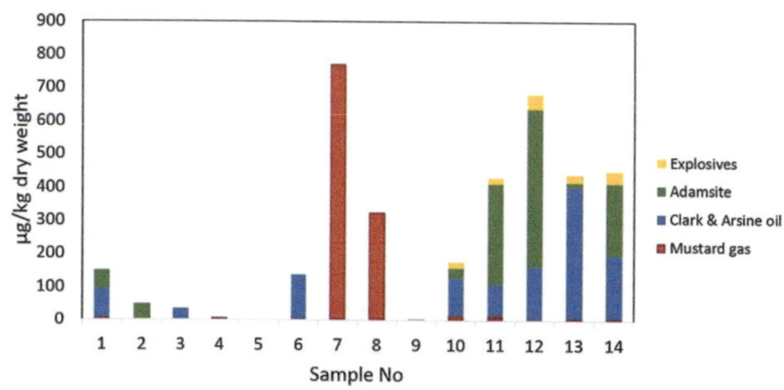

Fig. 4.13 Results of the analysis of the samples taken near the wreck in Bornholm Deep. Sampling points are shown in Fig. 4.12

Table 4.2 Results of the analysis of monitoring samples taken in Baltic dumpsites

Samples	Concentration (µg/kg *dw*)												
	Mustard					Adamsite	Clark		Arsine oil				CN
	Mustard	1,4-Dithiane	1,4-Oxathiane	1,4,5-Oxadithiepane	1,2,5-Trithiepane	5,10-Dihydrophen-arsazin-10-ol 10-oxide	Diphenylarsinic acid	Diphenyl-propylthioarsine	Triphenylarsine	Triphenylarsine *(after oxidation)*	Phenylarsonic acid	Dipropyl phenylarsonodithioite	α-Chloroaceto-phenone
Bornholm 1		660	7.9	880	1100		*			*			
Bornholm 2							*			*			
Bornholm 3		0.37				*	32			5.8	15		
Bornholm 4	1.3	1.1			11	3.4	*	84	1900	*	120	400	4.2
Bornholm 5		1.3		2.0	5.7		*		4.2	*			
Bornholm 6		0.60								*			
Bornholm 7		0.31											
Bornholm 8						4.8	*						
Gotland 1							*			*			
Gotland 2							*			*			
Gotland 3													
Gdańsk 1										*			7.7
Gdańsk 2													
Gdańsk 3													
Gdańsk 4							*			*			
Gdańsk 5													
Gdańsk 6													
Gdańsk 7													

The presented concentrations are in µg/kg of dry sediment. Findings marked with asterisks (*) are below the lowest calibration standards, but the identifications fulfill identification criteria grade sulphur mustard. Typically, no intact agents are seen in sediment samples. Some of the samples contain target chemicals from several types of munitions – e.g. the fourth sample from Bornholm – contain chemicals related to sulphur mustard, Adamsite, Clark, arsine oil and CN. As Clark and arsine oil can be related to winter-grade mustard, contamination from at least three types of munitions/containers is seen this sample.

Although, the concentrations in some Bornholm Deep samples are high, these is in line with earlier results. The sulfur mustard degradation product concentrations in the first Bornholm sample are the highest detected so far.

Results from Gotland Deep are lower than expected. In this dumpsite, the munitions/containers are more randomly distributed than e.g. in Bornholm Deep. With low number of samples, in is possible to find almost empty areas within the dumpsite.

The finding of arsenic-containing chemicals in the Gdańsk Deep are in line with previous results. However, CN has not been previously found in Gdańsk Deep.

4.4.3 High-Resolution Mass Spectrometry Analysis

The extracts of the sediments contain a large number of chemicals. The methods used for the analysis of the sediment extracts have traditionally been targeted methods – either methods based on selected reaction monitoring (SRM) using triple quadrupole instruments as discussed above or on selected ion monitoring (SIM) on quadrupole instruments (e.g. Tørnes 2006). Due to the high number of background chemicals, there are typically signals present at retention times corresponding to the target chemicals. Therefore, strict identification criteria have been applied on retention times and ion ratios to verify that only correct chemicals are being identified.

For the identification of trace amounts of contaminates high resolution mass spectrometry (HRMS) is a techniques that provide higher degree of evidence as compared to tandem mass spectrometry. This can be of importance if new constituents are suspected from e.g. unexpected dumped chemicals. HRMS is a tool for exact mass determination of a molecular ion allowing the prediction of its elemental composition. The development of technology has led to user-friendly robust instrumentation in combination with separation on liquid chromatography and gas chromatography that allows the high-resolution analysis to been done on a routine basis. The combination with ultra-high performance liquid chromatograph-high resolution mass spectrometry (UHPLC–HRMS) has drastically increased the separation speed and efficiency and it has become an extreme powerful tool for rapid and accurate analysis of complex samples. With state of the art instrument UHPLC–HRMS, data can be collected with full scan analysis with a sensitivity and selectivity matching trace analysis with tandem mass spectrometry. Furthermore, the full scan analysis contains the accurate mass information for all ionisable constituents in the sample. However, the ability for ionisation is dependent on the molecular properties and has to be optimised for each class of analytes with selection of UHPLC-eluent (basic or acid) and mode of ionisation, electrospray (ESI) or

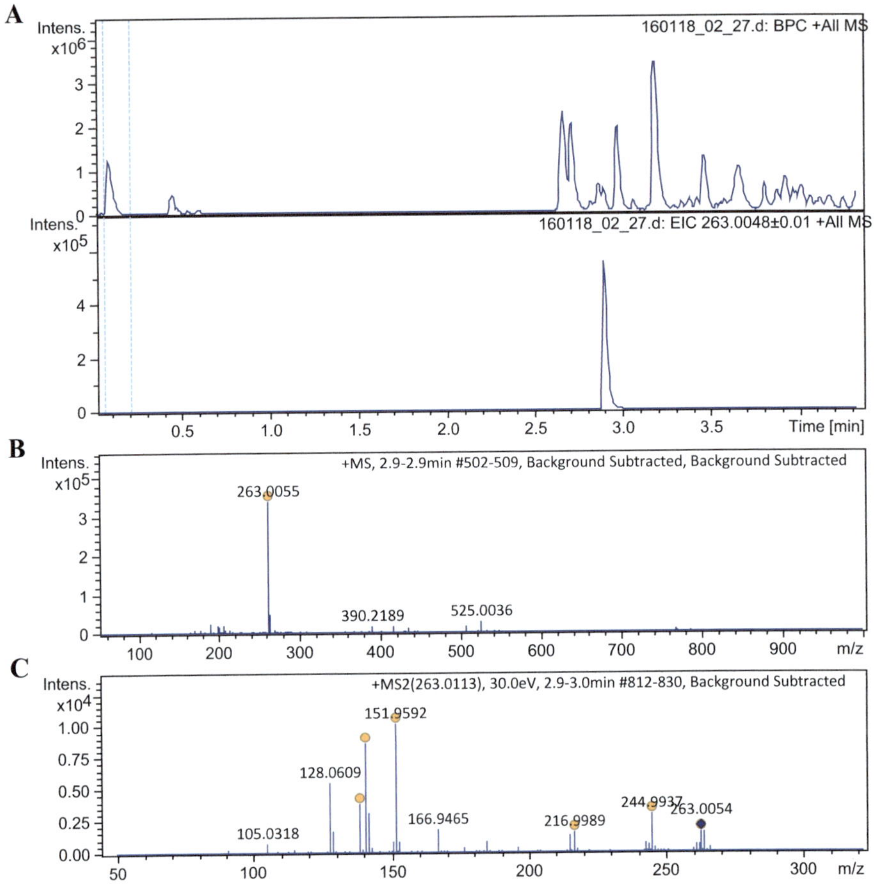

Fig. 4.14 LC–ESI/MS/HRMS analysis of sediment extract showing presence of diphenylarsinic acid: (**a**) full scan chromatograms – top: base peak chromatogram, bottom: extracted ion chromatogram for *m/z* 263.0048 (ion $[M + H]^+$, mass window ±0.01 u), (**b**) background subtracted mass spectra of the peak at 2.9 min corresponding to a molecular composition of $C_{12}H_{12}AsO_2$, (**c**) product ion spectrum for the peak at 2.9 min using precursor ion *m/z* 263

atmospheric pressure ionisation (APCI). Identified molecular ions of interest may be subjected to tandem mass spectrometry which will provide accurate mass determination of each individual fragment that will strengthen the fragmentation interpretation (Zedda and Zwiener 2012, Meyer and Maurer 2016). This mass spectrometry analysis combination with determination of the molecular ion combined with daughter ion spectra should be combined with the analysis of an authentic standard to obtain an unambiguous identification if new degradation products or unexpected dumped chemicals are indicated. An example of HRMS analysis of oxidized Adamsite – diphenylarsinic acid – is shown in Fig. 4.14.

4.4.4 Total Arsenic Measurements

X-Ray Fluorescence Spectrometry (XRF) is an analytical technique which utilizes X-ray fluorescent technology, which is a non-destructive analytical technique used to determine the elemental composition of materials.

The XRF technique consists of irradiating a solid or a liquid sample with high energy X-rays from a controlled X-ray tube, which results in the emission of a fluorescent (or secondary) X-ray emitted from a sample when it is excited by a primary X-ray source. This fluorescence is unique to the elemental composition of the sample. Because each element has its own characteristic spectra related to atoms excitation, therefore XRF can measure exactly what elements are in the sample and in what quantity. Handheld XRF analyzers are designed for simple point-and-shoot operation without complex sample preparation, what makes it possible to be used on-board.

A portable X-ray Fluorescence Spectrometer (XRF) S1 Titan 600, from BRUKER Polska, with designed initial GeoCHEM calibration was used for elemental analysis of sediment samples collected in chemical munition dumpsites during the MODUM project. Calibration of S1 Titan 600 allows to achieve the results with the limit of detection (3 μg/g) much lower than the geochemical background values and with resolution of ± 3 μg/g, which is accurate for detection of elevated arsenic concentrations. As the biogeochemical background values of arsenic concentration in muddy sediments of Baltic Sea is equal to 20 μg/g even rough estimation of actual sample concentration exceeding the background values is a valuable information in terms of additional arsenic source into the sediments from arsenic containing CWA. Results obtained with XRF were compared with results from Atomic Absorption Spectrometry technique which is vastly used in laboratories – mean values of arsenic concentration results from both methods didn't show significant differences (Fig. 4.15) although the standard deviation of triplicate measurements is considerable, therefore the obtained results show relevance of using the portable technique. Method is also reliable in terms of demonstrating the spatial differences in arsenic concentration.

Arsenic concentration varied from 9 to 24 μg/g within Danish strait (with mean value around 18 μg/g), from 16 to 24 μg/g in the area of Bornholm Deep (mean value: 20 μg/g) and from 13 to 23 μg/g within the Gdańsk Deep (mean value 19 μg/g) – according to the conducted study the Bornholm Deep is the biggest chemical munition dumping area monitored within MODUM project and is also considered the most CWA polluted area in the Southern Baltic.

Pairs of samples collected close and far from the suspicious objects concerned to be sunken munitions were analyzed. In samples from Bornholm Deep and Gdańsk Deep slight differences were observed – higher arsenic concentrations were measured in samples collected "close to objects" than "far from objects" (Fig. 4.16). However, the number of comparable samples pairs was too small and the differences are not always statistically significant, it is worth to further monitor the areas and continue the study for more samples as the visible variability of concentrations is likely.

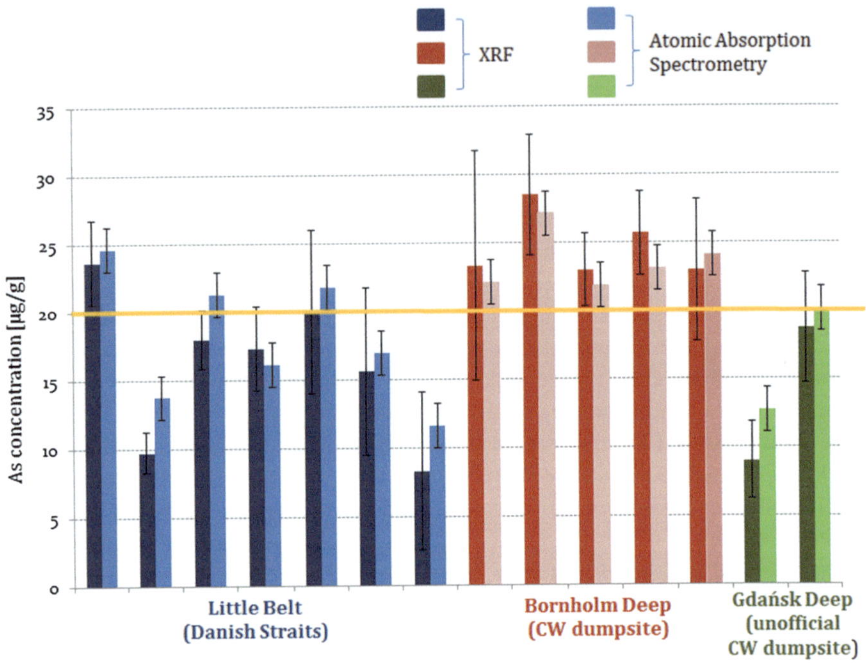

Fig. 4.15 Comparison of analytical results obtained by AAS and XRF (yellow line stands for the geochemical background value of arsenic concentrations in Baltic Sea muddy sediments)

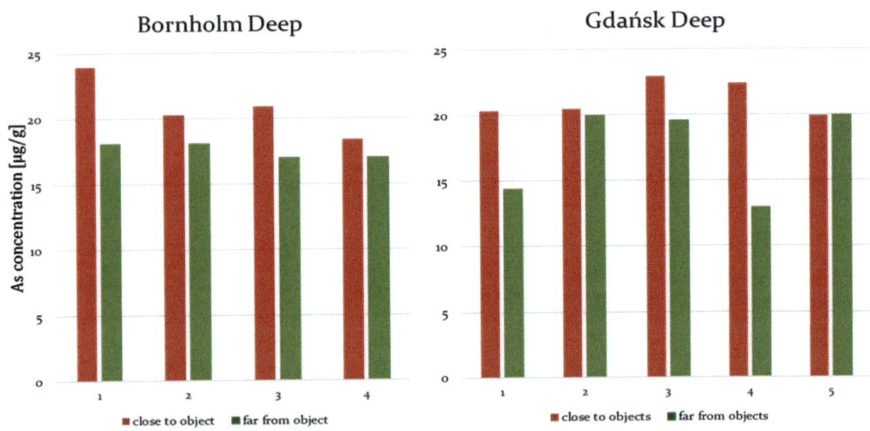

Fig. 4.16 Total arsenic concentrations in samples collected close and far from objects suspected to be dumped munitions

Arsenic Speciation Measurements An effort was given to implement new method for arsenic speciation in sediments for estimation of arsenic source to the sediments. Aim of this study was to develop and alternative technique for laboratories without authorization for measuring CWA and their derivatives, to estimate the possible source of arsenic in the sediments from chemical munition dumpsites.

The technique of ion chromatography separation of arsenic species (chromatographic column CF-Kit-As35, Elemental Scientific Instruments) was successfully installed to the Inductively Coupled Plasm Mass Spectrometer – ICPMS (Perkin Elmer, ELAN 9000) detector already used for years in The Marine Geotoxicology Laboratory, Institute of Oceanology Polish Academy of Sciences. This method allows to separate 5 species of arsenic from the liquid samples: As(III), As(V), arsenobetaine (AsB), monomethylarsine (MMA) and dimethylarsine (DMA). Differences in ratios between particular arsenic species concentrations were considered to indicate if elevated arsenic levels could be related to arsenic-containing CWA or occur from natural sources.

A method was developed to use this technique for environmental samples of marine origin. In the first step eluents were chosen and their pH was adjusted for better division of the peaks assigned to particular arsenic species. For this purpose and for determination of calibration curves, standard solutions of five arsenic species in different concentration were measured. Adjustment of chromatography parameters (time, eluents type, concentrations, pH, flow etc.) allowed to achieve well divided peaks for each arsenic species (Fig. 4.17). Calibration curves were established for each arsenic species.

Also extraction method was established for sediment samples based on available literature. Based on Fauser et al. (2013) and Elwood and Maher (2003), two extraction

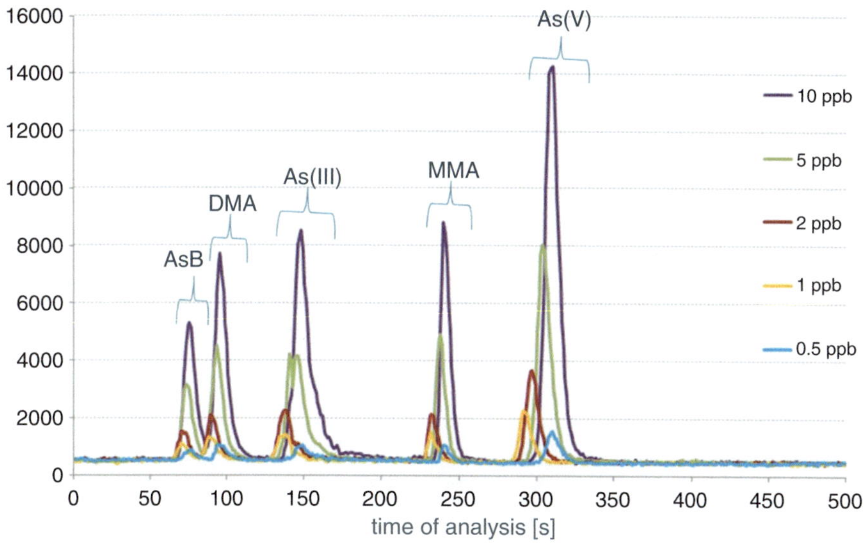

Fig. 4.17 Example chromatogram of five arsenic species in increasing concentrations

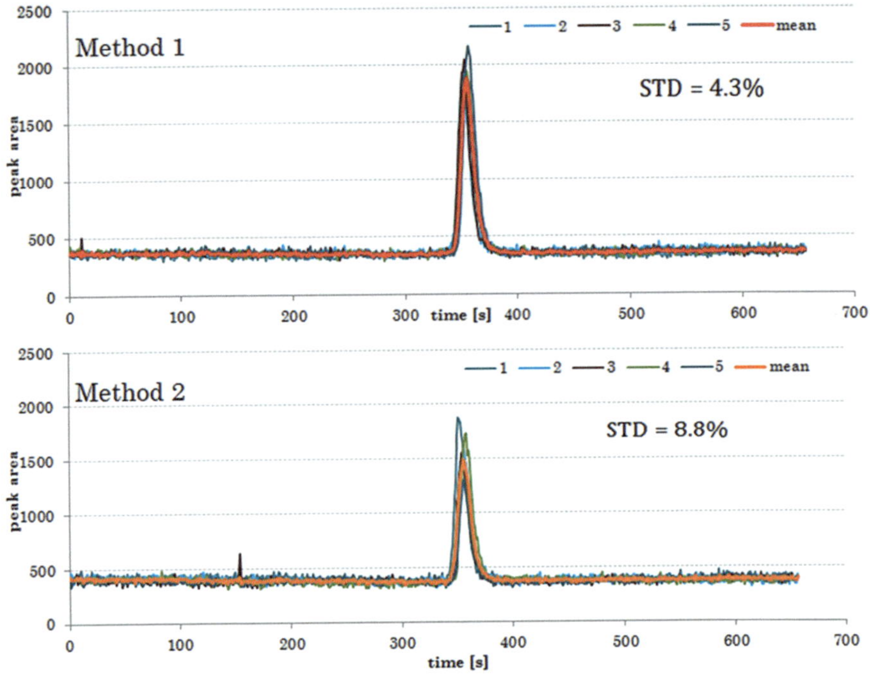

Fig. 4.18 Comparison of extraction methods

methods were chosen and compared. Certified Reference Material (CRM) JMS-1 (marine sediment with known and certified value of total arsenic concentration only) was used for recovery estimation. The CRM sediment was extracted with the use of both methods in five replicates to estimate the deviation of extraction procedure. Method 1 (Fauser et al. 2013) was chosen because of the lower STD value (4.3%) and higher recovery rate (56%) (Fig. 4.18).

Only As(V) was found in the JMS-1 sediments extract. JMS-1 CRM sediment is designed for heavy metals analysis in marine sediment samples and only total arsenic concentration is given. This sediment was used due to unavailability of arsenic species CRM in matrix similar to marine sediments or soil. The recovery is sufficient for this kind of analytical techniques so it was assumed that only As(V) was present in the CRM. Yet also imperfection of sample extraction should be taken into consideration.

Environmental samples were successfully analyzed after method adjustment. In total 18 sediment samples were analyzed with described arsenic speciation technique. Samples were collected in the area of Bornholm Deep and Gdańsk Deep. In all samples only peaks for As(V) were observed – this result may indicate that (1) only As(V) occurs in collected samples – in oxygenated environments As(V) occurs naturally and dominates while organic arsenic species occur in trace quantities if ever (Flora 2015), (2) extraction technique may be insufficient or (3) chemical reactions

Table 4.3 Concentrations of total arsenic and arsenic (V) in sediment samples

Sample	Total As [μg/g]	As(V) [μg/g]	% As(V)
Bornholm 1	17.0	12.1	71.6
Bornholm 2	20.2	13.2	65.0
Bornholm 3	24.0	12.3	51.3
Bornholm 4	18.4	19.1	103.8
Bornholm 5	20.9	27.0	129.4
Bornholm 6	18.0	23.7	131.1
Gdańsk 1	20.4	11.8	57.6
Gdańsk 2	19.5	19.4	99.3
Gdańsk 3	22.4	0.0	0.0
Gdańsk 4	20.5	2.9	14.0
Gdańsk 5	20.3	0.2	0.9
Gdańsk 6	19.9	<LOQ	–
Gdańsk 7	14.4	15.8	109.7
Gdańsk 8	13.0	11.8	91.1
Gdańsk 9	22.9	22.3	97.1
Gdańsk 10	20.0	41.5	207.4

occur between sample collection and extraction carrying out (the samples undergo freezing after collecting, freeze drying and homogenization before analysis). In some samples the peak from As(V) was visible however the calculated result remained under the quantification limit. When compared to total arsenic concentration analyzed in the same samples percentage of As(V) ranged from 0.01% to more than 100% (Table 4.3). There was no clear pattern within the samples from different areas with different As(V) proportion in total arsenic measured in the sample.

This draws to the conclusion that the method should be further developed and adjusted. First of all more samples should be analyzed with a special focus on samples collected from reference areas.

Few pore water samples were analyzed for arsenic speciation. Also in this case only the peak from As(V) occurred in chromatogram. The measured concentrations of As(V) were lower than in sediments samples. In most cases the result was below the quantification limit of the method.

4.5 Conclusions/Summary

Investigations of sea dumped CWA require the use of research vessels for both survey of munition locations and sampling. Modern investigation methods such as use of ROVs allow better identification of relevant objects on the sea floor. However, most objects are difficult to classify and require sampling for reliable identification.

If the analysis of the samples is done only afterwards at the reach back laboratory, the results of the analysis cannot help sampling but only help in planning of the next cruise. The on-going cruise can only follow the sampling plan made before the cruise.

The cruise plan can be made more flexible if tools can be introduced for on board sample analysis. The results can be used during the cruise to shift the focus of the sampling activities.

Two major types of CWA are being analyzed from the Baltic Sea sediments: mustards and arsenicals. Chemical not related to these two types are also analyzed, but they have a minor role in these investigations. In order to have good on board capabilities, both types of chemicals should be includes in on board analysis. In the MODUM project the first type, mustards, has been addressed.

The approach using deployable headspace GC–MS instrument has been successfully used for finding sulfur mustard related cyclic degradation products in collected sediment samples. The findings have been confirmed with analysis at the reach back laboratory. It seems that on the basis of this method it is possible to reliably identify sulfur mustard-related objects.

The other on board sample analysis method using capillary electrophoresis was tested successfully in laboratory, but would require further development before it could be used effectively during on board analysis.

Future method development for on board analysis techniques should address the analysis of arsenicals. If this capability could be developed, both major types of agents could be covered.

Reach-back laboratory analyses are still the main source for data on chemicals to be used for subsequent estimation of the environmental effects of the CWAs. The mass spectrometric analysis methods for identification and quantitation of the CWAs have been already established in several international projects, e.g. MERCW, CHEMSEA and MODUM. They still need to be further developed to fulfil any requirements of future projects and to make use of the developments in analytical instrumentation. One constantly developing and very promising analysis method, already applied in MODUM project, is high resolution mass spectrometry. This method allows more reliable identification of the known chemicals and search for previously unknown new chemicals possibly present in the samples.

The reach back laboratory analysis performed in MODUM project continue the survey of Baltic dump sites started in previous projects. The results for Bornholm and Gotland Deeps were in line with the previous results. The results for Gdańsk Deep further confirm it as a CWA dumpsite. Additionally, a new chemical, α-chloroacetophenone (CN), previously not found in the area, was identified.

The previous research projects have also included reach back laboratory analysis of arsenic in the sediment samples. This has mainly included analysis of total arsenic while efforts have been made to distinguish between organic and inorganic arsenic. In MODUM project, new methodology has been introduced for speciation of arsenic. This will allow better estimation of the effect of the arsenic levels in the sediment and possibly allow detection of CWA-related arsenic in future projects.

References

Aleksenko SS, Gareil P, Timerbajev AR (2011) Analysis of degradation products of chemical warfare agents using capillary electrophoresis. Analyst 136:4103–4118

Arison HL III (2013) European disposal operations: the sea disposal of chemical weapons.. CreateSpace

Bełdowski J, Szubska M, Emelyanov E, Garnaga G, Drzewińska A, Bełdowska M, Vanninen P, Östin A, Fabisiak J (2016a) Arsenic concentrations in Baltic Sea sediments close to chemical munitions dumpsites. Deep Sea Research Part II 128:114–122

Bełdowski J, Klusek Z, Szubska M, Turja R, Bulczak AI, Rak D, Brenner M, Lang T, Kotwicki L, Grzelak K, Jakacki J, Fricke N, Östin A, Olsson U, Fabisiak J, Garnaga G, Rattfelt Nyholm J, Majewski P, Broeg K, Söderström M, Vanninen P, Popiel S, Nawała J, Lehtonen K, Berglind R, Schmidt B (2016b) Chemical Munitions Search & Assessment-an evaluation of the dumped munitions problem in the Baltic Sea. Deep-Sea Research II 128:85–95

Bissen M, Frimmel FH (2003) Arsenic – a Review. Part I: occurrence, toxicity, speciation, mobility. Acta Hydrochim Hydrobiol 31:9–18

Borg H, Jonsson P (1996) Large-scale metal distribution in Baltic Sea sediments. Mar Pollut Bull 32:8–21

BSH, Bundesamt für Seeschifffahrt und Hydrographie (1993) Chemical munitions in the southern and western Baltic Sea – report by a federal/Länder government working group (in German) and cited references. Hamburg, Germany

Coleman K (2005) A history of chemical warfare. Palgrave Macmillan, Houndmills, Basingstoke, Hampshire

Ding Y, Rogers K (2008) Measurement of nitrogen mustard degradation products by poly(dimethylsiloxane) microchip electrophoresis with contactless conductivity detection. Electroanalysis 20:2192–2198

Ellwood MJ, Maher WA (2003) Measurement of arsenic species in marine sediments by high-performance liquid chromatography-inductively coupled plasma mass spectrometry. Anal Chim Acta 477:279–291

Emelyanov EM (2007) The geochemical and geoecological situation of the Gotland Basin in the Baltic Sea where chemical munitionss were dumped. Geologija 60:10–26

Emelyanov E, Kravtsov V, Savin Y, Paka V, Khalikov I (2010) Influence of chemical weapons and warfare agents on the metal contents in sediments in the Bornholm Basin, the Baltic Sea. Baltica 23:77–90

Fauser P, Sanderson H, Hedegaard RV, Sloth JJ, Larsen MM, Krongaard T, Bossi R, Larsen JB (2013) Occurence and sorption properties of arsenicals in marine sediments. Environ Monit Assess 185:4679–4691

Fish JT, Stout RN, Wallace E (2010) Practical crime scene investigations for hot zones. CRC Press, Boca Raton

Flora SJS (ed) (2015) Handbook on arsenic toxicology. Elsevier, London

Garnaga G, Wyse E, Azemard S, Stankevicius A, de Mora S (2006) Arsenic in sediments from the southeastern Baltic Sea. Environ Pollut 144:855–861

Hammes TX (2004) The sling and the stone: on war in the 21st century. Zenith, Grand Rapids, MI

Hanaoka S, Nomura K, Wada T (2006) Determination of mustard and lewisite related compounds in abandoned chemical weapons (yellow shells) from sources in China and Japan. J Chromatogr A 1101:268–277

Harkabusová V, Macharáčková B, Čelechovská O, Vitulová E (2009) Determination of arsenic in the rainbow trout muscle and rice samples. Czech J Food Sci 27:404–406

Heleg-Shabtai V, Gratziany N, Liron Z (2006) Separation and detection of VX and its methylphosphonic acid degradation products on a microchip using indirect laser-induced fluorescence. Electrophoresis 27:1996–2001

IMO, International Maritime Organization (1972) Convention on the prevention of marine pollution by dumping of wastes and other matter http://www.imo.org/en/About/conventions/listofconventions/pages/convention-on-the-prevention-of-marine-pollution-by-dumping-of-wastes-and-other-matter.aspx. Accessed 12 April 2017

IMO, International Maritime Organization (1996) Protocol to the convention on the prevention of marine pollution by dumping of wastes and other matter, 1972. http://www.imo.org/en/OurWork/Environment/PollutionPrevention/Pages/1996-Protocol-to-the-Convention-on-the-Prevention-of-Marine-Pollution-by-Dumping-of-Wastes-and-Other-Matter,-1972.aspx. Accessed 12 April 2017

Jõul P, Lees H, Vaher M, Kobrin EG, Kaljurand M, Kuhtinskaja M (2015) Development of a capillary electrophoresis method with direct UV detection for the analysis of thiodiglycol and its oxidation products. Electrophoresis 36:1202–1207

Knobloch T, Bełdowski J, Böttcher C, Söderström M, Rühl N-P, Sternheim J (2013) Chemical munitions dumped in the Baltic Sea. Report of the ad hoc expert group to update and review the existing information on dumped chemical munitions in the Baltic Sea (HELCOM MUNI), Baltic Sea environment proceeding (BSEP) no. 142, Baltic marine environment protection commission (HELCOM), 128 pages

Kubáň P, Seiman A, Makarõtševa N, Vaher M, Kaljurand M (2011) In situ determination of nerve agents in various matrices by portable capillary electropherograph with contactless conductivity detection. J Chromatogr A 1218:2618–2625

Li SFY (1992) Capillary electrophoresis: principles, practice, and applications, Journal of Chromatography Library, vol 52. Elsevier Science Publishers, The Netherlands

Magnusson R, Nordlander T, Östin A (2016) Development of a dynamic headspace gas chromatography-mass spectrometry method for on-site analysis of sulfur mustard degradation products in sediments. J Chromatogr A 1429:40–52

Makarõtševa N, Seiman A, Vaher M, Kaljurand M (2010) Analysis of the degradation products of chemical warfare agents using a portable capillary electrophoresis instrument with various sample injection devices. Procedia Chem 2:20–25

Mazurek M, Witkiewicz Z, Popiel S, Sliwakowski M (2001) Capillary gas chromatography-atomic emission spectroscopy-mass spectrometry analysis of sulphur mustard and transformation products in a block recovered from the Baltic Sea. J Chromatogr A 919:133–145

MERCW, Modelling of Ecological Risk Related to Sea-Dumped Chemical Weapons, (2006) Synthesis paper on available data. http://www.mercw.org/images/stories/pdf/synthesis_d2.1.pdf. Accessed 12 April 2017

Meyer M, Maurer H (2016) Review: LC coupled to low- and high-resolution mass spectrometry for new psychoactive substance screening in biological matrices – where do we stand today? Anal Chim Acta 927:13–20

Munro N, Talmage SS, Waters LC, Guy AB, Griffin D, Watson P, King JF, Hauschild V (1999) The sources, fate, and toxicity of chemical warfare agent degradation products. Environ Health Perspect 107:933–974

Nicholas DR, Ramamoorthy S, Palace V, Spring S, Moore JN, Rosenzweig RF (2003) Biogeochemical transformations of arsenic in circumneutral freshwater sediments. Biodegradation 14:123–137

Niedzielski P, Siepak M, Siepak J (2000) Występowanie i zawartość arsenu, antymonu i selenu w wodach i innych elementach środowiska. Rocznik Ochrona Środowiska 2:317–341

OPCW, Organisation for Prohibition of the Chemical Weapons (1993) Convention on the Prohibition of the development production, stockpiling and use of chemical weapons and on their destruction. https://www.opcw.org/fileadmin/OPCW/CWC/CWC_en.pdf. Accessed 12 April 2017

Östin A (2013) Review of analytical methods for the analysis of agents related to dumped chemical weapons for the CHEMSEA project, CHEMSEA Project

Pumera M (2006) Analysis of nerve agents using capillary electrophoresis and laboratory-on-a-chip technology. J Chromatogr A 1113:5–13

Røen BT, Unneberg E, Tørnes JA, Lundanes E (2010) Headspace-trap gas chromatography-mass spectrometry for determination of sulphur mustard and related compounds in soil. J Chromatogr A 1217:2171–2178

Rohrbaugh D, Yang Y (1997) Liquid chromatography/electrospray mass spectrometry of mustard-related sulfonium ions. J Mass Spectrom 32:1247–1252

Sáiz J, Mai TD, Hauser PC, García-Ruiz C (2013) Determination of nitrogen mustard degradation products in water samples using a portable capillary electrophoresis instrument. Electrophoresis 34:2078–2084

Sanderson H, Fauser P, Thomsen M, Sorensen PB (2008) Screening level fish community risk assessment of chemical warfare agents in the Baltic Sea. J Hazard Mater 154:846–857

Seiman A, Jaanus M, Vaher M, Kaljurand M (2009) A portable capillary electropherograph equipped with a cross-sampler and a contactless-conductivity detector for the detection of the degradation products of chemical warfare agents in soil extracts. Electrophoresis 30:507–514

Söderström M (2014) Summary of chemical analysis of sediment samples, CHEMSEA Project

Tørnes JA, Opstad AM, Johnsen BA (2006) Determination of organoarsenic warfare agents in sediment samples from Skagerrak by gas chromatography-mass spectrometry. Sci Total Environ 356:235–246

Uścinowicz S (2011) Geochemistry of Baltic Sea surface sediments. Polish Geological Institute, Poland

VanderNoot V, Ferko S, Van De Vreugde J, Patel K, Volponi J, Morrissey K, Forrest L, Horton J, Haroldsen B (2012) On-line monitoring system for chemical warfare agents using automated capillary micellar electrokinetic chromatography. J Chromatogr A 1249:233–240

Wagner G, Maciver B, Rohrbaugh D, Yang Y (1999) Thermal degradation of bis(2-chloroethyl) sulfide (mustard gas). Phosphorus Sulfur Silicon 152:65–76

Wang J, Chatrathi MP, Mulchandani A, Chen W (2001) Capillary electrophoresis microchips for separation and detection of organophosphate nerve agents. Anal Chem 73:1804–1808

Wang J, Pumera M, Collins GE, Mulchandani A (2002) Measurements of chemical warfare agent degradation products using an electrophoresis microchip with contactless conductivity detector. Anal Chem 74:6121–6125

Wang J, Zima J, Lawrence NS, Chatrathi MP, Mulchandani A, Collins GE (2004) Microchip capillary electrophoresis with electrochemical detection of thiol-containing degradation products of V-type nerve agents. Anal Chem 76:4721–4726

Zedda M, Zwiener C (2012) Is nontarget screening of emerging contaminants by LC-HRMS successful? A plea for compound libraries and computer tools. Anal Bioanal Chem 403:2493–2502

Chapter 5
Environmental Toxicity of CWAs and Their Metabolites

Morten Swayne Storgaard, Ilias Christensen, and Hans Sanderson

Abstract This chapter reviews the environmental toxicity of CWAs and their metabolites as well as mixtures of CWAs. We used Microtox™ to generate EC_{50} value for 11 compounds. We observed hormetic effects for two compounds namely Triphenylarsine and Triphenylarsine oxide. None of the mixtures tested show sign of synergism. Two compounds can be characterized as *very toxic* as both α-chloroacetophenone (EC_{50} = 11.20 μg L^{-1}) and 2-chlorovinylarsinic acid (EC_{50} = 31.20 μg L^{-1}) demonstrated EC_{50} values below 1000 μg L^{-1}. Several compounds can be characterized as *toxic* as 1,2,5-trithiepane (EC_{50} = 1170 μg L^{-1}), 1,4,5-oxadithiepane (EC_{50} = 1700 μg L^{-1}), phenarsazinic acid (EC_{50} = 5330 μg L^{-1}) and 1,4-dithiane (EC_{50} = 9970 μg L^{-1}) as these compounds demonstrated EC_{50} values between 1000 μg L^{-1} and 10,000 μg L^{-1}. An *D. magna* acute LC_{50} for, the compound most frequently detected compound (DPA [ox]), was determined to be 100,000 μg L^{-1}. A chronic *D. magna* $LC50_{19days}$ of 640 μg L^{-1} was derived for the compound. A 14-day locomotor behaviour test on adult male Zebrafish (*Danio rerio*) revealed altered behaviour when exposed to concentrations of 1,4,5-oxadithiepane down to 40.3 ± 2.9 μg L^{-1}. A $NOEC_{weight}$ and $NOEC_{mortality}$ greater than 1533 μg L^{-1} was determined for 1,4,5-oxadithiepane.

M.S. Storgaard • H. Sanderson (✉)
Department of Environmental Science, Aarhus University, Roskilde, Denmark
e-mail: hasa@envs.dmu.dk

I. Christensen
Department of Environmental Science, Aarhus University, Roskilde, Denmark

Department of Environmental Engineering, Technical University of Denmark, Lyngby, Denmark

J. Bełdowski et al. (eds.), *Towards the Monitoring of Dumped Munitions Threat (MODUM)*, NATO Science for Peace and Security Series C: Environmental Security, DOI 10.1007/978-94-024-1153-9_5

5.1 Introduction

The environmental toxicity of Chemical warfare agents (CWAs) has in recent years received more attention (Della Torre et al. 2013; Christensen et al. 2016; Sanderson et al. 2014) than previously, where it was mainly the intended and unintended adverse effects on humans that have been studied. The fact that the CWAs are illicit weapons has hampered the development of standardized environmental results from the broader environmental toxicological community. Hence, Quantitative Structure-Activity Relationship (QSAR) based assessments have been shown to be a suitable first tier of assessing their acute environmental toxicities. The predicted toxicities for all known CWAs and dissipation products ranged from 1380 (Clark I (CAS# 712-48-1)) to >9.8 10^7 µg L^{-1} (N-mustard dissipation product (CAS# 54060-15-0)). Dissipation products were found in general to be less toxic than the parent compound – with organophosphate CWAs as the exception (VX dissipation product CAS# 5842-97-9 at 74 µg L^{-1}) (Sanderson et al. 2007). Parent CWAs have not been detected in the water phase in the recent projects in the Baltic Sea are rarely found in the sediments – it is mainly CWA dissipation products that are found in sediments and sometimes in the pore water. Those were narrowed down to 14 compounds in this project, whereas only three were parent compounds, sulphur mustard, triphenylarsine (TPA) and α-chloroacetophenone (Söderström 2014). The rest was arsenicals and degradation products of sulphur mustard. Sanderson and colleagues (Sanderson et al. 2007, 2008, 2010) have modelled the properties of CWAs as well as their toxicity and risk towards the community based upon data gathered in the FP6 project “Modelling of Ecological Risks Related to Sea-Dumped Chemical Weapons” (MERCW) from 2005 to 2009 (http://www.mercw.org/) and NordStream Projects (2008–2012).

The aim of this chapter is to perform a hazard assessment of observed CWAs and obtain novel ecotoxicity data to fill the data gaps regarding ecotoxicity hereby providing reliable data for a future risk assessment in the MODUM project. Firstly, we will present the desk-top work where we update the literature review; the results based on QSAR assessments as QSAR previously has been a suitable first tier of assessing CWA acute toxicity (Sanderson et al. 2007) and new environmentally relevant dissipation products have been found; and lastly the toxicogenomic responses of CWAs. In addition, the field of toxicogenomics will be included with the aim to identify toxicant-gene interactions at the earliest stage as toxicogenomics is an emerging field of research in ecotoxicology and environmental risk assessment. From the experimental side we will address the acute and chronic toxicity of the available detected compounds by the following tests.

1. We addressed the acute toxicity using the marine bacteria *Aliivibiro fischeri* in Microtox™ where we tested all the soluble compounds individually as well as in four different mixture compositions to determine any risk of synergistic effects. Additionally, two mustard heel compounds were recognized and synthesized, they are not included in Christensen et al. (2016), and are yet to be assigned names and CAS no. Hence, they will be named after the features which

characterize them from each other, namely a chloride group and a hydroxyl group, respectively. Consequently, we propose the names *mustard heel-chloride* and *mustard heel-hydroxy*.

2. We moreover determined the acute and chronic toxicity of the most frequently detected CWA related compound (Diphenylarsinic acid (DPA [ox]) towards zooplankton (*Daphnia magna*).
3. We also determined the impact of one of the mustard gas dissipation products (1,4,5-oxadithiepane) on fish behaviour, fitness and survival.

5.2 Methods

5.2.1 Literature Review

The literature review consists of two different information searches. The first is a search for new ecotoxicity values published in ecotoxicity databases which were not included in the 2007 Persistence, Bioaccumulation and Toxicity (PBT)-screening (Sanderson et al. 2007). The second part of the review is a search for new literature and papers from 2007 and forward which may relate to other aspects than simply EC_X-values. Below is described how the searches were conducted. In regard to *ecotoxicity* the underlying meaning in this study is *aquatic toxicity*. The focus has therefore been on aquatic species which are included in the standards of testing ecotoxicity which are fish, algae and daphnia. The databases used are:

1. Hazardous Substance Data Bank (HSDB) via TOXNET;
2. The United Stated Environmental Protection Agency's (US EPA) ECOTOX database; and
3. SciFinder
4. Web of Science
5. ChemidPlus

In regard to the identification of new literature and papers, the search for articles and reports published post 2007 (incl.) was conducted using:

1. SciFinder;
2. Web of Science; and
3. Google Scholar

The keywords used in various combinations were: Chemical warfare agents; toxicity; ecotoxicity; aquatic toxicity; marine; Baltic Sea; and chemical munitions. Relevant articles were also identified from the cited references in key articles (Christensen 2015).

5.2.2 Computational Toxicogenomics

Toxicogenomics is an emerging field of research in the area of environmental risk assessment in support of deriving Adverse Outcome Pathways (AOPs). In order to find the data needed to form a hypothesis about what effects could be caused and measured due to molecular interactions with the CWAs, two databases will be used: The Comparative Toxicogenomics Database (www.ctdbase.org) and ToxCast™ (http://actor.epa.gov/actor/faces/ToxCastDB/Home.jsp). The Comparative Toxicogenomics Database (CTD) has collected over 2.5 million toxicogenomic disease associated relationships (Davis et al. 2012). The site provides an intuitive interface where the only input needed is a CAS number. The result is a list of interacting genes, what diseases are related to the genes and an inference score which indicates the likelihood of connectivity between chemical and gene. CTD contains human toxicogenomic relationships. ToxCast™ contains animal toxicogenomics which will add to the findings in CTD. The findings from CTD and ToxCast™ will be cross-checked in genomic databases for whether the affected genes are present in fish. The genome of zebrafish (*Danio rerio*) is well studied and used for research of gene function and human genetic diseases as approximately 70% of the human genome has a homologous gene sequence in zebrafish [3]. The zebrafish genome is accessed through the American National Center for Biotechnology Information (NCBI) (www.ncbi.nlm.nih.gov) using the search string ' "*Danio rerio*" AND "[gene name]" '. In addition to NCBI, the Zebrafish Gene Collection (ZGC) (http://www.zgc.nci.nih.gov) was used to cross-check NCBI findings. The ambition was to qualify the potential Molecular initiating Event (MiE) in the AOP for the impact of the compounds.

5.2.3 Microtox™

The Microtox™ kit was bought from Aboatox including both freeze-dried bacteria (*Aliivibrio fischeri*) and reconstitution solution. All detected CWA dissipation products were assessed in the test. All chemicals used was of purity >97% except for Bis(2-chlorovinyl)arsinic acid and 2-chlorovinylarsonic acid with purities of ≥90% and ≥91%, respectively. α-chloroacetophenone, 1,4-oxathiane, 1,4-dithiane, thiodiglycolic acid, triphenylarsine, triphenylarsine oxide and 3,5-dichlorophenol (reference compound) were purchased from Sigma-Aldrich Denmark ApS. Phenylarsonic acid was purchased from abcr GmbH. The remaining compounds thiodiglycol sulfoxide, diphenylarsinic acid, phenarzasinic acid, 1,2,5-trithiepane and 1,4,5-oxadithiepane were synthesised by Envilytix GmbH. Bis(2-chlorovinyl) arsinic acid and 2-chlorovinylarsonic acid were synthesized by VERIFIN. The procedure of the Microtox™ test has in detail been described in (Christensen et al. 2016). Additionally, both of the mustard heel compounds were synthesized by FOI and both was of 93% purity. Mustard heel-hydroxy was dissolved in 2% NaCl water and a few drops of methanol and mustard heel-chloride was dissolved in 2% NaCl water. The luminescence was measured in the same pace as the pace in which the chemicals was added ensuring consequent and quantifiable readings. Luminescence measurements were captured and stored on a coupled pc using Microtox Omni v1.18. The assay was conducted in accordance to the ISO guideline (International Organization for Standardization 2007).

5.2.4 *Mixture Toxicity (Microtox™)*

Specific compounds were combined in four different mixtures including a mixture of all detected sulphur mustard degradation products in order to mimic a leaking sulphur mustard bomb. This description will help identifying the type of toxicity, which will be tested in the mixture toxicity tests. Toxicity mixtures can in general be described by three different types of interactions; namely Addition, Synergism and Antagonism. Addition is the type of interaction where each compound present contributes to the total toxicity of the solution as if the compound was tested in a single-compound test. Usually it is predicted that the compounds in the mixtures will act in addition if they have the same mode of action (MoA) and the same site of action (Walker et al. 2012). One can utilize different models to describe the above-mentioned situations. We will use the concentration addition (CA) model. The concept of CA will be used to predict the mixture toxicity as this model has previously shown excellent predictive power for Microtox™ when assessing compounds with similar MoA (Altenburger et al. 2000).

The CA model, also called the Loewe equation, originates from the early works of Loewe (1953), below adopted by Faust and colleagues (2001):

$$ECx_{mix} = \sum_{i=1}^{n}\left(\frac{p_i}{ECx_i}\right)^{-1}$$

In which p_i is the fraction $p_i = \frac{\mathrm{EC}x_i}{\sum_{i=1}^{i=n} ECx_i}$ which is the ratio between the toxicity of (single) chemical and mixture.

The deviation between the observed toxicity (Microtox™ test) and CA predicted toxicity is expressed as the model deviation ratio:

$$MDR = \frac{EC_{50\,pred}}{EC_{50\,obs}}$$

This study will present the following mixtures:

1. A mixture mimicking the composition of a sulphur mustard bomb by mixing all detected degradation products of sulphur mustard;
2. A mixture of the three most toxic sulphur mustard degradation products;

3. All organoarsenic compounds; and
4. All soluble compounds.

5.2.5 Daphnia magna – *Acute*

The acute toxicity test on the freshwater crustacean *Daphnia magna* is a 48 h test using pure solutions of a selected compound. In this test, *Daphnia magna* was exposed to six different concentrations of DPA [ox] (CAS: 4656-80-8; <99%, Sigma Aldrich Denmark ApS). Effect curves/EC-values were calculated based on the amount of immobilised/dead daphnia observed after 24 and 48 hours compared to the controls. Solutions were prepared in M7 media (artificial lake water prepared according to the ISO standard) and each dilution was prepared in four replicates incl. The control solution. The solutions were prepared in 100 mL flasks so that each test vial contained 25 mL solution. Each vial contained five daphnia. The test followed the OECD test guideline 202 (Christensen 2015).

5.2.6 Daphnia magna – *Chronic*

The chronic toxicity test lasts for 21 days with the main endpoint being inhibition of reproduction. Deaths will also be noted as well as daphnia size at the end of the test. All endpoints are compared to the control group after which an effect curve and EC-values are derived. Ten daphnia pr. concentration (at least five concentrations) are held individually in 50 mL solution. At the start of the test, parent daphnia are less than 24 h old but after approximately 10 days they will produce offspring which are counted daily and disposed of. The test will be semi-static meaning that the media containing the solutions will be renewed three times a week. The daphnia are fed thrice a week according to the change of media using concentrated *Raphidocelis subcapitata* (formerly known as *Pseudokirchneriella subcapitata*) suspensions. These algae are cultivated in the lab (also in M7 water) and the concentration of algae added is measured beforehand using a Coulter counter. pH and oxygen levels are measured in the highest concentration and in the control at every change of media. The growth of the parent daphnia is measured at day 0 and 21 using microscope photography. The test followed the OECD test guideline 211 (Christensen 2015).

5.2.7 *Zebrafish (*Danio rerio*) Subchronic Test*

The adult male fish zebrafish (*Danio rerio*) were exposed to nominal concentrations of 1,4,5-oxadithiepane (CAS: 3886-40-6; 99% purity, Envilytix A/S, Germany) in three groups of 1930, 193 and 19.3 μgL^{-1} and one solvent-control (C) group receiving the carrier-solvent acetonitrile. The nominal concentration of acetonitrile was 26 μL L^{-1} in all exposure tanks, and hence, in accordance with OECD guidance document 23, did not exceed 100 μL L^{-1} (OECD 2000). The actual concentrations of 1,4,5-oxadithiepane were determined using gas chromatography tandem mass spectrometry (GC-MS/MS) (Bełdowski et al. 2016) to 40.3 ± 2.9 μg L^{-1}, 133.3 ± 6.7 μg L^{-1} and 1533 ± 266.7 μg L^{-1}. The low concentration is an environmental relevant concentration previously detected in the pore water in the CHEMSEA project (Söderström 2014). The stock, consisting of 300 male and female Zebrafish (*Danio rerio*), was obtained from Credo Fish (Aalborg, Denmark). The stock was acclimated, fed several times daily with Tetra-Min (Tetra Werke Melle, Germany) and observed for health 2 weeks prior to experimental start. Ten morphological male zebrafish were placed in each of the twelve 38 L seamless glass exposure tanks (46 cm × 28 cm × 28 cm) (length × width × height) (Struers KEBO Laboratory, Copenhagen, Denmark). The adult male Zebrafish (*Danio rerio*) were fed daily with Tetra-Min (Tetra Werke Melle, Germany). The experimental set-up and behaviour measurements follows the procedure of Baatrup and Henriksen (2015) and Larsen et al. (Larsen et al. 2008) with modifications described here and in (Swayne Storgaard et al. 2016). Standard length of the fish measured from the tip of the snout to the posterior end of the last vertebrae, excluding the caudal fin was determined. The weight and standard length were used to calculate the Fulton's condition factor (100*weight(g)/length^3(cm) for each fish. Survival was also recorded in the test.

5.3 Results

5.3.1 Literature Review

The literature contains several studies of effects on (mainly) fish and mussels exposed to CWAs. The distribution of papers is visualized on Fig. 5.1. The literature review was performed in 2015, since then Greenberg et al. (2016) published a comprehensive review article in *Clinical Toxicology* on environmental risks of sea-dumped chemical munitions. In our review we focus on the environmental toxicological hazard information in the literature.

The studies we found typically focus on biomarkers and –indicators to find evidence that local biota is affected. Geno- and cytotoxicity is a frequently used endpoint in recent studies (Barsiene et al. 2014; Della Torre et al. 2010, 2013). Della Torre et al. conducted two studies in 2010 and 2013. In the first study, blackbelly rosefish and European conger eel were sampled from CWA dumpsite in the Mediterranean. No residues from sulphur mustard were found in the fish but elevated levels of arsenic were measured. Both species showed signs of chemical exposure such as lesions, ulcerations and DNA-damage as well as increased ethoxyresorufin-*O*-deethylase (EROD). A direct link between observed effects and

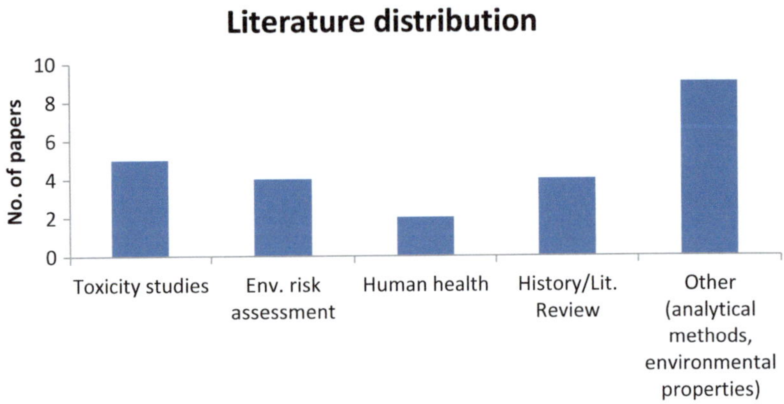

Fig. 5.1 Distribution of 24 papers found in the literature. The studies are categorised into toxicity studies, environmental risk assessments, human health risk assessments, historical papers, and finally other papers that are specific to e.g. compound detection

exposure to sulphur mustard still needs to be established, hence a toxicity study was carried out in 2013 using only eel as it showed to be the most sensitive species. The study included an in-vivo experiment where sulphur mustard was injected into eel and examined 24 and 48 h later. Elevated EROD and UDP-glucuronosyltransferase (UGT) activity was measured as well as an inflammatory reaction on the skin where sulphur mustard was injected. Again, neither sulphur mustard nor its degradation products could be detected in the injected fish. The question remains whether the detected CWAs cause chronic effects and it would therefore be better with a longer exposure. However, it is improbable that causality would be established in light of the acute 48 h test. Apart from geno−/cytotoxicity several fish health indicators have been measured: fitness conditions (condition factor, liver somatic index etc.), diseases and pathology (gross parasites, liver histopathology, etc.), neurotoxicity (Acetylcholinesterase (AChE) inhibition), and oxidative stress (antioxidant defence enzymes) (Lang et al. 2013). *In situ* tests carried out on caged mussels near hot spots seem to fail establishing causality (unreviewed data) (Turja and Lehtonen 2012).

There is a tendency to look for sub-lethal effects rather than determining the traditional EC_{50} of the CWAs. This shows that there is (understandably) more focus on the chronic effects rather than acute effects due to the assumedly yearlong exposure of CWAs in relatively low concentrations. The literature review revealed a trend that sub-lethal effects are tried detected by means of biochemical/enzymatic markers (Amato et al. 2006; Dabrowska et al. 2014; Della Torre et al. 2010, 2013; Kroening et al. 2009; Lang et al. 2013). We therefore further investigated the possible sub-lethal effects, the following section looks into the types of biomarkers and the effects that follow when they are "triggered". A common biomarker is ethoxyresorufin-*O*-deethylase (EROD) induction, e.g. used by Della Torre et al. (2010, 2013). EROD is increased as a function of xenobiotics binding to certain receptors triggering transformation of these xenobiotics (Whyte et al. 2000). The

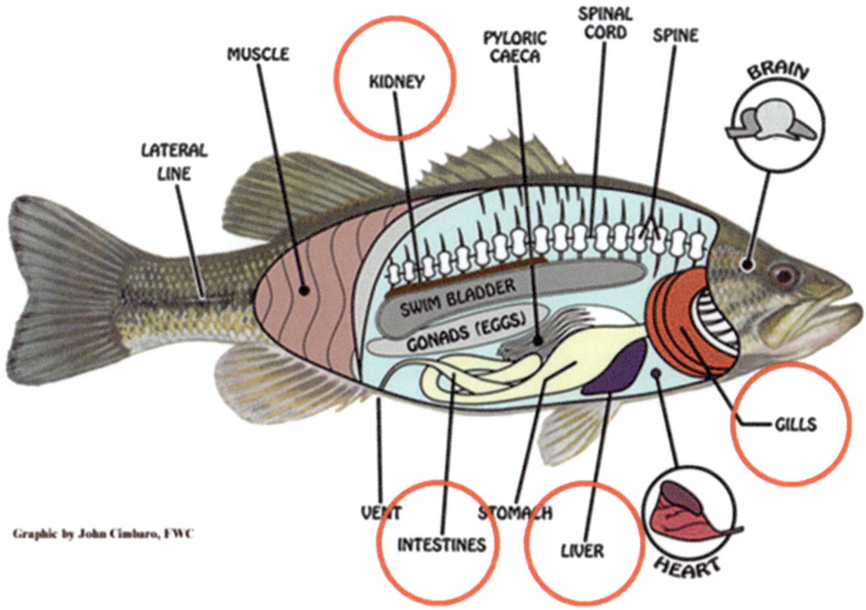

Fig. 5.2 Visualisation of areas affected by EROD (*red*) and AChE (*green*) activity. The figure also shows where tissue samples are taken to measure EROD or AChE activity (Source: John Cimbaro, FWC)

enzymes related to this transformation are present in the kidney, liver, intestines and gills which are therefore sampling areas for EROD-measurement (Della Torre et al. 2010; Whyte et al. 2000), see Fig. 5.2.

Research suggest that increased EROD activity does not only indicate chemical exposure but is directly involved in detrimental effects, but this theory needs more research (Whyte et al. 2000). Cholinesterases (ChE) are also common biomarkers of exposure and are used to tell whether compounds have a neurotoxic effect (Lang et al. 2013). A significant decrease in AChE interferes with the nervous system in a such degree that paralysis occurs (Fulton and Key 2001). It is estimated that the lethal reduction of AChE occurs at >70% (Sancho et al. 2000). Other biomarkers of exposure include Glutathione-S-transferase (GST) and UGT induction which are related to antioxidation and transformation of xenobiotics (van der Oost et al. 2003). No specific thresholds for lethal induction were found. The popular biomarkers in fish, EROD, AChE, GST, and UGT, were investigated in this report to find out whether they had any harmful outcome – if not, they are strictly speaking uninteresting in an ecotoxicological context as they will not cause changes to the size of a fish population. Only AChE could have a lethal outcome, which highlights the need to find more biomarkers for CWAs.

Table 5.1 below, contains a review of the detected compounds from the Baltic Sea in terms of their PBT properties, as well as key physical/chemical properties.

Table 5.1 PBT screening of CWAs included in Sanderson et al. (2007) as well as CWAs found in the Baltic Sea focusing on fish toxicity

Endpoint							QSAR derived toxicity data ECOSAR v. 1.11				Measured
Compound	CAS#	Log(K_{OW})	BCF	K_{OC}	Water solubility (μg L^{-1})	Biodeg	LC_{50} (fish) (μg L^{-1})	ECOSAR class	Chronic (fish) data (μg L^{-1})	ECOSAR class	Acute LC_{50} (μg L^{-1})
Parent compounds											
α-Chloroacetophenone (CAP)	532-27-4	1.93	0.8	98.9	$1.635 \cdot 10^6$	Not pers.	74	Halo ketone	14	Halo ketone	1050–1200 (F) [a], 1100 (F) [b]
Triphenylarsine (TPA)	603-32-7	5.97	7901	151,700	120	Not pers.	68	Neutral organics	10	Neutral organics	NA
Triphenylarsine oxide (TPA [ox])	1153-05-5	5.97	7901	151,700	7.3	Not pers.	72	Neutral organics	11	Neutral organics	NA
Clark I metabolites											
Diphenylarsinic acid (DPA[ox])	4656-80-8	2.8	32.6	269 22.30[b]	$1.629 \cdot 10^6$	Not pers.	41,300	Neutral organics	4410	Neutral organics	NA
Adamsite metabolites											
Phenarsazinic acid (DM[ox])	4733-19-1	2.33	16.1	148 4401 [b]	$3.338 \cdot 10^6$	Pers.[a]	113,590	Neutral organics	11,510	Neutral organics	NA
Sulphur mustard metabolites											
Mustard heel-hydroxy	NA	2.97	143.4	143	125,400	Not pers.	22,049	Neutral organics	2396	Neutral organics	NA
Mustard heel-chloride	NA	4.69	579	17,500	3422	Pers. [a]	689	Neutral organics	90	Neutral organics	NA
Thiodiglycol sulfoxide (TDG[ox])	3085-45-8	−2.76	3.16	0.047	$1 \cdot 10^9$	Not pers.	$2.15 \cdot 10^9$	Neutral organics	$1.559 \cdot 10^7$	Neutral organics	500,000 (A) [c] < $1 \cdot 10^6$ (F) [d], $1 \cdot 10^7$ (F) [c]

Thiodiglycolic acid	123-93-3	1.16	3.16	0.327	$4 \cdot 10^{8}$	Not pers.	$8.47 \cdot 10^{8}$	Neutral organics	$1.651 \cdot 10^{7}$	Neutral organics	NA
1,4-Dithiane	505-29-3	0.77	3.16	4.66	$3.000 \cdot 10^{6}$	Not pers.	352,760	Neutral organics	32,980	Neutral organics	23.2 (D), 24.0 (B) [e]
1,4-Oxathiane	15980-15-1	0.53	3.16	13.4	$3.9880 \cdot 10^{7}$	Not pers.	1,780,230	Neutral organics	98,410	Neutral organics	NA
1,4,5-Oxadithiepane	3886-40-6	1.49	4.44	35.7	$1.6277 \cdot 10^{7}$	Not pers.	323,620	Neutral organics	29,910	Neutral organics	NA
1,2,5-Trithiepane	6576-93-8	2.34	16.2	107	769,000	Not pers.	62,100	Neutral organics	6300	Neutral organics	NA
Lewisite metabolites											
2-Chlorovinylarsonic acid (L1[ox])	64038-44-4	−0,472	3.162	−0,40	$1.296 \cdot 10^{8}$	Pers.[a]	6,421,930	Vinyl/allyl halide	1.90010^{7}	Neutral organics	NA
Bis(2-chlorovinyl)arsinic acid (L2[ox])	157184-21-9	1.79	1.79	35.74 307[b]	$3.8204 \cdot 10^{7}$	Pers.[a]	50,850	Vinyl/allyl halide	28,010	Neutral organics	NA
Organoarsenicals											
Phenylarsonic acid (PDCA[ox])	98-05-5	0.03	3.16	1.13 1001[b]	$1 \cdot 10^{9}$	Not pers.	9,712,340	Neutral organics	394,410	Neutral organics	$>1 \cdot 10^{6}$ and $7.10 \cdot 10^{5}$, $9.00 \cdot 10^{5}$, $4.20 \cdot 10^{5}$, $6.00 \cdot 10^{5}$ (F)[f]

[a]BIOWIN 3: Weeks-Months. (A) Algae, (B) Bacteria (D) Daphnia & (F) Fish. [a] US EPA ECOTOX [b] Summerfelt and Lewis (1967) [c] HSBD [d] Munro et al. (1999), [e] Galli et al. (1994) [f] Tsuji et al. (1986). The K_{OC} chosen are the lowest value (most conservative) of the two modes in Epi Suite, molecular connectivity index (MCI) and $Log(K_{OC})$ based

In order to evaluate the persistence of the compounds the EU TGD (2003) recommends using the three BIOWIN models 2,3 and 5 with the benchmark values: (non-linear model(<0.5 biodegradation probability = persistent)) or MITI non-linear model (<0.5) and ultimate biodegradation ≥ months, respectively). The compounds must comply with all the requirements in order to be categorized as potential persistent (Pers.) which only serves as an indicator for further evaluation. The following chemicals meet the requirements: DM [ox], L1 [ox], L2 [ox] and Mustard heel-chloride with the note that the BIOWIN 3 module deemed the ultimate degradation to occur between weeks to months for the compounds marked with an ^{x}. All of the other degradation products are not persistent. However, TPA and TPA [ox] failed one module (BIOWIN 2). The evaluation of persistent must be deemed conservative as marine settings is influenced by specific parameters such as temperature and salinity. Two of the agents showed considerable potential to bio accumulate (BCF > 2000), namely TPA and its degradation product TPA [ox]. Both agents showing the highest bio accumulation potential also have the greatest Log(K_{ow}), both 5.97, of all compounds in this study and in (Sanderson et al. 2007) followed by the mustard heel-chloride (Log(K_{ow} = 4.69)) and mustard heel-hydroxy (Log(K_{ow} = 2.97)). Mustard heel-chloride and mustard heel-hydroxy showed a BCF = 579 and BCF = 143.4, respectively. The majority of the compounds (12) have BCF < 50. Due to the salting out effect, the solubility of a contaminant in marine settings is typically reduced by a factor of 1.36(EU TGD 2003). Table 5.1 summarizes the predicted LC_{50} ($\mu g\ L^{-1}$) values from Epi Suite for fish. The two parent compounds TPA and CAP is predicted to demonstrate the highest toxicity towards fish. Noteworthy, the degradation product of TPA, TPA [ox], demonstrating similar toxicity.

5.3.2 Toxicogenomic Read-Across Analysis

The Computation Toxicogenomic Database (CTD) was screened for all CWAs detected in the Baltic Sea. Only two compounds, PDCA[ox] and CAP, resulted in hits showing various chemical gene interactions. These genes were cross-checked with NCBI and ZGC to leave out genes not existing in Zebrafish (*D. rerio*). The genes, diseases and inference scores are listed in Table 5.2. Many more diseases than the ones listed can be triggered but only the three highest scoring diseases are included. PDCA[ox] can, among others, cause effects related to the digestion system (gastroparesis), nervous system (adrenoleukodystrophy) and skin (mastocytosis) Even though PDCA[ox] only interacts with one gene, HMOX1, this gene is related to 56 diseases. The inference score for the three diseases is between 5.96 and 5.84 which is considered a low value since inference scores of 130 < was observed for other (unrelated) compounds. CAP revealed slightly higher inference scores of up to 12.98 and was found to interact with four genes existing in the Zebrafish genome: CALCA; CYB5A; TAC1; and TAC2. The top three gene-related diseases were related to the nervous system (migraine and neurogenic inflammation) and

Table 5.2 Possible gene interactions and diseases (human) caused by three of the CWAs detected in the Baltic Sea (MiEs)

Compound	Interaction (human)	Diseases (human)	Likelihood in fish[a]	Inference score
Phenylarsonic acid PDCA[ox]	Increased expression of HMOX1 mRNA	Gastroparesis	Likely	5.96
		Adrenoleukodystrophy	Likely	5.84
		Systemic mastocytosis	Likely	5.84
α-Chloroacetophenone CAP	Increased expression of CALCA protein	Migraine without Aura	Less likely	12.98
	Increased expression of CYB5A protein	Neurogenic inflammation	Likely	12.37
	Increased expression of TAC1 protein	Arthritis	Less likely	10.96
	Increased expression of TAC2 protein			
Diphenylarsinic acid DPA[ox]	Increased expression of BCL2 mRNA	–	–	NA
	Increased activity of CASP3 protein			
	Increased cleavage of GCLC protein			
	Increased expression of CCT2			
	Increased expression of CYP1B1 mRNA			
	Decreased expression of CYP3A2 mRNA			
	Decreased expression of DLG1 protein			
	Increased expression of GCLC protein			
	Decreases expression and increases degradation of GLS			
	Increased expression of HMOX1 protein			
	Effect on the localization and increased activity of NFE2L2 protein			
	Increased expression of PTGS2 mRNA			
	Increased expression of RPLP0 protein			

Only the three most probable diseases (high inference score) are shown
[a]Personal estimation

joint disorder (arthritis). In Table 5.2 “Likelihood in fish” addresses whether the diseases triggered in humans could possibly occur in fish. Effects limited to cells are assessed to be likely whereas diseases on a higher systemic level as arthritis are deemed unlikely. As the TOXNET database lacks inference scores to indicate most likely gene-interactions, only the interacting genes are listed in Table 5.2. DPA[ox] interacts with 15 genes including HMOX1 which interacted with PDCA[ox].

The possible AOP would hence be that the weight gain, fitness and behaviour of the fish would be impacted – which could hence be a hypothesis for further experimental elucidation.

5.3.3 *Microtox*™

The three most toxic compounds found in the test were α-chloroacetophenone, 2-chlorovinylarsonic acid and 1,2,5-trithiepane with EC_{50}-values of 11.2, 31.2 and 1170 μg L^{-1}, respectively. Thiodiglycol sulfoxide and triphenylarsine did not induce an effect at the concentration level tested. Triphenylarsine were tested up to 200,000 μg L^{-1} and if tested in higher concentrations, the concentration of solvent would either be too high or volatilize. As almost no inhibition occurred, a complete dose-response curve was not found for triphenylarsine. The values listed as EC_{50} for these two compounds (200,000 and 74,250,000 μg L^{-1} for triphenylarsine and thiodiglycol sulfoxide, respectively) are the maximum concentration at which they were tested. The solution of 2-chlorovinylarsonic acid was clear, but contained very little crystallization particles. Therefore, we suggest that the toxicity presented here should be seen as conservative. Triphenylarsine oxide demonstrates hormesis at around 50,000 μg L^{-1}. Hormesis can be defined as the biphasic dose response curve with a stimulatory and inhibitory part (Calabrese and Baldwin 2002). The hormesis observed for triphenylarsine oxide was also seen in the organoarsenicals mixture. The parent compound, triphenylarsine also demonstrated similar pattern of hormesis at 100,000 μg L^{-1} (Christensen et al. 2016) (Table 5.3).

5.3.4 *Microtox*™ *Mixture*

Of the different compounds detected in the Baltic Sea, four different mixtures were assessed for toxicity: One including the mustard gas degradation products, one including the organoarsenicals, a combination of all detected compounds and finally a combination of the three most toxic sulphur mustard compounds. The CA model was applied to each mixture to predict an EC_{50} value in order to make suitable mixtures in the range of the predicted EC_{50} value of the mixture. The mixture tests demonstrated lower toxicities than predicted by CA except the organoarsenical mixture as shown in Table 5.4. Despite the slight differences the observed toxicity and the predictive toxicity lies within a factor of 1.5–2.5 (except for the sulphur mustard mixture).

Table 5.3 Toxicity screening of mixtures of CWA residues found in the Baltic Sea

Mixture/compound	CAS #	Highest concentration tested [μg L^{-1}]	EC_{50} (CA-predicted) [μg L^{-1}]	EC_{50} *A. fischeri* (95% conf. int.) [μg L^{-1}]
Sulphur mustard gas bomb:				
TDG-acid	505-29-3	55,000	16,500	82,000 (63,600–106,000)
1,4-dithiane	15980-15-1	18,900		
1,4-oxathiane	3886-40-6	94,700		
1,4,5-oxadithiepane	6576-93-8	3400		
1,2,5-trithiepane	3085-45-8	2200		
TDG [ox]	123-93-3	211,000		
Three most toxic sulphur mustard gas compounds:				
1,4-dithiane	505-29-3	18,900	42,100	9710 (7800–12,100)
1,4,5-oxadithiepane	3886-40-6	1700		
1,2,5-trithiepane	6576-93-8	2200		
Organoarsenic mixture[a]:				
Phenylarsonic acid	98-05-5	97,100	95,400	62,400 (52,900–73,600)
Phenarsazinic acid	4656-80-8	5330		
Diphenylarsinic acid	4733-19-1	127,000		
Triphenylarsine oxide	1153-05-5	155,000		
All tested compounds[a]:				
TDG-acid	505-29-3	55,000	61,600	94,800 (88,900–101,000)
1,4-dithiane	15980-15-1	18,900		
1,4-oxathiane	3886-40-6	94,700		
1,4,5-oxadithiepane	6576-93-8	3400		
1,2,5-trithiepane	3085-45-8	2200		
TDG [ox]	123-93-3	211,000		
Phenylarsonic acid	98-05-5	97,100		
Phenarsazinic acid	4656-80-8	5330		
Diphenylarsinic acid	4733-19-1	127,000		
Triphenylarsine oxide	1153-05-5	155,000		
α-chloroacetophenone	532-27-4	22.4		

Redrawn from Christensen et al. (2016)
[a]Except for TPA, L1 [ox] as these compounds was excluded due to solubility problems

5.3.5 Daphnia magna *Acute*

A 48 h acute screening test was performed with *D. magna* exposed to DPA [ox] using a maximum concentration of 100,000 μg L^{-1}. This concentration was not high enough to obtain a complete effect curve and an EC_{50}. No significant mortality was observed in concentrations up to 40,000 μg L^{-1}. However, at a concentration level of 100,000 μg L^{-1} 55% of the daphnia died, as evident on Fig. 5.3.

Despite an incomplete dose-response curve in the acute daphnia test, it indicated an acute EC_{50} of around 100,000 μg L^{-1} when exposed to DPA [ox].

5.3.6 Daphnia magna – *Chronic*

An initial 21-days chronic test was carried out reaching concentrations of 30,000 and 100,000 μg L^{-1} where the mortality shortly after test start was too high according to the validity criteria exceeding 20% mortality in the control group. The second 21-days chronic test was hereafter carried out with a concentration range of 100–10,000 μg L^{-1} which lasted the whole period. This chronic test failed to fulfil the validity criteria as the mortality in the control group exceeded 20% at test day 19

Table 5.4 Toxicity screening of CWA residues found in the Baltic Sea

Compound	CAS#	LogKow	EC_{50} *A. fischeri* (95% conf.int.) [μg L^{-1}]
Triphenylarsine[1]	603-32-7	5.97	>200,000*
Triphenylarsine oxide	1153-05-5	5.97	155,000* (124,000–194,000)
α-chloroacetophenone[1]	532-27-4	1.93	11.2 (8.6–14.5)
Mustard heel-hydroxy	NA	2.97	349,710 (323,300–378,280)
Mustard heel-chloride	NA	4.69	100,850 (75,350–134,970)
1,4-dithiane[2]	505-29-3	0.77	9970* (8360–11,900)
1,4-oxathiane[2]	15980-15-1	0.53	47,400 (42,100–53,300)
1,4,5-oxadithiepane[2]	3886-40-6	1.49	1700* (1460–1990)
1,2,5-trithiepane[2]	6576-93-8	2.34	1170* (920–1500)
Thiodiglycol sulfoxide[2]	3085-45-8	−2.76	>74,250,000
Thiodiglycolic acid[2]	123-93-3	1.16	22, 500 (21,300–23,800)
Phenylarsonic acid[3]	98-05-5	0.03	97,100 (91,500–103,000)
Diphenylarsinic acid[4]	4656-80-8	2.8	124,000 (118,000–137,000)
Phenarzasinic acid[5]	4733-19-1	2.33	5330* (5020–5670)
2-chlorovinylarsonic acid[6]	64038-44-4	−0.472	31.2* (28.4–34,300)
Bis(2-chlorovinyl)arsinic acid[6]	157184-21-9	1.79	Not tested (insoluble)

The columns listing log Kow is based on QSAR findings

Modified from Christensen et al. (2016)

[1]Parent compound, [2]Sulphur mustard metabolite, [3]Phenyldichloroarsine (PDCA) metabolite, [4]Clark I metabolite, [5]Adamsite metabolite, [6]Lewisite metabolite *Solvent used

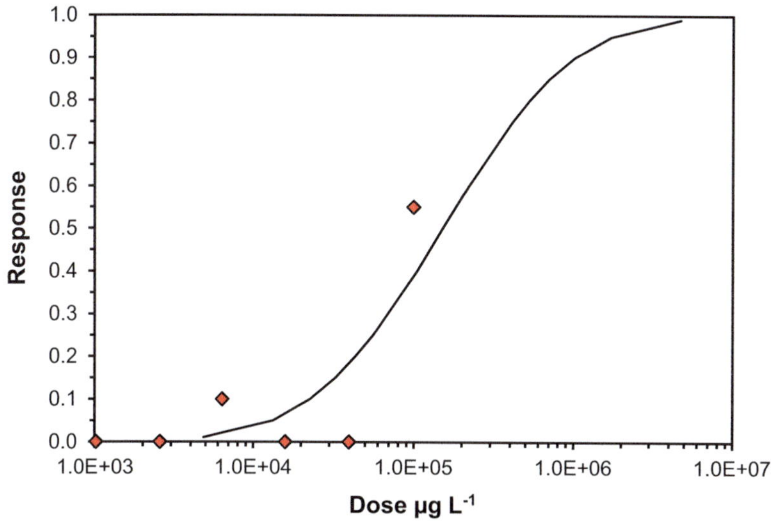

Fig. 5.3 Dose-response curve for acute toxicity test with DPA[ox] tested on *D. magna*

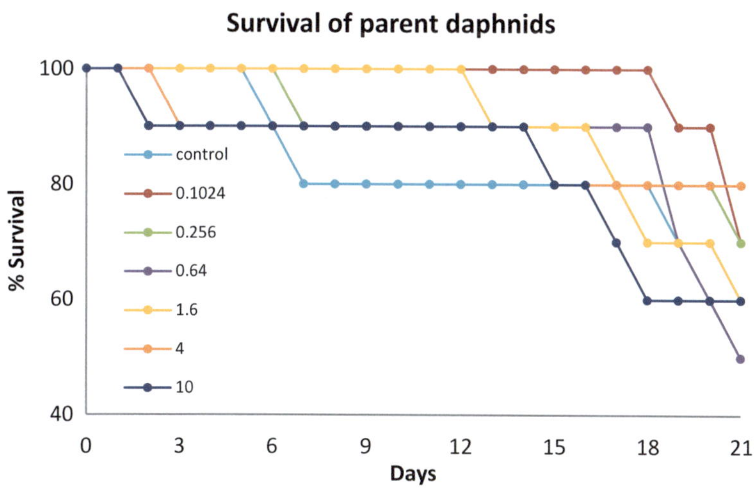

Fig. 5.4 This figure shows the survival of parent daphnia in the various concentrations. At test day 19 the survival in the control went under 80%

(Fig. 5.4). However, the data is still valid and included up until day 19. There does not seem to be a correlation between mortality and dosage. At day 21 the mortality in the control was higher than concentrations of 100, 260 and 4000 µg L^{-1} and actually equal to the mortality in the highest concentration, 10,000 µg L^{-1}. Highest mortality was observed at concentration 640 µg L^{-1} where only 50% of the parent

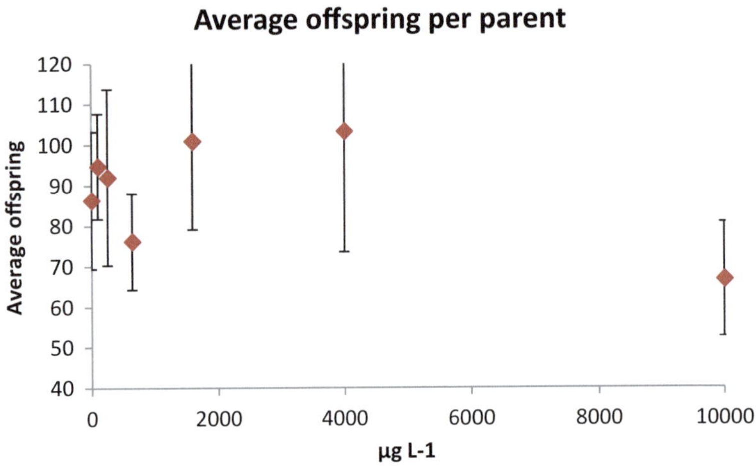

Fig. 5.5 Average offspring per parent in the various concentrations after 21 days. The bars indicate standard deviation (SD)

daphnia survived. A chronic $LC50_{19days}$ of 640 µg L^{-1} was therefore estimated. None of the other effect measures (brood size; time to first brood; parent size) produced statistically significant results at day 19 or 21 compared to the control.

The average offspring per daphnid per concentration was recorded and presented in Fig. 5.5. Even though the average offspring is lower in the highest concentration compared to the control, a two-sided T-test assuming unequal variances revealed no statistically significant difference ($p > 0.05$). Since the lowest concentration (100 µg L^{-1}) had a higher average offspring than the control, a T-test was performed showing a significant difference compared to the highest concentration ($p < 0.05$). The highest cumulative amount of offspring was seen at 1600 and 4000 µg L^{-1} reaching 127 and 161 daphnia respectively from a single parent.

The days until live offspring was born varied between 6 and 13 days. The average for the control and highest concentration was 6.7 and 7.6 days respectively. Standard deviations (SD) were high causing overlapping results in all concentrations, shown on Fig. 5.6 below. A T-test revealed no statistically significant difference between the control and highest concentration group ($p > 0.05$).

The length of the parent daphnia was measured at day 21 and the size averages of each concentration are presented in Fig. 5.7. A tendency of the average sizes is an increase from the control (4.05 mm) to 640 µg L^{-1} (4.21 mm) followed by a decrease to the lowest average in the highest concentration (3.92 mm). There was not any statistically significant difference ($p > 0.05$) between control and highest concentration when conducting a two tailed T-test. However, a significant difference ($p < 0.05$) was found for the between the lowest and highest concentration.

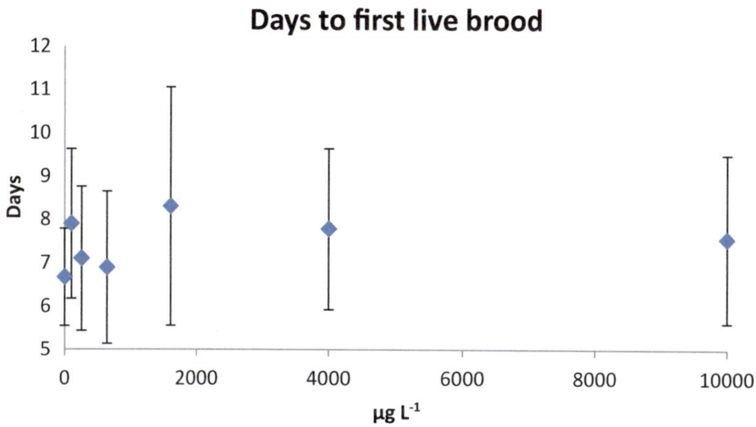

Fig. 5.6 Count of days until first live offspring was observed in the various concentrations. Bars indicate SD

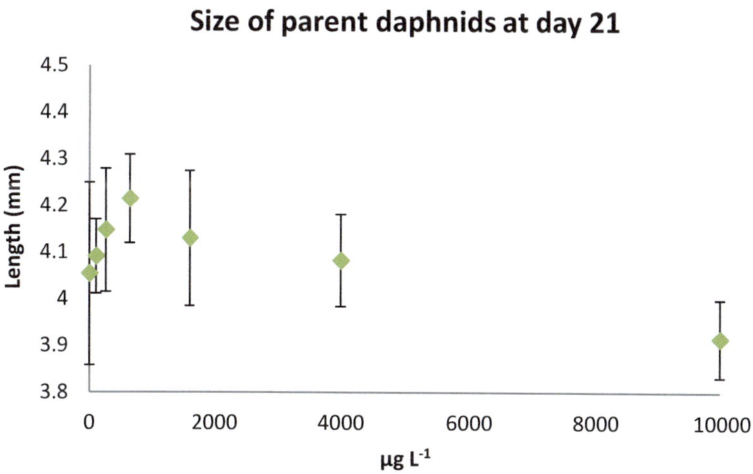

Fig. 5.7 Average of size measurements of the parent daphnia at the end of the test (day 21)

5.3.7 Zebrafish Sub-chronic Locomotor Behaviour Test

Only one fish died during the exposure period in the middle treatment group whereas no fish died in the highest treatment group. Accordingly, it seems unlikely that the single death can be ascribed to toxicity of 1,4,5-oxadithiepane. Fulton's condition factor was identical across all treatment groups and the control group (Kruskal-Wallis, $p = 0.559$). This demonstrates an equal weight relative to length across all treatment groups and the control group. This also suggests that 1,4,5-oxadithiepane at concentrations up to 1533 μg L^{-1} over 14 days is unlikely to affect the growth or weight of adult male Zebrafish (NOEC >1533 μg L^{-1}).

The general trend in this behaviour study is a decreased swimming activity in Zebrafish exposed to the environmentally realistic (low) concentration at 40.3 ± 2.9 μg L^{-1} of 1,4,5-oxadithiepane. At the middle treatment group, the activity resembled the controls whereas the highest treatment group again displayed a decreased swimming activity. The average velocity significantly decreased when Zebrafish were exposed to 40.3 ± 2.9 μg L^{-1} (ANOVA, Tukey Contrast, $p < 0.05$) compared to the control, whereas the average maximum swimming velocity decreased significantly when Zebrafish were exposed to 40.3 ± 2.9 μg L^{-1} compared to Zebrafish exposed to 133.3 ± 6.7 μg L^{-1} (ANOVA, Tukey Contrast, $p < 0.05$). The presented alterations in swimming behaviour should be seen in the context of the lacking dose-effect relationship resulting in a questionable causation of the treatment (Hill 1965) (Fig. 5.8).

In summary, the fish study provides novel ecotoxicity data on 1,4,5-oxadithiepane finding no effects after 14 days at 1533 μg L^{-1} flow-through exposure. Behavioural alterations when adult male Zebrafish was exposed to 1,4,5-oxadithiepane was also shown, however the causation of these require further studies (Swayne Storgaard et al. 2016).

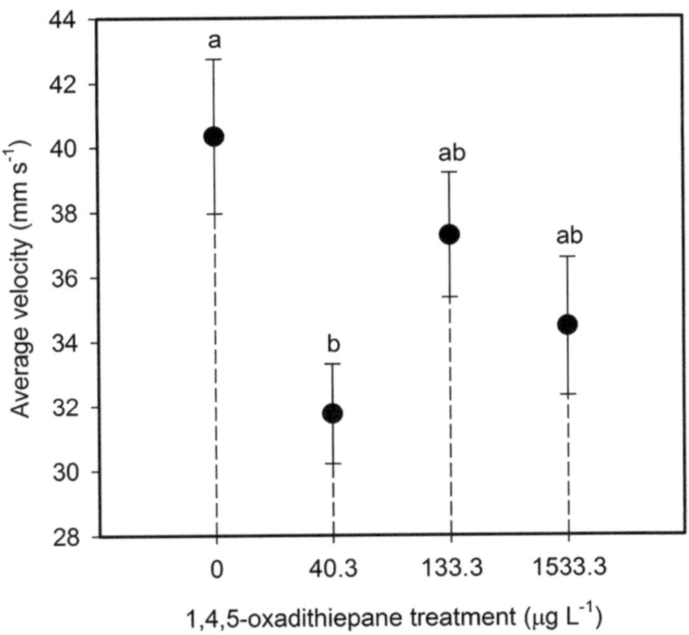

Fig. 5.8 Average velocity (mm s^{-1}) of adult male Zebrafish exposed to 1,4,5-oxadithiepane at 40.3 μg L^{-1} (n = 26), 133.3 μg L^{-1} (n = 26), 1533.3 μg L^{-1} (n = 27) and a solvent control (n = 27). during 45 minutes recording. Entries are presented as mean values ± SEM and the letters "a" and "b" indicate statistical significance. Modified and redrawn from Swayne Storgaard et al. (2016)

5.4 Discussion and Conclusion

During the MODUM project, novel ecotoxicity data has been gathered. These data can be useful for risk assessment and in general, for the scientific community. Contrary, these data also highlight the research needs and uncertainties which will end this discussion. Beginning this section, the toxicity of the compounds will be evaluated in the context of the Globally Harmonized System (GHS) of Classification and Labelling of Chemicals most applicable to the Microtox™ test. The GHS system of classification (UN 2011) has been developed by the UN by an international agreement in 1992. The classification system includes hazard labels etc. allowing to characterize/classify and communicate the acute and chronic toxicity of the specific compounds. An acute toxicity of a compound can be assigned a label of very toxic, toxic, harmful or not toxic. The boundaries of each label has been set as the following:

- Very toxic if $EC_{50} < 1000$ μg L^{-1},
- Toxic if $EC_{50} > 1000$ μg L^{-1} but $\leq 10{,}000$ μg L^{-1},
- Harmful if $EC_{50} \geq 10{,}000$ μg L^{-1} but $\leq 100{,}000$ μg L^{-1}
- Not toxic if $EC_{50} > 100{,}000$ μg L^{-1}.

The compounds tested in Microtox™ fits each of the different categories. Two compounds can be characterized as *very toxic* as both α-chloroacetophenone ($EC_{50} = 11.20$ μg L^{-1}) and 2-chlorovinylarsinic acid ($EC_{50} = 31.20$ μg L^{-1}) demonstrated EC_{50} values below 1000 μg L^{-1}. Several compounds can be characterized as *toxic* as 1,2,5-trithiepane ($EC_{50} = 1170$ μg L^{-1}), 1,4,5-oxadithiepane ($EC_{50} = 1700$ μg L^{-1}), phenarsazinic acid ($EC_{50} = 5330$ μg L^{-1}) and 1,4-dithiane ($EC_{50} = 9970$ μg L^{-1}) as these compounds demonstrated EC_{50} values between 1000 μg L^{-1} and 10,000 μg L^{-1}. In the acute toxicity test on *D. magna*, DPA [ox] can be characterized as *harmful/not toxic* as a $LC_{50} = 100{,}000$ μg L^{-1} was found. The mustard heel compounds can be characterized as *not toxic* as both mustard heel-hydroxy ($EC_{50} = 349{,}710$ μg L^{-1}) and mustard heel-chloride ($EC_{50} = 101{,}850$ μg L^{-1}) demonstrated little toxicity on *A. fisheri*. TPA and TPA [ox] are both not toxic. Prospectively, this can be used in a context of a general environmental risk assessments and more specifically in the MODUM project and subsequently for the dimensioning of the monitoring programme of CWAs in the Baltic Sea (Swayne Storgaard 2016).

The QSAR assessment predicts different MoAs for the compounds (Table 5.1). Predominantly, a narcotic (base-line toxicity) MoA was predicted. However, TPA and TPA [ox] demonstrated hormesis in the Microtox™ test. This might suggest an upregulation of the transcription of the genes associated with light production. Contrary to the predicted narcotic MoA, the upregulating effects are unlikely to be ascribed the narcotic MoA. The popular biomarkers in fish; EROD, ChE, GST, and UGT, were investigated to find out whether they had any harmful outcome when their expression was increased/decreased to the size of a fish population. Only ChE could potentially have a lethal outcome, whereas the others have a more subtle impact, which highlights the need to find more organismal biomarkers that could predict impacts to the ecosystem, e.g. in an AOP context. The study suggested that three of the detected CWAs (DPA[ox], CAP and PDCA[ox]) could potentially interact with up to 20 genes present in Zebrafish. Future studies of sub-lethal effects could investigate the molecular initiating events proposed, as specific genes known to be affected by CWAs are actually present in fish.

Different mixtures were also assessed in the Microtox™ test. The toxicity of the mixtures follows the CA model with MDRs ranging from 0.21–1.53. However, a MDR of 0.21 is almost a 5fold deviation but should not be regarded as an antagonistic mixture. It was decided that TDG [ox] must be present in the mixture despite an absent EC_{50}-value (EC_{50} > 74,250,000 μg L^{-1}). For the two mixture where this compound was a component, the MDR shows some deviation from CA (MDR = 0.21 and MDR = 0.65). This would definitely be more equal to CA with an exact EC_{50} to take into account. CA has previously shown excellent predictive power for similarly acting chemicals (Altenburger et al. 2000).

The toxicological studies performed during MODUM and other studies have solemnly been focusing on species inhabiting the water column, marine bacteria and different studies on fish species (Barsiene et al. 2014, 2016; Della Torre et al. 2010, 2013; Christensen et al. 2016; Swayne Storgaard et al. 2016). Since the munitions are dumped in the sediment and the highest concentration found of, generally, all compounds occur in the sediment, the toxicological hazard assessment will be closer to completion if testing on sediment-dwelling species will be included. MODUM finishes the first sound toxicity data study on the acute toxicity towards marine species and the first sound sub-chronic exposure of legacy compound towards fish. However, the authors will be plea for chronic toxicity studies and studies on different species inhabitant the sediment. This will definitely give a better perspective on toxicity of the compounds. Due to the behaviour of sulphur mustard lumps covered with an insoluble polymeric skin (Mazurek et al. 2001), it would be valuable to analyse those lumps for compounds presents and mimicking environmental realistic conditions of the exposure of those lumps.

To conclude this section, the results will briefly be summarized. By using the first tier screening test, Microtox™, we managed to gather 11 novel EC_{50} value (Table 5.4) and found hormesis from two compounds, Triphenylarsine and Triphenylarsine oxide. None of the mixtures tested show sign of synergism. An acute LC_{50} for, the compound most detected and found highest concentration, DPA [ox] was determined to be around 100,000 μg L^{-1} on *D. magna* whereas the 21-day chronic toxicity on *D. magna* failed the validity criteria of more than 20% mortality. However, a chronic $LC50_{19days}$ of 640 μg L^{-1} can be derived. A 14-day locomotor behaviour test on adult male Zebrafish (*Danio rerio*) revealed altered behaviour when exposed to concentrations of 1,4,5-oxadithiepane down to 40.3 ± 2.9 μg L^{-1}. However, The presented alterations in swimming behaviour should be seen in the context of the lacking dose-effect relationship resulting in a questionable causation of the treatment (Hill 1965). A $NOEC_{weight}$ and $NOEC_{mortality}$ greater than 1533 μg L^{-1}.

Acknowledgements NATO Science for Peace project #984589 (MODUM) for funding.

References

Altenburger R, Backhaus T, Boedeker W, Faust M, Scholze M, Grimme LH (2000) Predictability of the toxicity of multiple chemical mixtures to *Vibrio fischeri*: mixtures composed of similarly acting chemicals. Environ Toxicol Chem 19(9):2341–2347

Amato E, Alcaro L, Corsi I, Della Torre C, Farchi C, Focardi S, Marino G, Tursi A (2006) An integrated ecotoxicological approach to assess the effects of pollutants released by unexploded chemical ordnance dumped in the southern Adriatic (Mediterranean Sea). Mar Biol 149(1):17–23

Baatrup E, Henriksen PG (2015) Disrupted reproductive behavior in unexposed female zebrafish (*Danio rerio*) paired with males exposed to low concentrations of 17alpha-ethinylestradiol (EE2). Aquat Toxicol 160:197–204

Barsiene J, Butrimaviciene L, Grygiel W, Lang T, Michailovas A, Jackunas T (2014) Environmental genotoxicity and cytotoxicity in flounder (*Platichthys flesus*), herring (*Clupea harengus*) and Atlantic cod (*Gadus morhua*) from chemical munitions dumping zones in the southern Baltic Sea. Mar Environ Res 96:56–67

Barsiene J, Butrimaviciene L, Grygiel W, Stunzenas V, Valskiene R, Greiciunaite J, Stankeviciute M (2016) Environmental genotoxicity assessment along the transport routes of chemical munitions leading to the dumping areas in the Baltic Sea. Mar Pollut Bull 103(1–2):45–53

Bełdowski J, Klusek Z, Szubska M, Turja R, Bulczak AI, Rak D, Brenner M, Lang T, Kotwicki L, Grzelak K, Jakacki J, Fricke N, Östin A, Olsson U, Fabisiak J, Garnaga G, Nyholm JR, Majewski P, Broeg K, Söderström M, Vanninen P, Popiel S, Nawała J, Lehtonen K, Berglind R, Schmidt B (2016) Chemical Munitions Search & Assessment—an evaluation of the dumped munitions problem in the Baltic Sea. Deep-Sea Res II Top Stud Oceanogr 128:85–95

Calabrese EJ, Baldwin LA (2002) Defining hormesis. Hum Exp Toxicol 21(2):91–97

Christensen IMA (2015) Toxicity and risks of CWAs found in the Baltic Sea. Aarhus University, Roskilde

Christensen IMA, Swayne Storgaard M, Fauser P, Foss Hansen S, Baatrup E, Sanderson H (2016) Acute toxicity of sea-dumped chemical munitions: luminating the environmental toxicity of legacy compounds. Glob Secur Health Sci Policy 1(1):39–50

Dabrowska H, Kopko O, Gora A, Waszak I, Walkusz-Miotk J (2014) DNA damage, EROD activity, condition indices, and their linkages with contaminants in female flounder (*Platichthys flesus*) from the southern Baltic Sea. Sci Total Environ 496:488–498

Davis AP, Wiegers TC, Rosenstein MC, Mattingly CJ (2012) MEDIC: a practical disease vocabulary used at the comparative Toxicogenomics database. Database (Oxford) 2012:bar065

Della Torre C, Petochi T, Corsi I, Dinardo MM, Baroni D, Alcaro L, Focardi S, Tursi A, Marino G, Frigeri A, Amato E (2010) DNA damage, severe organ lesions and high muscle levels of as and hg in two benthic fish species from a chemical warfare agent dumping site in the Mediterranean Sea. Sci Total Environ 408(9):2136–2145

Della Torre C, Petochi T, Farchi C, Corsi I, Dinardo MM, Sammarini V, Alcaro L, Mechelli L, Focardi S, Tursi A, Marino G, Amato E (2013) Environmental hazard of yperite released at sea: sublethal toxic effects on fish. J Hazard Mater 248-249:246–253

Faust M, Altenburger R, Backhaus T, Blanck H, Boedeker W, Gramatica P, Hamer V, Scholze M, Vighi M, Grimme LH (2001) Predicting the joint algal toxicity of multi-component s-triazine mixtures at low-effect concentrations of individual toxicants. Aquat Toxicol 56(1):13–32

Fulton MH, Key PB (2001) Acetylcholinesterase inhibition in estuarine fish and invertebrates as an indicator of organophosphorus insecticide exposure and effects. Environ Toxicol Chem 20(1):37–45

Galli R, Rich HW, Scholtz R (1994) Toxicity of organophosphate insecticides and their metabolites to the water flea *Daphnia magna*, the Microtox test and an acetylcholinesterase inhibition test. Aquat Toxicol 30(3):259–269

Greenberg MI, Sexton KJ, Vearrier D (2016) Sea-dumped chemical weapons: environmental risk, occupational hazard. Clin Toxicol 54(2):79–91

Hill AB (1965) Environment and disease – association or causation. Proc R Soc Med Lond 58(5):295–300

International Organization for Standardization, I (2007) Water Quality – determination of the inhibitory effect of water samples on the light emission of *Vibrio fischeri* (Luminescent bacteria test)

Kroening KK, Solivio MJV, García-López M, Puga A, Caruso JA (2009) Cytotoxicity of arsenic-containing chemical warfare agent degradation products with metallomic approaches for metabolite analysis. Metallomics 1(1):59–66

Lang T, Fricke N, Broeg K, Baude R, Brenner M, Lehtonen K, Turja R, Barsiene J (2013) Health status of cod (*Gadus morhua*) at dumpsites for chemical warfare agents in the Baltic Sea, (2013)

Larsen MG, Hansen KB, Henriksen PG, Baatrup E (2008) Male zebrafish (*Danio rerio*) courtship behaviour resists the feminising effects of 17alpha-ethinyloestradiol--morphological sexual characteristics do not. Aquat Toxicol 87(4):234–244

Loewe S (1953) The problem of synergism and antagonism of combined drugs. Arzneimittelforschung 3:285–290

Mazurek M, Witkiewicz Z, Popiel S et al (2001) Capillary gas chromatography-atomic emission spectroscopy-mass spectrometry analysis of sulphur mustard and transformation products in a block recovered from the Baltic Sea. J Chromatogr A 919:133–145

Munro NB, Talmage SS, Griffin GD, Waters LC, Watson AP, King JF, Hauschild V (1999) The sources, fate, and toxicity of chemical warfare agent degradation products. Environ Health Perspect 107(12):933–974

OECD (2000) Guidance document on aquatic toxicity testing of difficult substances and mixtures. OECD (ed) OECD, OECD Environment Directorate, Paris, p 53

van der Oost R, Beyer J, Vermeulen NPE (2003) Fish bioaccumulation and biomarkers in environmental risk assessment: a review. Environ Toxicol Pharmacol 13(2):57–149

Sancho E, Ceron JJ, Ferrando MD (2000) Cholinesterase activity and hematological parameters as biomarkers of sublethal molinate exposure in Anguilla anguilla. Ecotoxicol Environ Saf 46(1):81–86

Sanderson H, Fauser P, Thomsen M, Sørensen P (2007) PBT screening profile of chemical warfare agents (CWAs). J Hazard Mater 148(1–2):210–215

Sanderson H, Fauser P, Thomsen M, Sørensen PB (2008) Screening level fish community risk assessment of chemical warfare agents in the Baltic Sea. J Hazard Mater 154(1–3):846–857

Sanderson H, Fauser P, Thomsen M, Vanninen P, Söderström M, Savin Y, Khalikov I, Hirvonen A, Niiranen S, Missaen T, Gress A, Borodin P, Medvedeva N, Polyak Y, Paka V, Zhurbas V, Feller P (2010) Environmental hazards of sea-dumped chemical weapons. Environ Sci Technol 44(12):4389–4394

Sanderson H, Fauser P, Rahbek M, Larsen JB (2014) Review of environmental exposure concentrations of chemical warfare agent residues and associated the fish community risk following the construction and completion of the Nord stream gas pipeline between Russia and Germany. J Hazard Mater 279:518–526

Söderström M (2014) Summary of chemicals analysis of sediment samples. CHEMSEA (ed)

Summerfelt RC, Lewis WM (1967) Repulsion of green sunfish by certain chemicals. J Water Pollut Control Fed 39(12):2030–2038

Swayne Storgaard M (2016) The environmental toxicity of chemical warfare agents and their degradation products found in the Baltic Sea. Aarhus University, Aarhus

Swayne Storgaard M, Sanderson H, Henriksen PG, Fauser P, Östin A, Baatrup E (2016) Suppressed swimming activity in Zebrafish (*Danio rerio*) exposed to 1,4,5-oxadithiepane, a sulphur mustard degradation product. Submitted

Tsuji S, Tonogai Y, Ito Y, Kanoh S (1986) The influence of rearing temperatures on the toxicity of various environmental pollutants for killifish (*Oryzias latipes*). Jpn J Toxicol Environ Health 32:46–53

Turja R, Lehtonen K (2012) Biological effects measured on caged mussels and cod, Helsinki

UN (2011) Globally harmonized system of classification and labelling of chemicals (GHS). Nations U (ed) United Nations, New York/Geneva, pp 215–241

Walker CH, Sibly RM, Hopkin SP, Peakall DB (2012) Principles of ecotoxicology. CRC Press, Boca Raton, pp 163–172

Whyte JJ, Jung RE, Schmitt CJ, Tillitt DE (2000) Ethoxyresorufin-O-deethylase (EROD) activity in fish as a biomarker of chemical exposure. Crit Rev Toxicol 30(4):347–570

Chapter 6
The Health Status of Fish and Benthos Communities in Chemical Munitions Dumpsites in the Baltic Sea

Thomas Lang, Lech Kotwicki, Michał Czub, Katarzyna Grzelak, Lina Weirup, and Katharina Straumer

Abstract The environmental characteristics of the deep basins in the Baltic Sea and their impact on the occurrence of selected biota – benthos and fish communities – are described in chemical munitions dumping site areas. Results of the NATO-funded SfP project MODUM "Towards the Monitoring of Dumped Munitions Threat" (2013–2016) and other related previous activities regarding the impact of chemical warfare agents (CWA) on biodiversity and status of benthic fauna and regarding the health status of Baltic cod (*Gadus morhua*) are presented and discussed in the light of requirements for monitoring ecological risks associated with dumped CWA.

6.1 General Introduction

The Baltic Sea was formed approximately 8000 years ago as a result of the last glaciation and is one of the largest brackish ecosystems of the world. Its water exchange with the open ocean is limited by the narrow and shallow Danish Straits. The Baltic Sea is characterised by conspicuous north–south and east–west gradients in climate and large differences in the most important abiotic conditions, including salinity (Ojaveer and Kalejs 2008). The distribution of living organisms in the Baltic Sea is controlled by these strong environmental gradients. Salinity, temperature and other physical factors directly affect the occurrence and survival of biota. Overall, salinity induces the strongest horizontal environmental gradient effects, as salinity

T. Lang (✉) • L. Weirup • K. Straumer
Thünen-Institut für Fischereiökologie, Deichstraße 12, 27472 Cuxhaven, Germany
e-mail: thomas.lang@thuenen.de

L. Kotwicki • M. Czub • K. Grzelak
Institute of Oceanology Polish Academy of Sciences,
Powstancow Warszawy 55, 81-712 Sopot, Poland

J. Bełdowski et al. (eds.), *Towards the Monitoring of Dumped Munitions Threat (MODUM)*, NATO Science for Peace and Security Series C: Environmental Security, DOI 10.1007/978-94-024-1153-9_6

is decreasing from approx. 20 PSU in the western Baltic to approx. 1 PSU in the north eastern Baltic, with a mean value in the range of approx. 7 PSU.

The biodiversity of the Baltic Sea fauna is limited and encompasses unique communities of both marine and freshwater origin that show some marked spatial distribution patterns (Ojaveer and Lehtonen 2001; HELCOM 2002; Ojaveer and Kalejs 2005, 2008; Zettler et al. 2008). The diversity of most groups of organisms in the Baltic Sea is much lower than that in typical marine or freshwater areas, because, in the brackish environment, most species live at the border of their physiological salinity limit, as they are either of freshwater or marine origin (Ojaveer and Lehtonen 2001).

Salinity gradients in the Baltic Sea do not only occur at the horizontal, but also at the vertical scale. In deep areas, such as the Bornholm and Gotland Basins as well as the Gdansk Deep, a permanent halocline is observed throughout the year, thereby preventing water mixing from the surface to the bottom and creating a surface layer with salinities around 7–8 PSU and more saline heavier bottom waters with salinities of 12–14 PSU. The halocline in the deep basins of the Baltic is located at depths of about 60–80 m and prevents vertical mixing of the water column and the delivery of well-oxygenated water from the surface to the bottom, thus creating hypoxia and oxygen stress for the organisms living there. The salinity-caused stratification and its effects on oxygen concentrations have become perhaps the strongest factor influencing the biodiversity of fish and benthic communities within the Baltic deeps and explain the patterns in their distribution (Ojaveer and Lehtonen 2001; Laine 2003; Bonsdorff 2006).

The Baltic Sea ecosystem provides multiple natural services for a large human population living in the catchment area. It is also highly sensitive to many forms of human impact (Elmgren 2001; Korpinen et al. 2012; Andersen et al. 2015). For instance, benthic areas have increasingly been facing severe degradation of living conditions, directly linked with the human-induced nutrient overload. Eutrophication due to the heavy use of fertilizers for the agricultural production in the catchment zone is considered responsible for the observed dropdown of dissolved oxygen (DO) concentrations in bottom waters and the creation of the so called „benthic desert" occurring below the halocline (Diaz and Rosenberg 2008). According to Conley et al. (2009), oxygen deficiency in the Baltic Sea as a result of human activities has increased drastically during the last century.

Currently, environmental conditions, particularly in the deep basins of the Baltic Sea mostly depend on inflows of saline and oxygen-rich water from the North Sea. Significant events, so called Major Baltic Inflows (MBI), have a direct positive impact on both sediment and bottom water quality (Carstensen et al. 2014). The MBI observed in December 2014 was the first large-scale water exchange between the Baltic and North Seas during the last decade. It was estimated to have brought 198 km^3 of dense and oxygen-rich water to the Baltic Sea (Morholz et al. 2015). However, the sustainability of its effects is not clear yet.

There can be no doubt that the Baltic Sea and its fauna are facing multiple anthropogenic pressures (e.g., associated with fishery, input of nutrients and organic matter as well as hazardous substances, construction of large wind farms and large-scale extraction of seabed resources), causing temporary or permanent cumulative environmental impact including effects on biodiversity (Korpinen et al. 2012; Andersen et al. 2015). From the above, there is evidence that especially the deep basins of the Baltic Sea with their specific hydrographic conditions constitute hostile and stressful habitats for many species and that anthropogenic impact is amongst the causal factors. For a long time, in particular toxic metals (e.g., mercury, cadmium, lead) and persistent organic pollutants (POPs) (e.g., PCBs, pesticides) have been regarded as hazardous substances exerting the greatest threats to the Baltic Sea environment (HELCOM 2010), but, largely due to the development of more sensitive analytical technologies, other pollutants have increasingly gained scientific interest. Amongst these, emissions from dumped conventional and chemical munitions and unexploded ordnance have caused considerable concern because of their known toxicity and potential effects on the marine environment. The majority of chemical munitions and chemical warfare agents disposed in the Baltic Sea after WW 2 by order of the allied forces was dumped in the deep areas, mainly of the Bornholm and Gotland Basins (Bełdowski et al. 2016) and may, thus, act as additional stressor with impact on the biota inhabiting the areas. As a result of the CHEMSEA project, the Gdansk Deep was confirmed as additional dumpsite (Bełdowski et al. 2016).

Amongst the aims of the CHEMSEA and MODUM projects were the analysis and assessment of ecological risks of CWA posed to major biota in the deep basins used for dumping purposes (Bełdowski et al. 2016). In the following, results of CHEMSEA and MODUM studies on effects of CWA on the biodiversity of benthic invertebrates and on the health status of Baltic cod (*Gadus morhua*) in the deep basins of the Baltic Sea are addressed.

6.2 Benthos Communities in the Deep Basins

Since the Baltic Sea is a large brackish water body, it contains unique benthos communities of both marine and freshwater origin (Zettler et al. 2008). The species diversity is much lower than that in typical marine area since, in the brackish environment, most species live at the border of their physiological salinity limit as they are either of freshwater or marine origin. The distribution of benthic species in the Baltic Sea is controlled by strong environmental gradients. Besides temperature and other physical factors, salinity induces the strongest environmental impact, directly affecting the occurrence and survival of biota. Salinity together with oxygen concentrations best circumscribe the Baltic Sea diversity of benthic fauna and explain the patterns in its distribution (Laine 2003; Bonsdorff 2006). The permanent halocline observed in the waters of the deep basins throughout the year prevents water mixing from the surface to the bottom and, thus, delivery of well-oxygenated waters to the bottom and its benthic fauna, creating oxygen stress.

The deep soft sediments are inhabited by zoobenthos, which is divided, based on the size structure, into two main groups: larger organisms (≥1 mm) being referred to as the macrozoobenthos (macrofauna) and smaller (≥30 μm) ones constituting the meiobenthos (meiofauna). The macrofauna inhabiting deep soft bottom areas comprises typically only a few species (Bonsdorff and Pearson 1999; Laine 2003; Villnäs and Norkko 2011). The soft seabed of the deep basins is mainly inhabited by several glacial relicts such as *Monoporeia affinis* or *Saduria entomon*. Macrobenthic studies conducted in Gotland Deep showed that this community is species-poor with low densities and never reach maturity (Laine et al. 1997; Olenin 1997). As a result of Major Baltic Inflows (MBI), Baltic deeps can be recolonized by more nectobenthic species – *Pontoporeia femorata* (Crustacea) and *Bylgides sarsii* (Polychaeta) – which spend more time swimming in the water column and can even be passively transported by currents from more oxygenated waters (Olafsson and Limen 2002).

Most of the areas formerly used for the dumping of warfare materials in the Baltic Sea are deep basins, which serve as sediment traps with poorly oxygenated sediments. For all of the Baltic deeps, the mean oxygen concentration showed almost anoxic conditions, with values not exceeding 2 mg/l (see Fig. 6.1). During the CHEMSEA and MODUM projects, no macrofaunal organisms were found at 37 stations located in the investigated Baltic deeps (Kotwicki et al. 2016) (see Fig. 6.2). However, during the ROV inspections of Bornholm Deep in March 2015, numerous

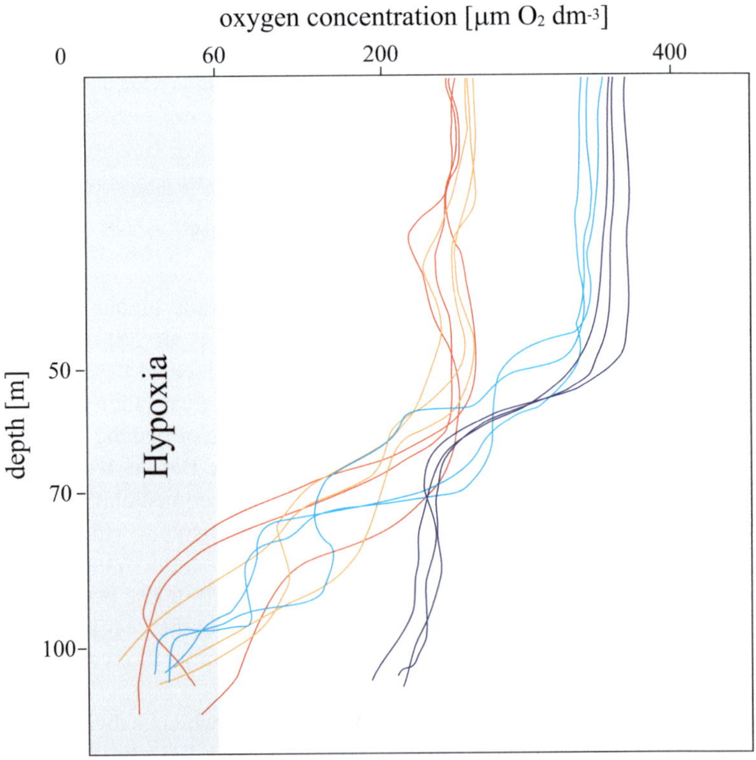

Fig. 6.1 Vertical profile of oxygen concentrations in relation to water depth in the Bornholm Deep, Gotland Deep and Gdańsk Deep in 2012–2015

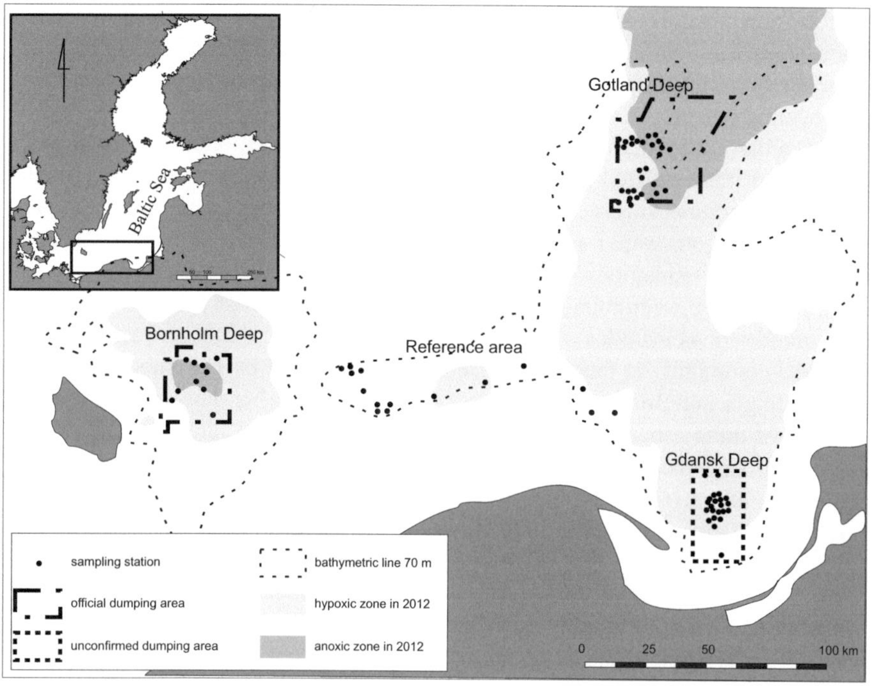

Fig. 6.2 Dumpsites of chemical warfare agents (CWA) in the deep basins of the Baltic Sea used for studies on benthos communities, with 70 m water depth conture and illustration of hypoxic and anoxic bottom water zones

active individuals of benthic amphipods (*Monoporeia* ssp) were observed, at depths exceeding 90 m, very likely as a result of the MBI happening in 2014. The permanent absence of macrozoobenthic organisms was first discovered in the Bornholm Deep in 1964–1965. The same was found at the bottom of the Gotland Deep and Gdansk Deep in the early 1970s. In later years, the seabed areas with heavily reduced macrofauna or devoid of benthic invertebrates increased in spatial extent, such areas occurring even at shallower depths of just 60–80 m in the Baltic Sea proper (Villnäs and Norkko 2011). Species simple display a wide range of sensitivity to hypoxia, anoxia, and the presence of hydrogen sulphide (Diaz and Rosenberg 1995).

As offshore soft bottom macrozoobenthic community shows a clear response to the physicochemical environmental state, particularly oxygen deficiency, meiofaunal organisms were used in the CHEMSEA and MODUM projects as a key group to explore the faunal communities inhabiting dump sites for chemical warfare agents in the Baltic deeps. The meiobenthos of the deep soft seabed of the Baltic Sea is rather poorly known. Investigations of meiofauna in the Baltic Sea were restricted to some shallow areas only (Stockholm archipelago and southern Baltic Sea). There, nematodes always dominating in number are accompanied by others taxa: gastrotrichs, harpacticoid copepods, ostracods, turbellarians, halacarids, kinorhynchs and juvenile forms of macrozoobenthos. At larger depths in the deep, muddy Baltic Sea basins experiencing hypoxia and anoxia and devoid of the macrozoobenthos, the metazoans are represented exclusively by sparsely occurring meiobenthic nematodes (Elmgren et al. 1984; Radziejewska 1989). This was confirmed in the CHEMSEA and MODUM studies, where meiofauna recorded in the CWA dumpsites consisted exclusively of nematodes. Nematode densities were relatively low, with the lowest number found in the Bornholm Deep dumpsite, where most of the stations were not inhabited by any specimen. Nevertheless, a relatively high diversity of nematodes genera was found in the Baltic deeps with highly visible differences in their communities' composition between areas. Almost 80% of the total number of nematodes at Gotland Deep constituted of Sabatieira representatives, the community at the Gdansk Deep was dominated by Halomonhystera and Terschellingia, and Bornholm basin was characterised by Microlaimus and Sabatiera genera. All these groups have been found to be extremely tolerant to low oxygen concentrations, since they can withstand harsh environmental hypoxic and even anoxic conditions (Giere 2009). Sabatieria, Terschellingia and Halomonhystera are very resistant groups; they can inhabit even sulphidic sediments (Soetaert and Heip 1995). Sabatieria species are often dominant in species-poor assemblages of disturbed environments, and even tend to increase in density under such conditions (Steyaert et al. 2007). Therefore, in many studies, Sabatieria species are used as indicators of low oxygenated conditions.

Also, *Halomonhystera disjuncta*, dominant in the Gdańsk Deep and regularly noted in the Gotland Deep, is known to be resistant to environmental stress resulting from low oxygen or elevated heavy metals concentrations (Vranken et al. 1989; Grzelak and Kotwicki 2016). Twenty percent of the analysed *H. disjuncta* population carried living eggs and/or juveniles inside the uteri, indicating an ovoviviparous strategy of reproduction (Fig. 6.3). Typically, ovoviviparity can be observed upon exposure to environmental stress induced by pollutant toxicity, extreme temperatures or natural disturbance (Tahseen 2012). Under normal conditions, the eggs of most nematode species are released to the ambient environment. Such adaptation as

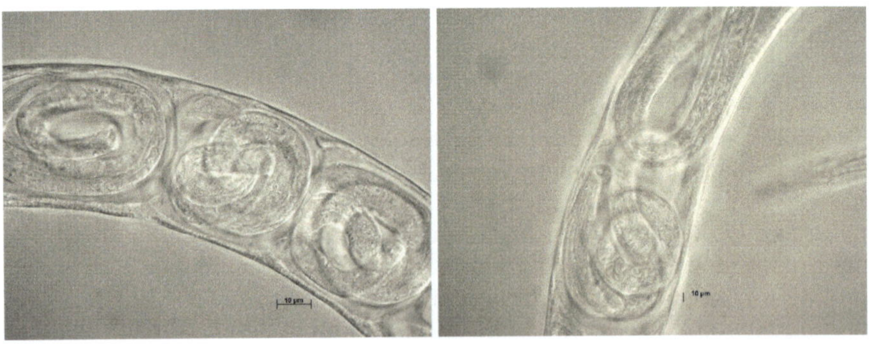

Fig. 6.3 *Halomonhystera disjuncta*, a common ovoviviparous nematode in the Gdańsk Deep and Gotland Deep

ovoviviparity is important for securing the survival and development of the brood under harsh environmental conditions. As a stress-resistant species, *H. disjuncta* may benefit from the harsh geochemical conditions occurring at the munitions disposal sites unoccupied by more sensitive species (Grzelak and Kotwicki 2016).

6.3 Fish Communities in the Deep Basins

The fish fauna of the brackish Baltic Sea encompasses both marine and freshwater species and their spatial distribution is largely driven by the pronounced west-northeast salinity gradient. Around 200 species have been recorded in the Baltic Sea including the Kattegat (Ojaveer et al. 2010), amongst which there is only a limited number of marine fish that has established successfully in the Baltic Sea and is able to reproduce in parts or the entire Baltic Sea (Ojaveer and Kalejs 2005).

The fish fauna of the Baltic Sea can be divided into pelagic species that mainly inhabit the water column and into demersal species that inhabit bottom-near waters. The pelagic fish fauna is dominated by two species, herring (*Clupea harengus*) and sprat (*Sprattus sprattus*) (Axenrot and Hansson 2004). These are also the species that are most abundant in the pelagic zone of the deep basins. Additionally, but less abundant, pelagic species such as Atlantic salmon (*Salmo salar*), trout (*Salmo trutta*), smelt (*Osmerus eperlanus*), mackerel (*Scomber scombrus*), horse mackerel (*Trachurus trachurus*), garfish (*Belone belone*) and pilchard (*Engraulis encrasicolus*) occur. Sometimes, three-spined stickleback (*Gasterosteus aculeatus*) is abundant in pelagic catches, e.g., in the Gotland Basin (Lang, unpublished data). Some of the pelagic species are residents in the Baltic Sea, others are regarded as occasional visitors, spending only part of their life in the Baltic Sea (Ojaveer and Kalejs 2005).

The biodiversity of the demersal fish fauna in the deep basins is largely influenced by the hydrographic conditions of the bottom water, which is characterized, at least during periods of stagnation, by low oxygen concentrations (see above). These may affect the general well-being, metabolisms and spawning success of fish and the availability of suitable benthic food organisms (Eero et al. 2015).

Baltic species that are bound to benthic habitats, e.g., common and widely distributed flatfish species such as flounder (*Platichthys flesus*) and plaice (*Pleuronectes platessa*) are, thus, rare in the deep basins and are apparently largely restricted to the margins of the basins with lower water depths and more favourable oxygen and feeding conditions. However, sometimes they occur in pelagic catches even in the centre of the deep basins (Lang, unpublished data).

The Atlantic cod (*Gadus morhua*) is one of the most abundant demersal fish species in the Baltic Sea, inhabiting almost the entire area, including shallow areas as well as the deep basins. The majority of cod in the deep basins belongs to the eastern Baltic cod stock, which is abundant in areas east of the island of Bornholm, including the Bornholm Sea, Gdansk Bight and the Gotland Sea, while the western cod stock is common west of Bornholm in the Arkona Sea, Mecklenburg Bight and the Belt Sea (Aro 1989). Both stocks apparently overlap in the Arkona Sea (Aro 1989;

Schmidt 2000; Bleil and Oeberst 2005a, b), but can be separated based on population genetic markers (Schmidt 2000; Nielsen et al. 2003; O'Leary et al. 2007).

As a demersal species, the cod shows a preference to stay close to the sea floor, even in the deep basins, as long as environmental conditions (i.e., oxygen concentrations) allow. However, hydrographic conditions have been unfavourable in the last few decades and eutrophication has worsened the situation. As a result, the deep basins have been chronically hypoxic most of the time for the last 30 years and young cod have moved to shallow waters, whereas adult cod have largely resorted to living in the water column, above the most hypoxic water (Uzars 1994; Samuelsson 1996; Tomkiewicz et al. 1998; Chabot and Claireaux 2008). Interestingly, there is evidence from studies carried out in the Bornholm Basin that cod can undertake excursions to deep zones (Neuenfeldt et al. 2009; Schaber et al. 2009, 2012) even at surprisingly low oxygen concentrations (<1 ml/l), possibly in order to forage on sprat which can also tolerate low oxygen concentrations (Stepputtis 2006; Hinrichsen et al. 2011). If available, they may also feed on macrozoobenthos that normally is deeply burrowed in the sediment, but emerge from the sediment to the surface when oxygen concentrations are low and is, thus, easily accessible for foraging cod (Pihl 1994; Neuenfeldt et al. 2009). According to Chabot and Claireaux (2008), cod are able to tolerate low oxygen concentrations for short periods by using anaerobic metabolism for part of their energy requirements, which may explain the finding that cod dives even into anoxic zones in the deep basins. However, anaerobic metabolism is a pathway that cannot be sustained for long periods by most organisms, ultimately resulting in death by suffocation (Chabot and Claireaux 2008). However, since the main prey items of cod in the Baltic Sea are pelagic clupeids, cod can also be abundant in the pelagic zone of the deep basins, following schools of sprat and herring.

The deep basins are important spawning grounds of the eastern cod since only they provide suitable conditions required for a successful reproduction of the stock. Spawning used to take place from June to September in the Bornholm Sea, Gotland Basin, the Gdansk Deep and Slupsk Furrow (Bagge et al. 1994). However, current data indicate that a shift took place and that present successful spawning sites are located in the deepest parts of the Bornholm Basin and even in the Arkona Sea (Bleil and Oeberst 2000; Wieland et al. 2000; Bleil and Oeberst 2005a, b). During spawning time, cod in the Bornholm Basin were shown to be aggregated in an intermediate water layer about 15 m thick, with the halocline as the upper limit and with the oxygen content of the water defining the lower limit (Tomkiewicz et al. 1998).

As a marine species, spawning success of cod is greatly influenced by hydrographic conditions. Because of their specific density, spawned and fertilised eggs of Baltic cod require a minimum salinity of ≥11 PSU, which, in the eastern Baltic Sea, is only present in the deeper water layers of the basins, to prevent them from sinking to the bottom. Furthermore, a minimum oxygen concentration of ≥2 ml/l is required for survival and a successful development of embryos (Nissling et al. 1994; Wieland et al. 1994). However, during periods of hydrographic stagnation, salinity and oxygen concentrations in the deep basins are decreasing so that anoxic conditions occur and embryos cannot survive. Only if sustainable major inflows of oxygen-rich and saline waters from the North Sea into the deep basins of the Baltic Sea occur can the cod stock reproduce successfully over longer periods of time.

Cod, sprat, and herring have always been the commercially most important fish species in the central Baltic Sea (Sparholt 1994; Köster et al. 2003). Current estimated landings of cod have been in the range of 42,000 t (data from 2014: western stock 8000 t, eastern stock 34,000 t) (http://fischbestaende.portal-fischerei.de/Fischarten/?c=stock&a=detail&stock_id=606). Compared to the late 1970s and early 1980s with maximum landings of eastern cod of around 450,000 t (Anon 1991, 1994; Thurow 1993), the landings have significantly decreased, largely due to a stock collapse associated with a high fishing pressure and unfavourable environmental conditions (see above). However, largely due to management regulations, a positive stock development took place, and the eastern Baltic cod stock was considered as recovered and stable for the first time in 2010 (ICES 2015a, b). However, thereafter fishery was unable to achieve predicted quotas, and cod, especially around the island of Bornholm, were noticed to exhibit low condition indeces. The biomass of cod turned so low that 30% of the catch was not marketable (Weirup 2015; Zimmermann and Krumme 2015).

According to Eero et al. (2015), the development of the Baltic cod stock is subject to numerous ecological impact factors that are still poorly understood. Impacts (stressors) which may negatively influence the health of cod stocks, but which have so far been largely neglected in stock assessments, might be effects of contaminants fish are exposed to and/or the infection/infestation with diseases and parasites. To assess the impact of such environmental stressors, cod has been studied on a regular basis as part of national environmental monitoring programmes (Lang 2002). The majority of munitions containing chemical warfare agents (CWA) was dumped in the deep basins of the Baltic Sea, especially in the Bornholm Basin. This area is of utmost importance for the performance of the cod stock because it serves as major spawning ground (see above). However, the Bornholm Basin is characterised by hostile conditions for the fish, mainly linked to low oxygen concentrations and a lack of benthic food organisms. At the same time, the fish and its offspring may be exposed to CWA leaking from corroding munitions or dumped CWA storage containers. Hostile conditions, contaminant exposure, spawning stress and disease may exert combined effects on the general health status of cod in the areas that may explain the present poor condition of the eastern cod stock in the Baltic Sea.

6.4 Fish Health in the Deep Basins

Systematic studies on the health status of fish stocks in the Baltic Sea started in the early 1980s (Lang 2002) and countries currently carrying out regular fish disease survey with cod as one of the main target species are Germany, Poland, Russia and Sweden (ICES 2016). Methodologies applied in existing fish disease surveys are largely based on established guidelines developed by the International Council for the Exploration of the Sea (ICES), the fish disease component of the Biological Effects Quality Assurance in Monitoring programme (BEQUALM) and on experience made during the ICES/BSRP Sea-going Workshop on Fish Disease Monitoring in the Baltic Sea (WKFDM) (ICES 1989; Dethlefsen et al. 1986; Bucke et al. 1996; Lang and Mellergaard 1999; ICES 2006, www.bequalm.org).

The majority of studies on cod diseases in the Baltic Sea have addressed externally visible diseases and parasites (Dethlefsen and Watermann 1982; Draganik et al. 1994; Kosior et al. 1997; Mellergaard and Lang 1999; Weirup 2015). However, more currently, studies on internal pathologies and parasites have been conducted (Buchmann and Kania 2012; Nadolna and Podolska 2013; Faber 2014; Haarder et al. 2014; Horbowy et al. 2016). Table 6.1 and Plate 6.I provide information on the main target diseases and parasites recorded in the course of regular monitoring surveys.

In the course of the MODUM project, six fish disease surveys were carried out on board RV Walther Herwig III (WH III) in the period 2013–2015; in the course of the previous CHEMSEA project, five surveys were conducted in the period 2011–2013 (Table 6.2) in areas of the western and eastern Baltic Sea, including CWA dumpsites (known: Bornholm Basin, Gotland Deep; suspected: Gdansk Deep). The geographical location of study areas is shown in Fig. 6.4. Study areas were selected in a way that they covered areas representing the habitat of the western (west of the island of Bornholm) and of the eastern cod stock (east of Bornholm). Known ("official" and marked in the sea charts) or suspected CWA dumpsites studied were the major CWA dumpsite east of Bornholm (area B13), the dumpsite in the Gotland Basin (area B14) and the Gdansk Deep (area B15). The areas B10 and B11 served as western reference areas and area B09 as eastern reference area, all of which considered to be free of CWA dumping.

Table 6.1 Common diseases and parasites of cod (*Gadus morhua*) in the Baltic Sea used for monitoring purposes (Bucke et al. 1996; ICES 2006 with modifications) and their mean prevalence recorded during the MODUM and CHEMSEA projects 2011–2015 (all RV cruises and study areas combined)

Disease/parasite	Type of disease/causes	Mean prevalence
Skin ulcerations (acute/healing stages)	Bacterial	5.5%
Skin ulcer pre-stages (haemorrhagic ulcers)[a]	Bacterial	4.2%
Epidermal hyperplasia/ papilloma	Viral (likely)	2.1%
Fin rot/erosion (acute/healing stages)	Bacterial	3.1%
Skeletal deformities (lordosis, scoliosis, pugheadedness)	Multifactorial	2.1%
Pseudobranchial swelling/pseudotumour (X-cell disease)	Amoeba-like parasite (likely)	1.1%
Cryptocotyle lingua	Parasitic trematode (Digenea) (metacercariae)	15.3%
Lernaeocera branchialis	Parasitic copepod	1.6%
Loma morhua[a]	Parasitic microsporidia	>62.6%
Nematode larvae in the body cavity[a]	Parasitic nematode (Anisakidae)	33.9%

[a]Included because often monitored in combination with externally visible diseases

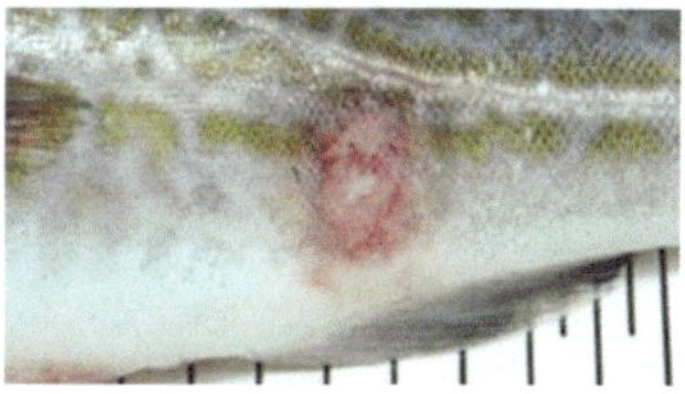	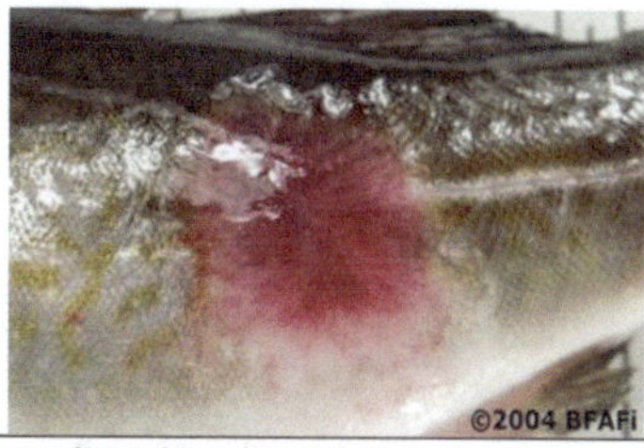
Acute skin ulceration	Healing skin ulceration
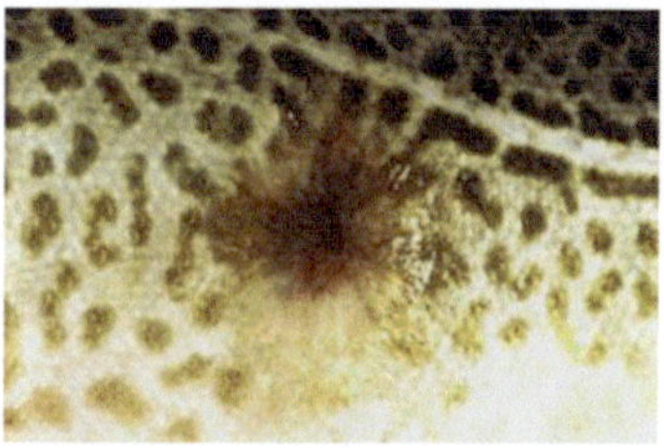	
Healed skin ulceration	Acute fin rot/erosion
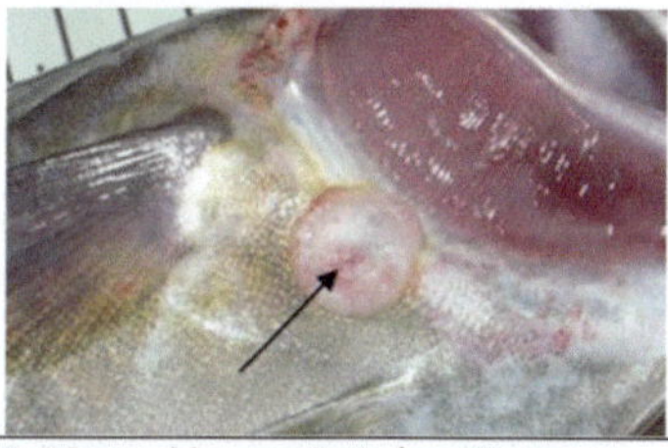	
Epidermal hyperplasia/papilloma on the skin	Skeletal deformity (compression of the spine)
	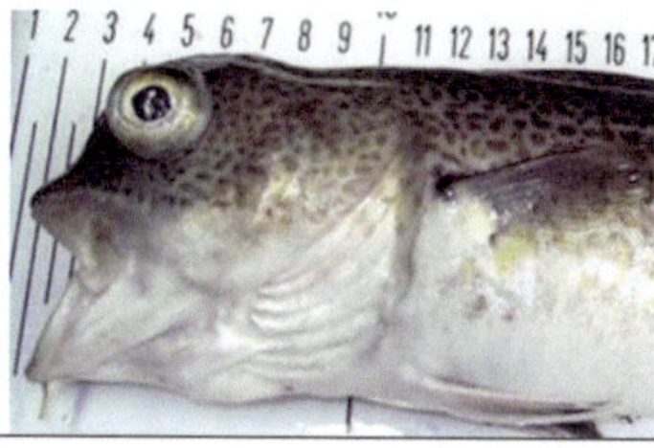
Skeletal deformity (lordosis and scoliosis)	Skeletal deformity (pugheadedness)

Plate 6.I Common grossly visible diseases and parasites of Baltic cod (*Gadus morhua*) recorded during the MODUM and CHEMSEA projects

Skeletal deformity (lordosis and scoliosis)	Skeletal deformity (pugheadedness)
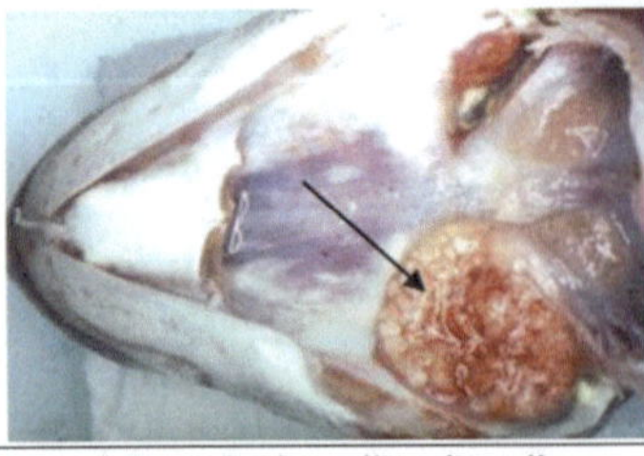	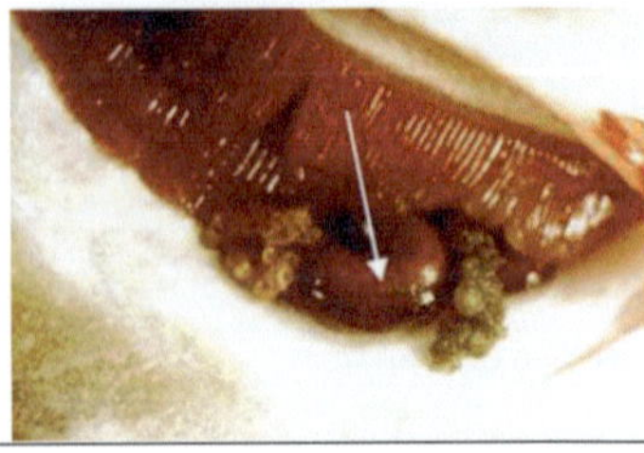
Pseudobranchial swelling (X-cell pseudotumour) in the oral cavity, upper jaw	*Lernaeocera branchialis* (Copepoda) in the gill chamber
	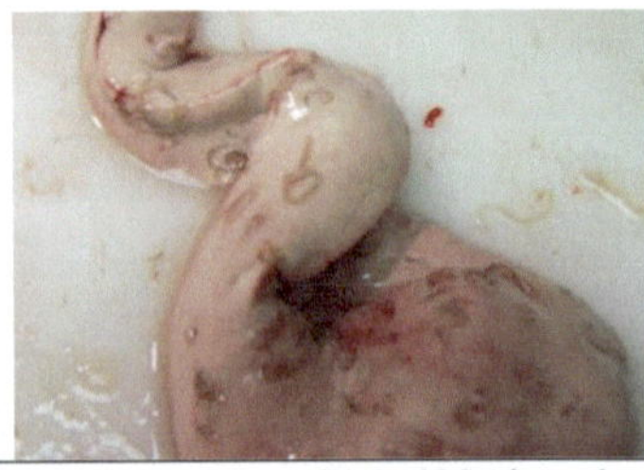
Cryptocotyle lingua (Digenea, metacercaria) in the skin	Nematode larvae (Anisakidae) on the liver surface
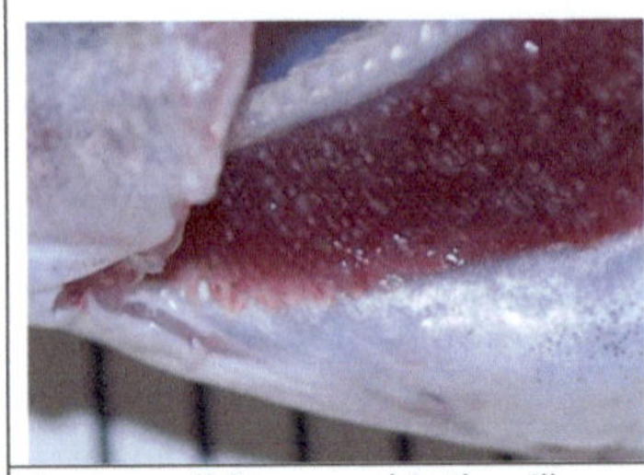	
Loma sp. (Microspora) in the gills	

Plate 6.I (continued)

In total, more than 18,000 individual cod sampled either using standard bottom or pelagic trawls (the latter in CWA dumpsites) were examined for the presence of externally visible disease and parasites. Out of these, more than 5,000 specimens were examined for gross internal diseases and selected parasites. In some fish, histopathological studies on liver lesions were carried out (Faber 2014).

Table 6.1 shows data on the mean prevalence of the disease conditions recorded over all RV cruises and study areas. From these, it can be seen that the disease conditions differ in prevalence. Prevalence of the parasitic infestations with *C. lingua*, *L. morhua* and larval nematodes was the highest. Among the infectious diseases with a bacterial or viral aetiology, acute/healing stages and pre-stages of skin

Table 6.2 RV cruise data for fish health studies during the CHEMSEA and MODUM projects in the period 2011–2015 (WH III: RV Walther Herwig III) (for location of areas see Fig. 6.1)

RV	Cruise No.	Dates	Areas	N cod examined	Project
WH III	349	02.–21.12.2011	B01, B11, B10, B13, B09, B15	1,484	CHEMSEA
WH III	354	02.–22.05.2012	B11, BHB, B13	196	CHEMSEA
WH III	357	23.08.–07.09.2012	B10	817	CHEMSEA
WH III	360	10.–20.12.2012	B01, B11, B13, B14	1,466	CHEMSEA
WH III	367	28.08.–12.09.2013	B01, B11, B10, B13, B09	2,238	CHEMSEA
WH III	370	30.11.–20.12.2013	B01, B11, B10, B13, B09	1,700	MODUM
WH III	374	14.05.–04.06.2014		271	MODUM
WH III	377	28.08.–17.09.2014	B01, B11, B10, BHB, B13, B09, B15, B14	3,974	MODUM
WH III	380	01.–19.12.2014	B01, B11, BHB, B13, B09	2,149	MODUM
WH III	387	28.08.–19.09.2015	B01, B11, B10, BHB, B13, B09, B15, B14	3,365	MODUM
WH III	390	04.–19.12.2015	B01, B11, B10, B13, B09	862	MODUM
			Sum	18,522	

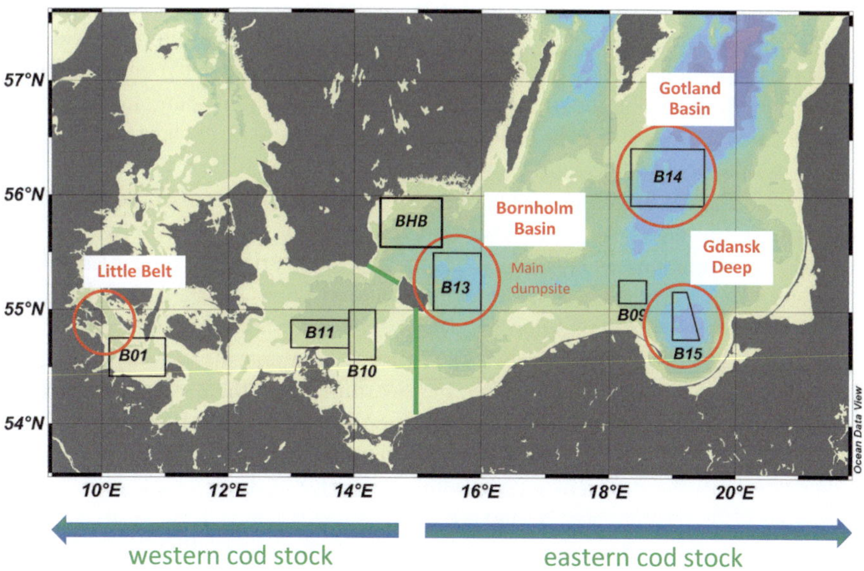

Fig. 6.4 Location of sampling areas used for fish disease studies in the course of the CHEMSEA and MODUM projects (borderline between western and eastern cod stock is marked; red circles: known or suspected dumpsites for chemical warfare agents)

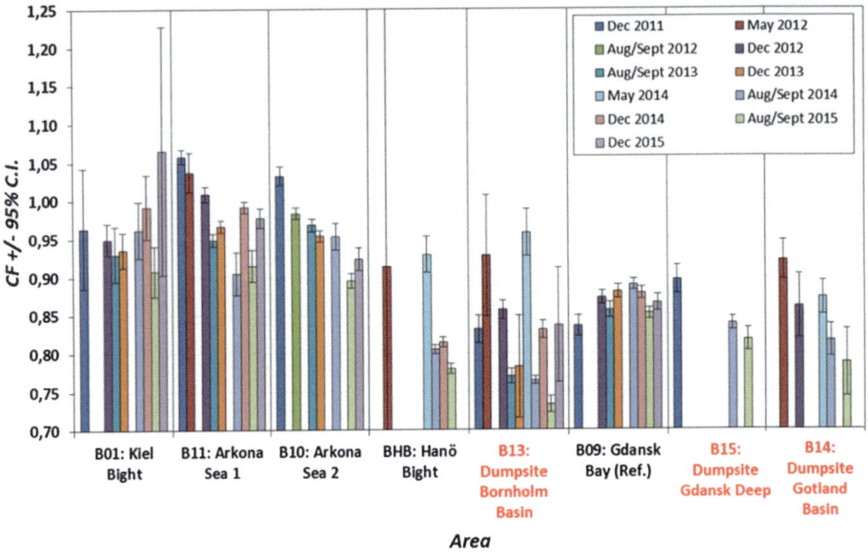

Fig. 6.5 Mean Condition Factor (based on ungutted weight) (CF and 95% confidence intervals) in cod (*Gadus morhua*) from CWA dumpsites and reference areas in the eastern and western Baltic Sea. Data from CHEMSEA and MODUM are shown for RV cruises carried out in the period 2011–2015 (for location of study areas see Fig. 6.4)

ulcerations were most prevalent, followed by fin rot/erosion. More details on the diseases can be found in Plate 6.I.

As addition to the examination for diseases and parasites, Fulton's condition factor K (CF = ungutted wet weight [g]*100/total length [cm]^3) was measured as an indicator of general fish health status. According to Ovegård et al. (2012) and references therein, simple condition indices, such as Fulton's K condition factor, are frequently used methods for measuring the energy reserves and general health status of fish.

Figure 6.5 provides the results of the calculations of the condition factor CF of cod for each study area and each cruise in the period 2011–2015. As a general observation, the data reveal that cod sampled in the reference areas west of the island of Bornholm (areas B01, B10, B11) were characterised by higher CF values compared to cod from the areas east of Bornholm (areas BHB, B13, B09, B14, B15), reflecting a worse condition in eastern than in western cod. In particular cod from areas BHB (Hanö Bight) and B13 (major CWA dumpsite east of Bornholm) had low CF values. The data indicate that in some areas there was partly strong variation in CF between cruises. This was pronounced in areas B13 (major CWA dumpsite), where the lowest CF values of all were recorded. This variation may have to do with the different seasons during which the studies took place (May, Aug./Sept., Dec.). Especially in May 2012 and May 2014, the CF values in area B13 were high, possibly reflecting preparation for spawning in the following summer in that area. In contrast, CF values in area B13 were low during Aug./Sept. 2013, 2014 and 2015, possibly reflecting post-spawning conditions. In areas B10, B11, B14 and B15 there seem to be a tendency for a decrease in CF over time, indicating that the nutritional status has worsened in the observation period. In the other areas, no such trend seems to exist.

In Fig. 6.6, temporal changes in the prevalence of diseases (incl. parasite infestation) are shown for three study areas: the major CWA dumpsite east of Bornholm (area B13), the eastern reference area outside the Gulf of Gdansk (area B09) and a

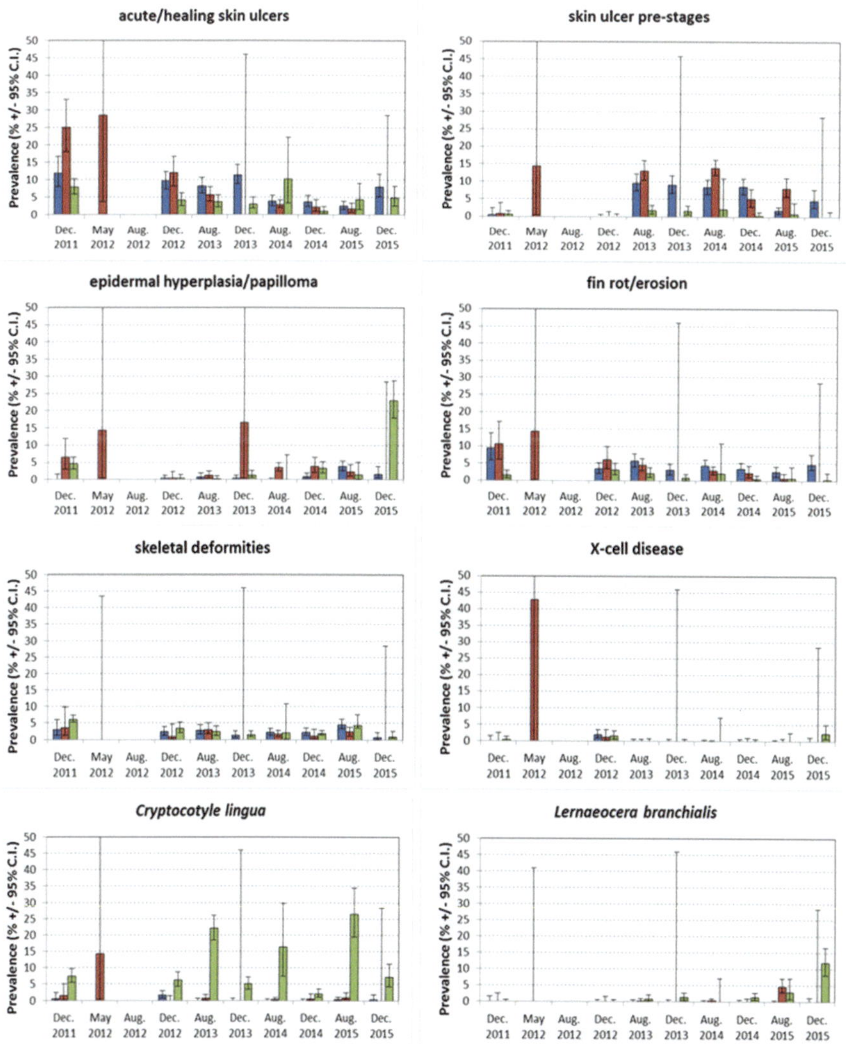

Fig. 6.6 Prevalence (% rate and 95% confidence intervals for binomial distribution) of eight common disease conditions (incl. two parasites) of Baltic cod recorded during 10 RV cruises (data from May 2014 not shown) in the period 2011–2015. Bars from left to right: reference area B09 outside the Gulf of Gdansk (blue), major CWA dumpsite B13 east of Bornholm (red), reference area B11 in the Arkona Sea (green) (data for study areas B01, B10, BHB, B14, B15 not shown) (MODUM and CHEMSEA data)

reference area in the Arkona Sea (area B11), the latter exemplifying the situation for the western cod stock (for location of areas see Fig. 6.4).

For none of the single diseases there is a clear either upward or downward temporal trend over the observation period 2011–2015 (see Fig. 6.5). However, acute/healing stages of skin ulcerations and fin rot/erosion seem to have decreased in area B13 (major CWA dumpsite), although this tendency was not statistically significant as can be seen from the overlapping 95% confidence intervals (the partly large confidence intervals were due to the sometimes low number of fish in the samples examined). For infestation with *L. branchialis*, the data indicate a slight increase (especially in area B11, the western reference area. Again, this trend was not significant, however.

Regarding consistent spatial differences, the only disease condition showing significant differences in prevalence between areas was the infestation with *C. lingua*, which was more prevalent in the western reference area B11 as well as in the other western study areas B01 and B10 (data not shown). Apart from this, no other disease condition was markedly higher or lower in prevalence than others. Overall, there is no indication of higher disease prevalences in cod from the CWA dumpsite in Bornholm Basin compared to the other study areas.

In order to assess the health status of cod as the sum of its disease conditions, the Fish Disease Index (FDI) approach for externally visible diseases was applied. FDI values were calculated for individual fish, based on the presence/absence of seven disease conditions (acute/healing skin ulcerations, acute/healing fin rot/erosion, epidermal hyperplasia/papilloma, skeletal deformities, pseudobranchial swelling, *L. branchialis*, *C. lingua*; see Table 6.2), their intensities (3 severity grades), their suspected impact on the host (weighting by expert judgment) as well as on adjustment factors compensating for effects of length and sampling season on the disease prevalence. The (individual) FDI summarizes the disease status of individual specimen by a single number which can be in the range of 0 (no disease present) to 100 (all seven diseases present with the highest severity stages) (for methodological details see Lang and Wosniok 2008; ICES 2012; Weirup 2015; Lang et al. 2017). Results of the FDI approach are shown in Fig. 6.7.

In general, the mean FDI values were low. Calculated over all cruises for the eight study areas, the mean FDI values ranged between 1.17 (Hanö Bight, area BHB) and 3.61 (CWA dumpsite Bornholm, area B13) (see Fig. 6.6). The maximum FDI value was found to be 5.58 in area B13 in Dec. 2011. Taking into account that the maximum individual FDI value can be as high as 100 if all diseases with the highest severity grades are present in one fish, the results indicate a general good average health status of the cod examined.

From the graphs, there is evidence for marked and partly statistically significant variation in FDI values between cruises. This may have to do with the different seasons during which the studies took place (May, Aug./Sept., Dec.). However, it may also be a reflection of a changing environment, e.g. associated to some strong inflows of saline and oxygen-rich water masses from the Kattegat and North Sea in the pasts winters that even reached the deep basins east of Bornholm.

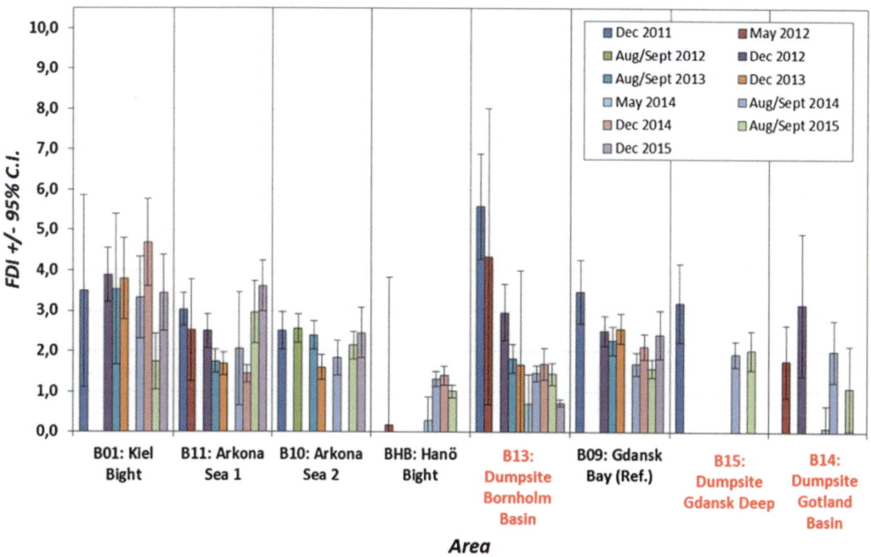

Fig. 6.7 Mean Fish Disease Index (FDI) (FDI and 95% confidence intervals) in cod (*Gadus morhua*) from CWA dumpsites and reference areas in the eastern and western Baltic Sea. Data are shown for RV cruises carried out in the period 2011–2015 (for location of study areas see Fig. 6.4)

Taking all data together and considering the variation in FDI, there is no clear indication of consistent differences in health status of cod from the various study areas. There is indication that cod from Kiel Bight (area B01) on average displayed the poorest health status and that cod from Hanö Bight (area BHB) were the healthiest. There was no big difference between eastern cod from the known or suspected CWA dumpsites in areas B13, B14, B15 and in the reference area B09. The average values in the western reference areas B10 and B11 were similar to the ones recorded in the eastern areas. However, when interpreting these finding, it has to be taken into account that some of the areas were only visited three to five times (the CWA dumpsites in Gotland Deep, area B14, and Gdansk Deep, area B15) in the period 2011–2015 while others were visited 11 times. This may have influenced the data to a certain degree.

Interestingly, there seems to be a decrease in FDI over time in the major CWA dumpsite east of Bornholm (area B13) and, to a lesser extent, also in the other eastern areas B09, B14 and B15, indicating that the health of eastern cod possibly has improved since 2011. However, taking into account the partly low number of fish examined and the resulting large confidence intervals for some of the mean FDI values, no firm conclusions can be drawn at this stage. Possible reasons for an improvement of health conditions has not yet been identified, but it cannot be excluded that there may be a link to improved oxygen conditions in this area caused by some Major Baltic Infow (MBI) events (see above).

There are two prevalent disease conditions that have so far not been incorporated in the FDI approach, namely infestations with nematodes in the body cavity and livers as well as infestations with the microsporidian parasite *Loma* sp. (likely *Loma morhua*) in the gills. As can be seen from Table 6.1, the mean prevalence of nematode infestation was 33.9% and the mean prevalence of infestation with *Loma* sp. was >62.6% (probably an underestimation of the true prevalence, since minor infestations of the gills are hard to detect because of the small size of the parasite cysts). Out of the disease conditions studied in MODUM and CHEMSEA, these two parasitic infestations were by far the most prevalent ones, and it cannot be excluded that they may have an impact on the overall health status and condition of Baltic cod, especially of the eastern stock, were the conditions are most prevalent. Therefore, future studies should focus on these two parasitoses and their population effects.

In case of nematode larvae, there is a clear trend revealing an increased prevalence in Baltic cod over time corresponding to a decrease in condition factor (Buchmann and Kania 2012; Nadolna and Podolska 2013; Haarder et al. 2014; Horbowy et al. 2016) which is also confirmed by data generated in the MODUM and CHEMSEA projects. Figure 6.8 shows the prevalence of liver nematodes in cod in the period 2013–2015, and from the data there is clear evidence that cod belonging to the eastern stock (areas BHB, B13, B09, B14, B15), characterised by low condition factors, have more frequently been infested than cod from the western stock (areas B01, B10, B11). For areas B10 and B11, there even is indication of an increase over the relatively short period 2013–2015.

According to Khan (2005), Powell et al. (2005) and Powell et al. (2006), also the infestation of fish species with *Loma* sp. may cause a reduced body condition, a decrease in energy stores (liver somatic index), mild anaemia and leukaemia and decreases in specific growth rate, with the metabolic cost of the disease being largely attributed to changes in branchial O_2 permeability due to the presence of parasite cysts and pathological tissue reactions (e.g., hyperplasia) of the gill cells. Since *Loma* sp. is very common in Baltic cod and sometimes occurs at high intensities/

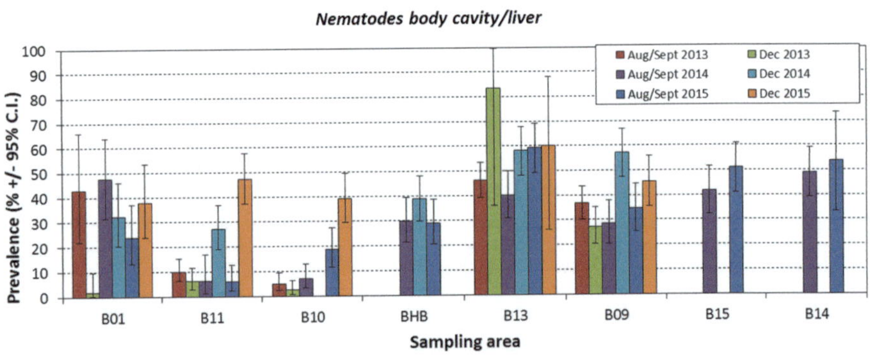

Fig. 6.8 Prevalence of infestation of Baltic cod (*Gadus morhua*) with larval nematodes (Anisakidae) in the body cavity in CWA dumpsites (B13, B14, B15) and reference areas (B01, B09, B10, B11) in the period 2013–2015

severity, it may well be causally involved in the decrease in body condition observed in eastern Baltic cod, in particular since oxygen uptake in regions with low oxygen concentrations in the water (e.g. in the Baltic deep basins) is a critical factor and affects body condition (Chabot and Claireaux 2008).

6.5 Summary and Conclusions

There can be no doubt that the deep basins of the Baltic Sea constitute specific and partly hostile habitats with characteristic and often adverse environmental conditions that affect the diversity and well-being of species in the areas. In particular salinity and oxygen concentrations are limiting factors, and there is increasing evidence that especially the latter not only affects the distribution of species, but also their general health status, condition and population performance.

In addition to suboptimal salinity and oxygen conditions, other potential and partly anthropogenic pressures may affect the health status of organisms. Besides many others, one such pressure may be dumped munitions and toxic emissions thereof, such as chemical warfare agents (CWA) and their degradation products which are still present in the deep basins in large quantities.

The results obtained through the MODUM and CHEMSEA studies on benthos communities and fish health and condition in CWA dumpsites can be considered to constitute a scientifically sound basis for future monitoring and assessment in that they provide baseline data both on spatial as well as temporal patterns of benthos biodiversity and of fitness and health indicators in fish. The results reveal that benthos and fish communities in the deep basins used as CWA dumpsites have to be regarded as stressed. However, the results did not provide clear-cut effects that can be directly linked to CWA exposure. One possible reason is that the reaction of organisms to external stressors often consist of non-specific responses and that, therefore, clear cause-effect relationships are only hard to establish. Nevertheless, some of the parameter responses measured in comparison to reference areas, such as the decrease of nematode abundance and diversity in CWA dumpsites and the low condition factors and partly higher prevalence of diseases/parasites found in cod from the main dumpsite east of Bornholm, should be taken as an alarm signal indicating that CWA exposure may have adverse ecological effects.

When interpreting the results, it has to be taken into account that cod, in contrast to benthic invertebrate organisms, is mobile and can undertake pronounced active horizontal and vertical migrations. Cod in the Baltic Sea belong to different stocks with different behaviour. For instance, western and eastern cod stocks have different horizontal migration patterns in term of timing and spatial preferences. Although these are fairly well known, it cannot be excluded that fish sampled at a given location have not stayed in the area before for longer periods and thus their biological characteristics do not reflect the conditions in the area. Furthermore, in the deep basins of the Baltic Sea (which are the major spawning areas of the eastern cod stock), cod displays pronounced vertical migration patterns (see above). During periods of stagnation, characterised by a thick bottom layers of deoxygenated water, cod moves from the bottom to the midwater, and, thus, the risk of CWA exposure at the sea bottom decreases. These factors need to be taken into account in the assessment of the results.

From the studies on diseases and parasites of Baltic cod carried out in MODUM and CHEMSEA by applying the Fish Disease Index (FDI) approach, there is some indication that the health status of cod has improved in the period 2011–2015 in some areas, e.g. the major CWA dumpsite east of Bornholm. It cannot be excluded that this was linked to some Major Baltic Inflow (MBI) events such as the one occurring in winter 2014. The associated increase in oxygen concentrations of the bottom water of the deep basins may have had some positive effects on the well-being of the cod and its resistance to diseases. However, it is by no means clear yet if this effect will be a longer lasting one. It has to be taken into account in this context that cod in the deep basins is facing a dilemma: if oxygen concentrations in the bottom waters increase to a level that lead to a recolonization of the sediments with benthos species which are suitable prey species for cod, the cod will increase its vertical migration from the midwater to the bottom for feeding purposes and will, thus, experience an increased risk of exposure to chemical warfare agents present at the sea floor. So, the positive effects of MBI may turn into negative effects through CWA exposure.

In conclusion, it seems appropriate to recommend that regular surveys on the health status of benthos and fish communities in munitions dumpsites and reference areas should be carried out, applying standardised and established methodologies. Even if cause-effect relationships are hard to establish, monitoring in dumpsites is required in order to identify ecological changes and risks. If significant changes occur, the results may serve as an "alarm bell", reflecting adverse ecological effects of environmental stressors. The surveys should, however, be part of an integrated monitoring programme, comprising chemical measurements of pollutants derived from dumped munitions and biological effects measurements at different levels of biological organisation (from the molecular to the individual and population level). Ideally, studies on specific biomarkers indicating exposure to and effects of toxic munitions compounds should be incorporated. These still need to be developed and validated as a future research priority. Whatever environmental monitoring strategies are being applied, they should always be accompanied by research activities in order to provide scientific background data required for the interpretation of monitoring data and to improve monitoring strategies and techniques, as appropriate.

References

Andersen JH, Halpern BS, Korpinen S, Murra C, Reker J (2015) Baltic Sea biodiversity status vs. cumulative human pressures. Estuar Coast Shelf Sci 161:88–92

Anon (1991) Report of the Working Group on the Assessment of Demersal Stocks in the Baltic. ICES C.M. 1991/Assess:16

Anon (1994) Report of the Working Group on the Assessment of Demersal Stocks in the Baltic. ICES. C.M. Assess:17

Aro E (1989) A review of fish migration patterns in the Baltic. Rapp P-v Réun Cons Int Explor Mer 190:72–96

Axenrot T, Hansson S (2004) Seasonal dynamics in pelagic fish abundance in a Baltic Sea coastal area. Estuar Coast Shelf Sci 60:541–547

Bagge O, Thurow F, Steffensen E, Bay J (1994) The Baltic cod. Dana 10:1–28

Bełdowski J, Klusek Z, Szubska M et al (2016) Chemical Munitions Search & Assessment—an evaluation of the dumped munitions problem in the Baltic Sea. Deep-Sea Res II Top Stud Oceanogr 136:1–132

Bleil M, Oeberst R (2000) Laichgebiete des Dorschs in der westlichen Ostsee. Inform Fischwirtsch Fischereiforsch 47(4):186–190

Bleil M, Oeberst R (2005a) Die Reproduktion von Dorschen (*Gadus morhua* L. und *Gadus morhua callarias* L.) in der Ostsee unter besonderer Berücksichtigung der Arkonasee: Teil 1: Allgemeiner Verlauf des jährlichen Reifeprozesses und der Laichaktivitäten in den verschiedenen Gebieten. Inform Fischwirtsch Fischereiforsch 52:74–82. doi:10.3220/Infn52_74-82_2005

Bleil M, Oeberst R (2005b) Die Reproduktion von Dorschen (*Gadus morhua* L. und *Gadus morhua callarias* L.) in der Ostsee unter besonderer Berücksichtigung der Arkonasee: Teil 2: Statistische Analysen zum Anteil reproduktiv aktiver Dorsche im Bezug auf gebietsspezifische Unterschiede und Gemeinsamkeiten, sowie deren mögliche Ursachen. Inform Fischwirtsch Fischereiforsch 52:83–90

Bonsdorff E (2006) Zoobenthic diversity-gradients in the Baltic Sea: continuous post-glacial succession in a stressed ecosystem. J Exp Mar Biol Ecol 330:383–391

Bonsdorff E, Pearson TH (1999) Variation in the sublittoral macrozoobenthos of the Baltic Sea along environmental gradients: a functional-group approach. Aust J Ecol 24:312–326

Buchmann K, Kania P (2012) Emerging *Pseudoterranova decipiens* (Krabbe, 1878) problems in Baltic cod, *Gadus morhua* L., associated with grey seal colonization of spawning grounds. J Fish Dis 35:861–866. doi:10.1111/j.1365-2761.2012.01409.x

Bucke D, Vethaak AD, Lang T, Mellergaard S (1996) Common diseases and parasites of fish in the North Atlantic: training guide for identification. ICES Tech Mar Environ Sci 19:27

Carstensen J, Jesper H, Gustafsson BG, Conley DJ (2014) Deoxygenation of the Baltic Sea during the last century. Proc Natl Acad Sci U S A 111(15):5628–5633

Chabot D, Claireaux G (2008) Environmental hypoxia as a metabolic constraint on fish: the case of Atlantic cod, *Gadus morhua*. Mar Pollut Bull 57:287–294

Conley DJ, Björck S, Bonsdorff E, Carstensen J, Destouni G et al (2009) Hypoxia-related processes in the Baltic Sea. Environ Sci Technol 43:3412–3420

Dethlefsen V, Watermann B (1982) Diseases of major fish species in western Baltic Sea. ICES CM 1982/E:19, p 20

Dethlefsen V, Egidius E, McVicar AH (1986) Methodology of fish disease surveys. Report of an ICES Sea-going Workshop held on RV 'Anton Dohrn' 3–12 January 1984. ICES Cooperative Research Report 140, p 33

Diaz RJ, Rosenberg R (1995) Marine benthic hypoxia: a review of its ecological effects and behavioural responses of benthic macrofauna. Oceanogr Mar Biol Annu Rev 33:245–303

Diaz RJ, Rosenberg R (2008) Spreading dead zones and consequences for marine ecosystems. Science 321:926–929

Draganik B, Grygiel W, Kuczynski J, Radtke K, Wyszynski M (1994) Results of the screening of fish diseases in the southern Baltic. ICES CM 1994/J:20, p 19

Eero M, Hjelm J, Behrens J, Buchmann K, Cardinale M, Casini M, Storr-Paulsen M (2015) Eastern Baltic cod in distress: biological changes and challenges for stock assessment. ICES J Mar Sci. doi:10.1093/icesjms/fsv109

Elmgren R (2001) Understanding human impact on the Baltic ecosystem: changing views in recent decades. Ambio 30:222–231

Elmgren R, Rosenberg R, Andersin A-B, Evans S, Kangas P, Lassig J, Leppäkoski E, Varmo R (1984) Benthic macro- and meiofauna in the Gulf of Bothnia (Northern Baltic). Finn Mar Res 250:3–18

Faber MN (2014) Studies of liver histopathology in cod (*Gadus morhua*) from chemical warfare agent dumpsites in the Baltic Sea. Master thesis, Humboldt University, Berlin, Germany, p 115

Giere O (2009) Meiobenthology. The microscopic motile Fauna of aquatic sediments, 2nd edn. Springer, Berlin

Grzelak K, Kotwicki L (2016) *Halomonhystera disjuncta* – a young-carrying nematode first observed for the Baltic Sea in deep basins with in chemical munitions disposal sites. Deep-Sea Res II 128:131–135

Haarder S, Kania PW, Galatius A, Buchmann K (2014) Increased *Contracaecum osculatum* infection in Baltic cod (*Gadus morhua*) livers (1982–2012) associated with increasing grey seal (*Halichoerus gryphus*) populations. J Wildl Dis 50(3):537–543

HELCOM (2002) Fourth Periodic Assessment of the State of the Marine Environment in the Baltic Sea Area, 1994–1998. Baltic Sea Environment Proceedings 82B

HELCOM (2010) Hazardous substances in the Baltic Sea – an integrated thematic assessement of hazardous substances in the Baltic Sea. Baltic Sea Environment Proceedings No. 120B

Hinrichsen H-H, Huwer B, Makarchouk A, Petereit C, Schaber M, Voss R (2011) Climate-driven long-term trends in Baltic Sea oxygen concentrations and the potential consequences for eastern Baltic cod (*Gadus morhua*). ICES J Mar Sci 68(10):2019–2028

Horbowy J, Podolska M, Nadolna-Altyn K (2016) Increasing occurrence of anisakid nematodes in the liver of cod (*Gadus morhua*) from the Baltic Sea: does infection affect the condition and mortality of fish? Fish Res 179:98–103

ICES (1989) Methodology of fish disease surveys. Report of an ICES Sea-going Workshop held on RV U/F 'Argos' 16–23 April 1988. ICES Cooperative Research Report 166, pp 33

ICES (2006) Report of the ICES/BSRP Sea-going Workshop on Fish Disease Monitoring in the Baltic Sea (WKFDM). ICES CM 2006, BCC:02, p 85

ICES (2012) Report of the Working Group on Pathology and Diseases of Marine Organisms. ICES CM 2012/SSGHIE:03, 48–61

ICES (2015a) Report of the Advisory Committee, 2015. Book 8. Baltic Sea. 8.3.3 Cod (*Gadus morhua*) eastern Baltic stock in Subdivisions 25–32 (Eastern Baltic Sea) and Subdivision 24

ICES (2015b) Report of the Advisory Committee, 2015. Book 8. Baltic Sea. 8.3.2 Cod (*Gadus morhua*) western Baltic stock in Subdivisions 22–24 (Western Baltic Sea)

ICES (2016) Interim Report of the Working Group on Pathology and Diseases of Marine Organisms (WGPDMO). ICES CM 2016/SSGEPI:07

Khan R (2005) Prevalence and influence of *Loma branchialis* (Microspora) on growth and mortality in Atlantic cod (*Gadus morhua*) in coastal Newfoundland. J Parasitol 91:1230–1232

Korpinen S, Meskia L, Andersen JH, Laamanena M (2012) Human pressures and their potential impact on the Baltic Sea ecosystem. Ecol Indic 15:105–114

Kosior M, Grygiel W, Kuczynski J, Radtke K, Wyszynski M (1997) Assessment of the health state of fish of the southern Baltic; observations of externally visible symptoms of diseases. Bull Sea Fish Inst Gdynia Poland 3:3–25

Köster FW, Möllmann C, Neuenfeldt S, Vinther M, St. John MA, Tomkiewicz J, Voss R, Hinrichsen H-H, Mac Kenzie B, Kraus G, Schnack D (2003) Fish stock development in the Central Baltic Sea (1974-1999) in relation to variability in the environment. ICES Mar Sci Symp 219:294–306

Kotwicki L, Grzelak K, Beldowski J (2016) Benthic communities in the chemical munitions dumping sites of the Baltic deeps with special focus on nematodes. Deep-Sea Res II 128:123–130

Laine AO (2003) Distribution of soft-bottom macrofauna in the deep open Baltic Sea in relation to environmental variability. Estuar Coast Shelf Sci 57:87–97

Laine AO, Sandler H, Andersin A-B, Stigzelius J (1997) Longterm changes of macrozoobenthos in the eastern Gotland basin and the Gulf of Finland (Baltic Sea) in relation to the hydrographical regime. J Sea Res 38:135–159

Lang T (2002) Fish disease surveys in environmental monitoring: the role of ICES. ICES Mar Sci Symp 215:202–212

Lang T, Mellergaard S (1999) The BMB/ICES Sea-going Workshop 'fish diseases and parasites in the Baltic Sea' – introduction and conclusions. ICES J Mar Sci 56:129–133

Lang T, Wosniok W (2008) The Fish Disease Index: a method to assess wild fish disease data in the context of marine environmental monitoring. ICES CM 2008/D:01, p 13

Lang T, Feist SW, Stentiford GD, Bignell JP, Vethaak AD, Wosniok W (2017) Diseases of dab (*Limanda limanda*): analysis and assessment of data on externally visible diseases, macroscopic liver neoplasms and liver histopathology in the North Sea, Baltic Sea and off Iceland. Mar Environ Res 124:61–69

Mellergaard S, Lang T (1999) Diseases and parasites of Baltic cod (*Gadus morhua*) from the Mecklenburg Bight to the Estonian coast. ICES J Mar Sci 56:164–168

Morholz V, Naumann M, Nausch G, Krüger S, Gräwe U (2015) Fresh oxygen for the Baltic Sea – an exceptional saline inflow after a decade of stagnation. J Mar Syst 148:152–166

Nadolna K, Podolska M (2013) Anisakid larvae in the liver of cod (*Gadus morhua*) L. from the southern Baltic Sea. J Helminthol. doi:10.1017/S0022149X13000096

Neuenfeldt S, Andersen KH, Hinrichsen H-H (2009) Some Atlantic cod *Gadus morhua* in the Baltic Sea visit hypoxic water briefly but often. J Fish Biol 75:290–294

Nielsen EE, Hansen MM, Ruzzante DE, Meldrup D, Gronkjaer P (2003) Evidence of a hybrid-zone in Atlantic cod (*Gadus morhua*) in the Baltic and the Danish Belt Sea revealed by individual admixture analysis. Mol Ecol 12:1497–1508

Nissling A, Kryvi H, Vallin L (1994) Variation in egg buoyancy of Baltic cod (*Gadus morhua*) and its implications for egg survival in prevailing conditions in the Baltic Sea. Mar Ecol Prog Ser 110:67–74

O'Leary DB, Coughlan J, Dillane E, McCarthy TV, Cross TF (2007) Microsatellite variation in cod (Gadus morhua) throughout its geographic range. J Fish Biol 70:310–335

Ojaveer E, Kalejs M (2005) The impact of climate changes on the adaptation of marine fish in the Baltic Sea. ICES J Mar Sci 62(7):1492–1500

Ojaveer E, Kalejs M (2008) On ecosystem-based regions in the Baltic Sea. J Mar Syst 74:672–685

Ojaveer E, Lehtonen H (2001) Fish stocks in the Baltic Sea: finite or infinite resource? Ambio 30:4–5

Ojaveer H, Jaanus A, Mac Kenzie BR, Martin G, Olenin S, Radziejewska T, Telesh I, Zettler ML, Zaiko A (2010) Status of biodiversity in the Baltic Sea. PLoS One 5(9):e12467. doi:10.1371/journal.pone.0012467

Olafsson E, Limen H (2002) Recovery of soft-bottoms after anoxic events: laboratory experiments with the amphipod Monoporeia affinis from the Baltic Sea. Ophelia 56(2):121–134

Olenin S (1997) Benthic zonation of the eastern Gotland Basin, Baltic Sea. Neth J Aquat Ecol 30:265–282

Ovegård M, Berndt K, Lunneryd S-G (2012) Condition indices of Atlantic cod (*Gadus morhua*) biased by capturing method. ICES J Mar Sci 69(10):1781–1788

Pihl L (1994) Changes in the diet of demersal fish due to eutrophication-induced hypoxia in the Kattegat, Sweden. Can J Fish Aquat Sci 51(2):321–336

Powell MD, Speare DJ, Daley J, Lovy J (2005) Differences in metabolic response to *Loma salmonae* infection in juvenile rainbow trout *Oncorhynchus mykiss* and brook trout *Salvelinus fontinalis*. Dis Aquat Org 67:233–237

Powell MD, Speare D, Becker JA (2006) Whole body net ion fluxes, plasma electrolyte concentrations and haematology during a *Loma salmonae* infection in juvenile rainbow trout, *Oncorhynchus mykiss* (Walbaum). J Fish Dis 29:727–735

Radziejewska T (1989) Large-scale spatial variability in the southern Baltic meiobenthos distribution as influenced by environmental factors. In: Styczyńska-Jurewicz E, Klekowski R (eds) Proceedings of the 21st EMBS, Polish Academy of Sciences, Institute of Oceanology, pp 403–412

Samuelsson M (1996) Interannual salinity variations in the Baltic Sea during the period 1954–1990. Cont Shelf Res 16:1463–1477

Schaber M, Hinrichsen H-H, Neuenfeldt S, Voss R (2009) Hydroacoustic tracking of individuals in environmental gradients—Baltic cod (Gadus morhua L.) vertical distribution during spawning. Mar Ecol Prog Ser 377:239–253

Schaber M, Hinrichsen H-H, Gröger J (2012) Seasonal changes in vertical distribution patterns of cod (*Gadus morhua*) in the Bornholm Basin, central Baltic Sea. Fish Oceanogr 21(1):33–43

Schmidt C (2000) Populationsgenetische Untersuchungen am Ostseedorsch (*Gadus morhua* L.). Dipl. thesis, University of Kiel

Soetaert K, Heip C (1995) Nematode assemblages of the deep sea and shelf break sites in the North Atlantic and Mediterranean Sea. Mar Ecol Prog Ser 125:171–183

Sparholt H (1994) Fish species interactions in the Baltic Sea. Dana 10:131–162

Stepputtis D (2006) Distribution patterns of Baltic sprat (Sprattus sprattus L.)—causes and consequences. PhD thesis, University of Kiel, p 153

Steyaert M, Moodley L, Nadong T (2007) Responses of inter-tidal nematodes to short-term anoxic events. J Exp Mar Biol Ecol 345:175–184

Tahseen Q (2012) Nematodes in aquatic environments: adaptations and survival strategies. Biodivers J 3:13–40

Thurow F (1993) Fish and fisheries in the Baltic Sea. ICES Coop Res Rep 186:20–36

Tomkiewicz J, Lehmann KM, St. John MA (1998) Oceanographic influences on the distribution of Baltic cod, *Gadus morhua*, during spawning in the Bornholm Basin of the Baltic Sea. Fish Oceanogr 7:48–62

Uzars D (1994) Feeding of cod (*Gadus morhua callarias* L.) in the Central Baltic in relation to environmental changes. ICES Mar Sci Symp 198:612–623

Villnäs A, Norkko A (2011) Benthic diversity gradients and shifting baselines: implications for assessing environmental status. Ecol Appl 21(6):2172–2186

Vranken G, Tiré C, Heip C (1989) Effect of temperature and food on hexavalent chromium toxicity to the marine nematode *Monhystera disjuncta*. Mar Environ Res 27:127–136

Weirup L (2015) Diseases and parasites of Baltic cod (*Gadus morhua* L.): Spatio-temporal patterns and host effects. Master thesis, University of Hamburg, Germany, p 83

Wieland K, Waller U, Schnack D (1994) Development of Baltic cod eggs at different levels of temperature and oxygen content. Dana 10:163–177

Wieland K, Jarre-Teichmann A, Horbowa K (2000) Changes in the timing of spawning of Baltic cod: possible causes and implications for recruitment. ICES J Mar Sci 57:452–464

Zettler ML, Schiedel D, Glockzin M (2008) Zoobenthos. In: Feistel R, Nausch G, Wasmund N (eds) State and evolution of the Baltic Sea, 1952–2005. A detailed 50-year survey of meteorology and climate, physics, chemistry, biology, and marine environment. Wiley, Haboken, pp 517–541

Zimmermann C, Krumme U (2015) Ostseedorsch am Tropf der Nordsee: Gut für die Umwelt ist nicht immer gut für die Fischbestände. Forschungsreport Ernähr Landwirtsch Verbrauchersch (1):40–43

Chapter 7
Estimation of Potential Leakage from Dumped Chemical Munitions in the Baltic Sea Based on Two Different Modelling Approaches

Jaromir Jakacki, Maria Golenko, and Victor Zhurbas

Abstract During the MODUM project two independent methods for estimation of potential leakage of dumped chemical munitions in the Baltic Sea have been developed. The first one is Lagrangian tracking of particles with random disturbance. The second one is using a passive tracer as a marker of potential leakage. The approaches have been developed in open source ocean models adapted for the Baltic Sea. But the models are quite different. The walking particles approach has been developed in the Princeton Ocean Model, which is nonlinear, free surface, hydrostatic, σ-coordinate, with an imbedded second and a half moment turbulence closure submodel. The passive tracer was implemented in the Parallel Ocean Program – a z-level coordinate, general circulation ocean model that solves 3-dimensional primitive equations for stratified fluid, using the hydrostatic and Boussinesq approximations. Because of many differences in our approaches we skipped a detailed comparison of the presented results (however, this will be the subject of the next stage in our work). Although the approaches and the models are quite different, the results are comparable.

7.1 Introduction

The Baltic Sea, with an area of ca. 377,000 km^2 and a volume of about 22,000 km^3, is one of the largest brackish water basins in the world. It spans between the latitude of 53°N and 66°N and the longitude of 10°E to 30°E, and it occupies a basin formed by glacial erosion. Connected to the North Atlantic via the Danish Straits, this shelf

J. Jakacki (✉)
Institute of Oceanology Polish Academy of Sciences, Sopot, Poland
e-mail: jjakacki@iopan.gda.pl

M. Golenko • V. Zhurbas
P.P. Shirshov Institute of Oceanology, Russian Academy of Sciences, Moscow, Russian Federation

J. Bełdowski et al. (eds.), *Towards the Monitoring of Dumped Munitions Threat (MODUM)*, NATO Science for Peace and Security Series C: Environmental Security, DOI 10.1007/978-94-024-1153-9_7

sea is enclosed by Denmark, Sweden, Finland, Russia, Estonia, Latvia, Lithuania, Poland and Germany. It is relatively shallow, with average depth of 52 m, and only 10% of its area is deeper than 100 m. The maximum depth is 459 m and the deepest point is located at the Landsort Deep near the Swedish coast. Due to its complex coastline and topography, the Baltic Sea can be divided into several separate regions: Gulf of Bothnia, Bay of Bothnia, Gulf of Finland, Gulf of Riga, Baltic Proper, Danish Straits and Kattegat. The largest part – the Baltic Proper – contains over 50% of the water volume. The Sea consists of many smaller topographical elements which are, despite their size, essential for hydrological properties of the basin. It is exposed to the influence of a number of changing environmental conditions, such as weather characteristics, freshwater discharge from rivers and inflows from the North Sea. Large number of rivers having their estuaries in the Baltic Sea bring a substantial supply of freshwater and nutrients. On the other hand, the exchange of water with the North Sea through the Danish Straits is limited. This leads to a very strong, permanent vertical stratification with respect to water density and salinity. Thermal stratification can also be observed. However, the depth of thermocline and the characteristics of stratification change significantly over a year. The average salinity in the Baltic Sea is ca. 7 PSU. However, in partly enclosed bays with major freshwater inflows and further away from the Danish Straits it is much lower. Salty and heavier seawater flowing from the North Sea is propagating near the bottom, thus salinity can reach much higher values in deeper parts of the Baltic Proper and near the Danish Straits. In winter a considerable part of the Baltic Sea is covered with ice. The average annual maximum coverage reaches ca. 45% of its freshwater as well as marine species, thus it is a unique area for in-situ experiments, remote measurements and modelling (Osinski 2007).

The processes of erosion and corrosion could cause leakage of the substances contained in chemical warfare agents (CWAs). If contents of the chemical weapon is for example sulphur mustard (or another dangerous chemical compound), it is dangerous for many forms of life, including mammals, birds and fishes. Dissolved dangerous material could be spread by natural and anthropogenic processes. Taking into consideration the surf zone effect, the common conclusion is that the horizontal relocation of large and heavy warfare materials is caused by anthropogenic activities. The main drivers that will have influence on scattering of the pollution are hydrodynamic forces in the dumped area. However, it could be disturbed by anthropogenic activities. We do not assume that the munitions could be moved from one place to another (however, it is possible for example to drag the shells or containers), we are thinking about anthropogenic processes that could have some influence on natural processes. For example, trawling fishing nets could have an important influence on seabed and bottom currents structure, and as a consequence the advection and diffusion processes will differ from undisturbed conditions. Natural processes that could have influence on spread of pollution are well known (however, the influence of each one of them could be difficult to estimate). Natural processes potentially causing spreading of chemical warfare agents can be divided into those which occur on a permanent basis, those which happen frequently and those which take place only occasionally (Knobloch et al. 2013):

1. Permanent – low force – diffusion from sources (e.g. resulting in contamination of adjacent sediments, pore water and water in the immediate vicinity of the leaking chemical munitions).
   Chemical warfare materials may be completely or partially buried in the sediment or they can be lying on the surface of the seafloor. Once the integrity of an encasement is breached, the contents of any object will spread due to the processes of advection and diffusion within the sediments and the water in the immediate vicinity of the point source. While advection is related to the movement of ambient media in duration and velocity, molecular diffusion follows any relative difference of concentration, which is a very slow process. It needs to be noted, however, that muddy sediments dominate the former dumping areas in the Baltic Sea and their permeability is so small that the process of dispersion caused by advection currents – either induced by density or resulting from the pressure gradient – can be disregarded. It is important to add that although diffusion processes are very slow, they could be accelerated by turbulences which can appear frequently or occasionally.
2. Frequently – low to medium force – horizontal currents of ordinary magnitude; disturbance by biota (bio-turbation); and vertical transportation with biodegradatively generated gas in the sediment and pore water from deeper layers, squeezed out due to the increasing weight of settling particles.
   The water of the Baltic Sea is circulating, which is most notably forced by wind but also by differences in water temperature and/or salinity and oxygenation levels. Strong forces may only occur in shallow waters and near to the shore. For instance, the maximum speed of currents in the Bornholm Basin has been measured at 20 cm/s at 5 m above the seabed and up to 40 cm/s at 40 m over the seafloor (Garnaga and Stankevičius 2005; Missiaen et al. 2010). In addition to horizontal movements, the water also undergoes vertical mixing. Findings made after the CHEMU Report (Paka and Spiridonov 2002) show that other noteworthy near-bottom turbulences occur in deeper waters as well. Two effects that might lead to the expulsion of contaminants from the seabed into the water column, and which could then be relocated by horizontal currents, have been described: settling sediments – the increasing weight of the growing and settling of the topmost sediment layer results in the expulsion of water from deeper sediment layers that might carry micro-particles/contaminants. Rising gas – generated by anaerobe biological degradation fermentation gases can form small bubbles rising upwards and causing microturbulences on the way, resulting in particles/contaminants being dragged along and ejected from the seabed. Diverse species of biota inhabit the seabed or visit it regularly to feed on benthic organisms. When animals dig burrows into the seafloor or scour the sediment for prey, the layers of particles are mixed and released into the near-bottom water. Contaminated sediments will also most likely respond to these disturbances of the seafloor or accompanying near-bottom water movements.

3. Occasionally – stronger force – extraordinary events like the inflow of cold, salty and oxygenated water from the North Sea into deep basins of the Baltic Sea; strong currents caused by storm surges and ice (in more shallow or coastal waters).
   Occasionally, oxygenated and salty water flows from the North Sea into the Baltic Sea. Due to its higher density, this water will replace the oxygen-deficient and less salty water of the deep basins from the deeper slopes towards the ridges. This salt water influx is recognized to have an important biological effect on the Baltic Sea's ecosystem. It used to occur on average every four to 5 years until the 1980s, but in recent decades it has become less frequent. The latest three inflows occurred in 1983, 1993 and 2003, suggesting a new inter-inflow period of about 10 years. Also, the dense waters are transported into the Baltic Proper (for example through the Słupsk Sill and the Słupsk Channel) and into northern parts of the Baltic Sea. The processes that have influence on the Baltic Sea inflows are complex and extreme inflows (the most important for the Baltic Sea ecosystem) are still not completely clear. But factors that play important role here are: water level in Kattegat and the Baltic Sea, salinity and temperature of the Kattegat water, the Baltic Sea oscillations and vertical structure of the water column. There are many papers that describe inflows from different points of view (Rak 2016; Mohrholz et al. 2015; Matthäus 2006). Other occasional strong force could be generated by strong wind-driven currents generated by storm surges. Also, mesoscale structures such as eddies could add another energy pack that could have influence on the level of pollution. Furthermore, ice cover could have indirect impact on the spreading process. Ice has strong influence on heat and momentum transfer into lower parts of water column, thus, it can block energy that could be transferred into the deeper part of the sea. The potential impact of these processes also depends on many local factors such as water depth; the depth of buried objects or point sources in the sediment; the composition of the upper layer of the seabed; and the temperature of the water.

7.2 Materials and Methods

In our work we present the results from different approaches that could be implemented as a pollution marker in the case of potential leakage of CWAs. One of them is based on analysing passive tracer implemented into z-coordinate model structure. The second one is Lagrangian tracking (LTR) with 'random' disturbance. Based on the consideration of the motion of individual particles due to mean currents, turbulence and gravitational settling (a random walk model) were applied to predict the propagation of substances in the sea. LTR was implemented into sigma coordinate model.

7.2.1 Passive Tracer

Under the MODUM project a hydrodynamic model, which is a part of a coupled ice-ocean model, was adapted for estimation of potential pollution in the case of chemical warfare agents (CWAs) leakage. The coupled ice-ocean model is a regional adaptation of the Community Earth System Model (CESM, Craig et al. 2012; Kay et al. 2015) for the Baltic Sea. Our regional adaptation consists of two active components: ocean model and ice model. Although we can assume that ice cover does not have direct influence on the bottom currents, it is an important part of heat and momentum budget, thus, it should not be ignored. The main part – the ocean model – is based on the Los Alamos National Laboratory (LANL) **P**arallel **O**cean **P**rogram (POP, Smith 2004), which evolved from the (Semtner 1974) global ocean model with added free surface formulation (Killworth et al. 1991). It is a z-level coordinate, general circulation ocean model (GCM) that solves 3-dimensional primitive equations for stratified fluid, using the hydrostatic and Boussinesq approximations. Numerically, the model computes spatial derivatives in the spherical coordinates using the finite difference technique. Placement of the model variables in the horizontal direction is according to an Arakawa B-grid (Arakawa and Lamb 1977). The barotropic equation is solved using a preconditioned conjugate gradient solver (PCG, Hu et al. 2013), the centred differencing represents the advection scheme. A biharmonic operator was chosen as a horizontal mixing parameterisation and a simple K-profile parameterisation to cover vertical mixing. We also used the equation of state introduced by (McDougall et al. 2003).

The ocean model is coupled through the "flux coupler" also called cpl7, (Craig et al. 2005) and (Craig et al. 2012) with the sea ice model (**C**ommunity **I**ce **C**od**E** – CICE model, Craig et al. 2014). The CICE uses an elastic-viscous-plastic ice rheology (Hunke and Dukowicz 1997; Hunke 2001). The Los Alamos CICE model is the result of an effort to develop a computationally efficient sea ice component for a fully coupled atmosphere-ice-ocean-land global climate model. It was designed to be compatible with the POP for use on massively parallel computers. CICE has several interacting components: a thermodynamic model (Bitz and Lipscomb 1999) that computes local growth rates of snow and ice due to vertical conductive, radiative and turbulent fluxes, along with snowfall; a model of ice dynamics which predicts the velocity field of the ice pack based on a model of the material strength of the ice; a transport model that describes advection of the ice area concentration, ice volumes and other variables of the state; and a ridging parameterisation that transfers ice among thickness categories based on energetic balances and rates of strain (Lipscomb et al. 2007). The CICE also has multiple thickness categories and ice thickness distribution evolves over time.

The horizontal resolutions of POP and CICE are identical and equal to ca. 2.3 km. The vertical resolution of the ocean model is 5 m for the whole Baltic Sea. Initial conditions of the hydrodynamic model have been prepared based on salinity and temperature from a climatological data set for the Baltic Sea (Jansen et al. 1999), the model bathymetry has been interpolated from high-resolution spherical grid topography (Seifert et al. 2001). The spinup time was 24 years (it was limited by the availability of sea level data for lateral boundary). The model integration is performed on 8 nodes (96 cores), and the simulation time depends on ice cover and it usually takes less than 1 h.

The model needs lateral boundary conditions to provide correct solutions for the areas where the model domain ends. The Baltic Sea is connected to the North Sea through Kattegat and Skagerrak. For proper representation of the flow through the Danish Straits (DS) it is sufficient to reproduce the sea level over the DS and it is also necessary to set salinity and temperature values at the boundary area (Stevens 1990; Meier and Kauker 2003). The sea level data from Goteborg have been assimilated in the Kattegat area. The data were provided by the Swedish Meteorological and Hydrological Institute (SMHI), however, all of the Baltic sea level stations are publicly available.

The presented results were obtained with the Baltic Sea model using atmospheric forcing from local forecast model UM run by the Interdisciplinary Modeling Centre at the University of Warsaw (ICM, approx. 4 km horizontal resolution, 2010–2014). Atmospheric forcing variables required by the model include:

- temperature at 2 m,
- specific humidity at 2 m,
- wind speed and direction at 10 m,
- mean sea level pressure,
- short and long wave radiation downward,
- total snow and rain precipitation.

Data from the atmospheric model have been interpolated into the Baltic Sea model domain outside of the model and the data provided as atmospheric forces have been entered from data files.

In its current configuration the model has horizontal resolution of ca. 2.3 km and 5 m thickness layers (vertical resolution) in most of the Baltic Sea area. Biharmonic operator represents the horizontal mixing parameterization, vertical mixing was implemented as a modified k-profile parameterization (KPP, (Large et al. 1994). KPP was designed for ocean models, but for shallow waters like the Baltic Sea the best vertical mixing parameterization was introduced by (Mellor and Yamada 1982) (level 2.5, MY), which is not available in the POP model (it is for sigma coordinates). To have something similar to MY, vertical viscosity function was modified. Also variable roughness was introduced in the bottom layer.

The main modification in the model was introducing the passive tracer that represents concentration of the dangerous gas/material in the water. The passive tracer is a prognostic variable in the model that has properties like salinity or temperature (active tracers), but it does not have influence on the density of the water. The passive tracer equation implemented in the model is presented as (7.1):

$$\frac{\partial C}{\partial t} + U\nabla_H C + (W - w)\frac{\partial C}{\partial z} = \frac{\partial}{\partial z}K_z\frac{\partial}{\partial z}C + \nabla_H K_H \nabla_H C + \varnothing \tag{7.1}$$

Where: C – represents tracer concentration, $\boldsymbol{U}$–horizontal velocity component (vector), W – vertical velocity component, w – settling velocity, K_H and K_z represent horizontal diffusion and vertical mixing coefficients, respectively, ∅ – sum of sources and sinks. Nobla is the horizontal gradient operator.

All simulations aimed at estimating contamination with use of the POP model have used the passive tracer based on Eq. (7.1). The passive tracer was initialized in the selected points shown on the map. The positions of the points are listed in Table 7.1. The points are also shown on the map in Fig. 7.1. The initial concentration was calculated based on the number of CWA found in one model cell (based on the data from the CHEMSEA project). It was assumed that one found point has a concentration equal to 10 μg/cm^3. The initial concentrations for all the considered points are included in Table 7.1. However, all the results based on the passive tracer will be recalculated for identical initial concentrations.

Table 7.1 Positions of the referenced points shown in Fig. 7.1

Station	Lon [deg]	Lat [deg]	Initial concentration	Depth [m]
Bornholm Deep (BM)	15.80	55.42	30	88
Slupsk_Furrow (SF)	17.02	55.18	50	73
Gotland_Deep (GT)	18.73	56.12	100	103
Gdansk_Deep (GD)	19.17	54.73	130	98

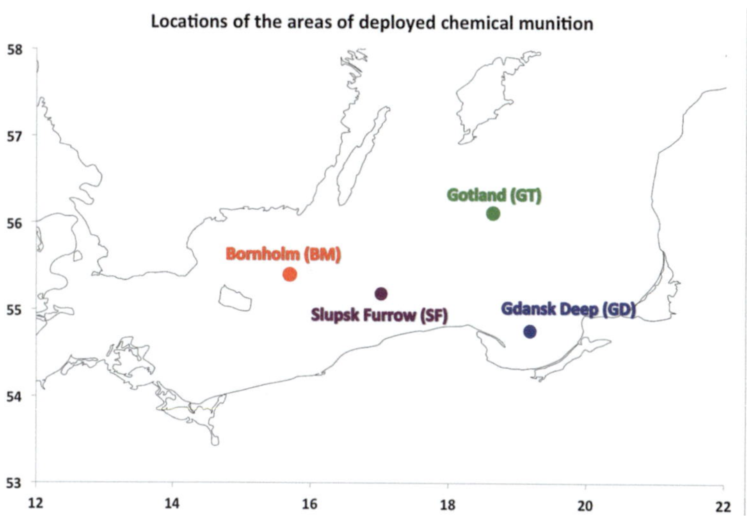

Fig. 7.1 Map of locations of dumpsites used as the referenced points in POP model simulations

7.2.2 Lagrangian Tracking

Pathways of suspended particles transport were calculated on the basis of current-velocity and eddy diffusivity data obtained from the Princeton Ocean Model (Blumberg and Mellor 1987; Mellor 1984). The model is a 3D, nonlinear, free surface, hydrostatic, σ-coordinate, with an imbedded second and a half moment turbulence closure sub-model (Mellor and Yamada 1982). The model domain comprises a wide area of the Baltic Sea closed at the Sounds (Kattegat) with the exclusion of the Gulf of Bothnia and the eastern part of the Gulf of Finland (Fig. 7.2); it has a horizontal resolution of ~1.8 km. The bottom topography was taken from a source (Seifert and Kayser 1995). There were 30 σ-layers specified vertically. Above the bottom boundary layer (BBL), the thickness of σ-layers was uniform, having a value of $0.038D$, where D is the total thickness of the water column. Since we are mostly interested in adequate resolution of flow in BBL, the thickness of σ-layers

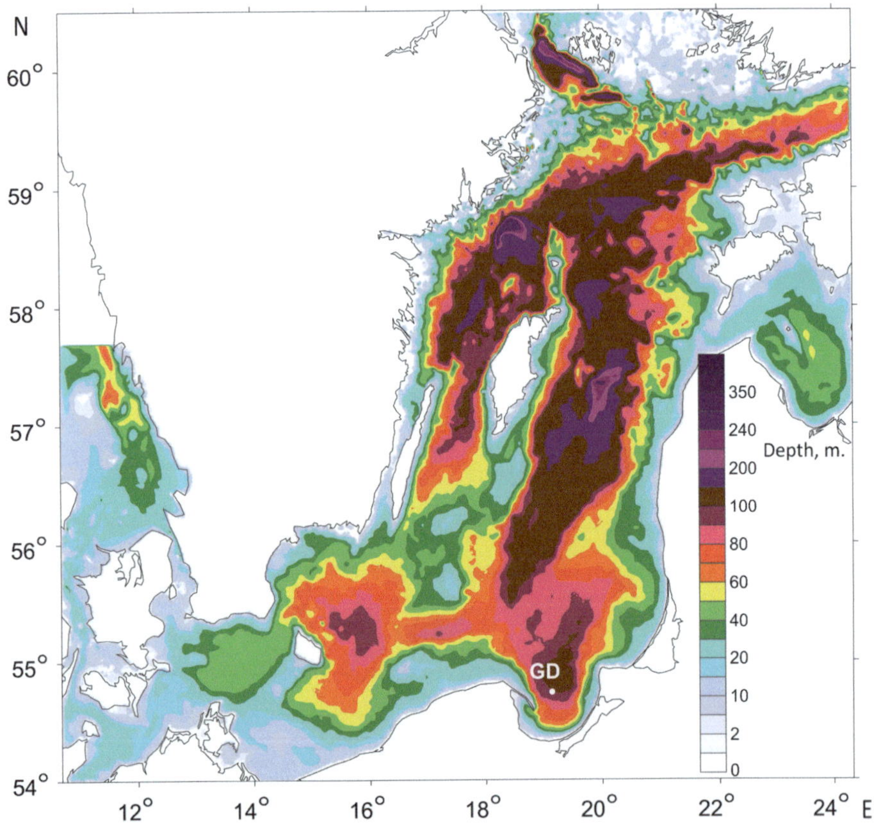

Fig. 7.2 Bathymetric map of the Baltic Sea, except the Gulf of Bothnia and the eastern part of the Gulf of Finland (the modeling area)

within BBL is taken logarithmically decreasing toward the bottom (Turner 1973), so that the σ-layer which is closest to the bottom is $0.002D$ thick. The period of calculations covers 4-month time-interval from the 01 January to 30 April 2012. The initial thermohaline stratification was taken from the HIROMB-BOOS (High Resolution Operational Model for the Baltic, Baltic Operational Oceanographic System) data for 01 January 2012 on an approximately 6 km horizontal grid. The output data from a version of HIRLAM (High Resolution Limited Area Model) applied at the Estonian Meteorological and Hydrological Institute (Männik and Merilain 2007) providing meteorological parameters on approximately 10 km horizontal grid with 3 h time step were used. The model used in the present study was already applied for investigation of different components of current velocity in the Baltic Sea: geostrophic, ageostrophic, as well as wave component related to quasi-inertial internal waves (Golenko and Golenko 2012). Modeling results showed a close correspondence to separate field data on thermohaline structure during upwelling (Golenko et al. 2015), ADCP velocity field data in separate points (Golenko and Golenko 2012) and satellite images (Zhurbas et al. 2004). The choice of the period of calculations is arbitrary. With reference to the above, the model data for this period were compared to satellite images (Golenko et al. 2015) and the ADCP current velocity data in the Gdańsk Deep for April 2012 (Bulczak et al. 2016), rather close correspondence to the sea surface temperature and horizontal current velocity in the whole water column (Fig. 7.3) was obtained. The 4-month period is considered to be long enough to give a picture of the dispersion of particles around the source, to reveal the areas of their transit and "capture", and to investigate the hydrodynamic processes affecting particles' propagation.

7.2.2.1 Particle Transport Model

Formulas for calculation of particles' displacement are described in detail in (Zhurbas et al. 2010). Suppose that at the time moment t_i a particle is located at the point of coordinates (x_i, y_i, z_i) where the velocity components and horizontal and vertical eddy diffusivities are (U_i, V_i, W_i), Kh_i, and Kv_i, respectively. The values of (U_i, V_i, W_i), Kh_i, and Kv_i are implied to have been found by means of interpolation of the circulation model results to the space-time point (x_i, y_i, z_i, t_i). Co-ordinates of the same particle $(x_{i+1}, y_{i+1}, z_{i+1})$ at a subsequent time moment $t_{i+1} = t_i + \tau$, where τ is a small time increment, can be expressed as:

$$\begin{aligned} x_{i+1} &= x_i + U_i\tau + x_i' \\ y_{i+1} &= y_i + V_i\tau + y_i' \\ z_{i+1} &= z_i + \left(W_i - w_S\right)\tau + z_i' \end{aligned} \tag{7.2}$$

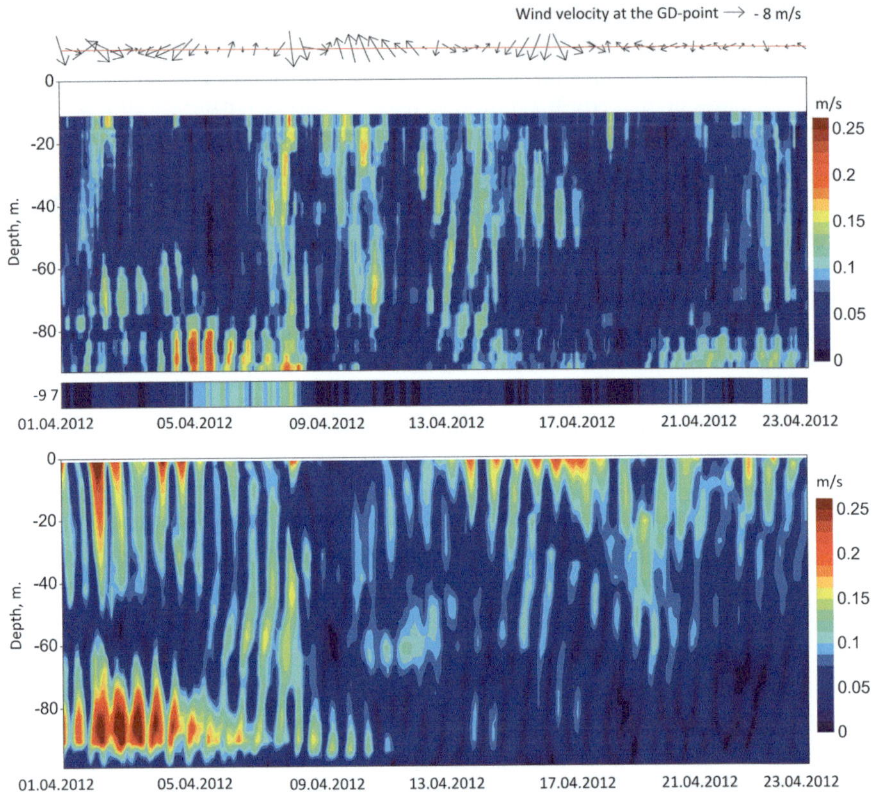

Fig. 7.3 The space-time variability of the velocity magnitude at the GD-point (see Fig. 7.2), measured by ADCP and Valeport in April 2012 (*top*), and obtained by modeling (*bottom*). The corresponding data of wind direction and magnitude are shown above. ADCP and Valeport data were obtained by D. Rak (IO PAN)

where $\left(x_i', y_i', z_i'\right)$ are the components of a random displacement of the particle due to the turbulent velocity fluctuations and w_S – is the settling velocity.

The random displacements $\left(x_i', y_i', z_i'\right)$ can be expressed through the eddy diffusivity and time increment; such an approach is typical for random walk models. The following formulas were used for the calculation of random displacements:

$$\left(x_i', y_i'\right) = \left(2 \cdot Kh_i \cdot \tau\right)^{1/2} A \tag{7.3}$$

$$z_i' = \frac{\partial Kv}{\partial z}\Big|_{z=z_i} \cdot \tau + \left\{2Kv\left[z_i + 0.5\frac{\partial Kv}{\partial z}\Big|_{z=z_i} \cdot \tau\right] \cdot \tau\right\}^{1/2} A \tag{7.4}$$

where Kh_i and Kv_i – are the values of Kh and Kv at the space-time point (x_i, y_i, z_i, t_i), A is the Gaussian random value with zero mean and unit variance. (Zhurbas et al. 2010) showed that the random-walk scheme (7.3) and (7.4) allows for non-uniform vertical profiles of the eddy diffusivity, characteristic for the real sea, and does not display unrealistic particles removal from highly turbulent BBL.

In the model run 10 particles with a given value of settling velocity 2 m/day are released at a given position just above the sea bottom and the trajectories of the particles are calculated using the random walk model with 3D velocity and diffusivity fields. This value of settling velocity corresponds to muds consisting of particles of predominant size <0.01 mm, which are typical sediment for dumpsites. The approach to calculation of Lagrangian particles' trajectories used in the present work is somewhat different than the one described in (Zhurbas et al. 2010). The difference is that the described calculations are performed using not stationary, the so-called "frozen" velocity field, corresponding to the time moment when the model reached quasi stationary regime under the constant in time, space-homogenous wind forcing, but taking into account 3D velocity and diffusivity fields varying with time, obtained under real wind forcing. During the calculations, the influence of quasi-inertial internal waves on particles' propagation was taken into account. (In calculations performed by (Zhurbas et al. 2010) near inertial oscillations were filtered out.)

7.2.2.2 Bottom Stress Analysis

The possibility that bottom sediments in the Bornholm, Gdańsk and Eastern Gotland Basin dumpsites can be re-suspended under the influence of near bottom currents was analysed. Some typical distributions of the normalized bottom stress (divided by density) [$N/m^2*m^3/kg = m^2/s^2$] in the Baltic Proper are shown in Fig. 7.4. Black polygons denote dumpsites under consideration. The variation of the normalized bottom stress in dumpsites shows that its values are, as a rule, more than an order of magnitude below the lower limit for the re-suspension threshold $\tau_0^{\min} / \rho = 0.0001\,\mathrm{m}^2 \cdot s^{-2}$ (this threshold value corresponds to muds consisting of particles of predominant size <0.01 mm). Rare exceptions are the dumpsites located in the Bornholm Deep and the South-Eastern periphery of the Eastern Gotland Basin.

The space-time variability of the horizontal model velocity magnitude in the points situated in the areas of dumpsites was also analysed (Fig. 7.5). Sometimes the velocity in the near bottom layer, corresponding to the upper border of the BBL, reached 10–15 sm/s. (The upper border of the model BBL is denoted with black solid lines.) These intensifications of the velocity were mainly due to near inertial internal oscillations. Since in the numerical models (namely in the POM) velocity descends to zero in the near bottom layer according to the logarithmic law, the model is not capable of reproducing some hydrodynamic processes, for example, reflection of internal waves from the bottom. Assuming that the energy of internal

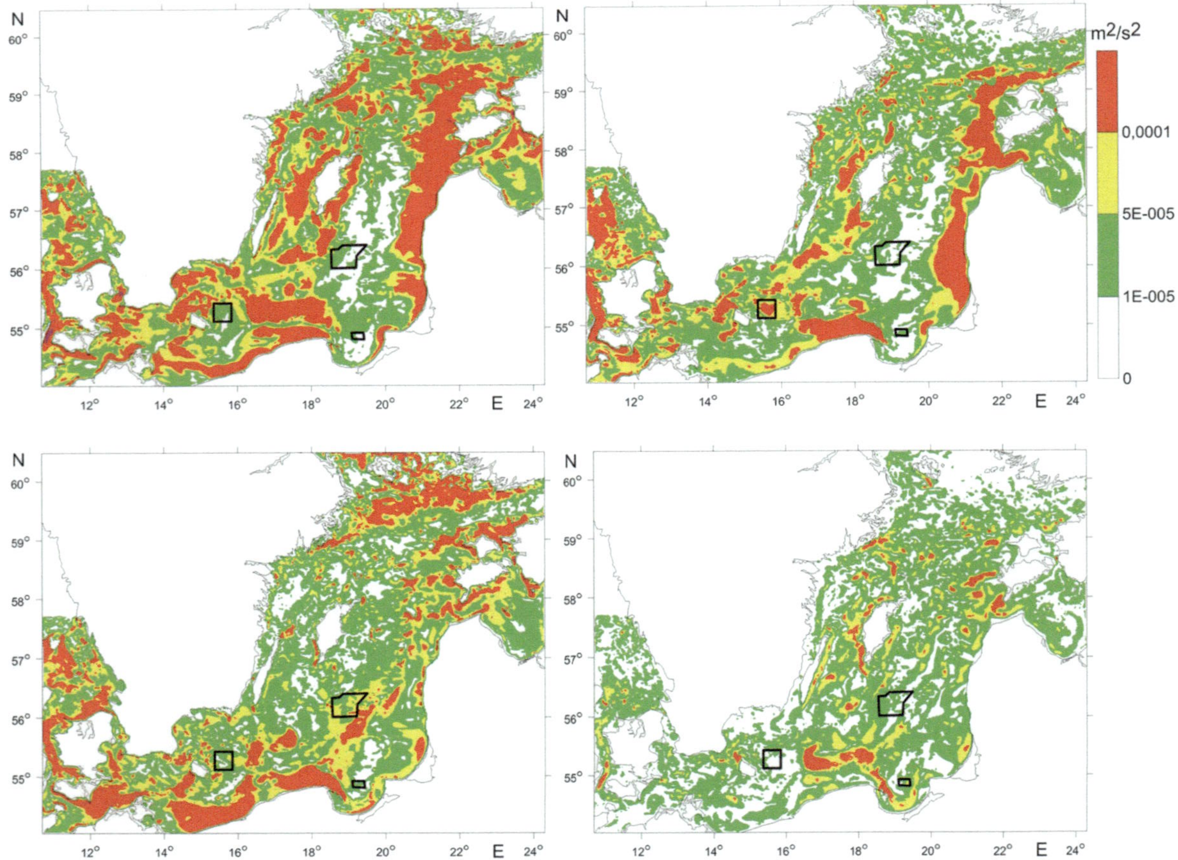

Fig. 7.4 Typical distributions of the normalized bottom stress [m^2/s^2] in the modeling area. *Black* polygons denote dumpsite areas under consideration. (Model calculations were performed for the period of time from 01.01 to 31.04 2012 under HIRLAM atmospheric forcing)

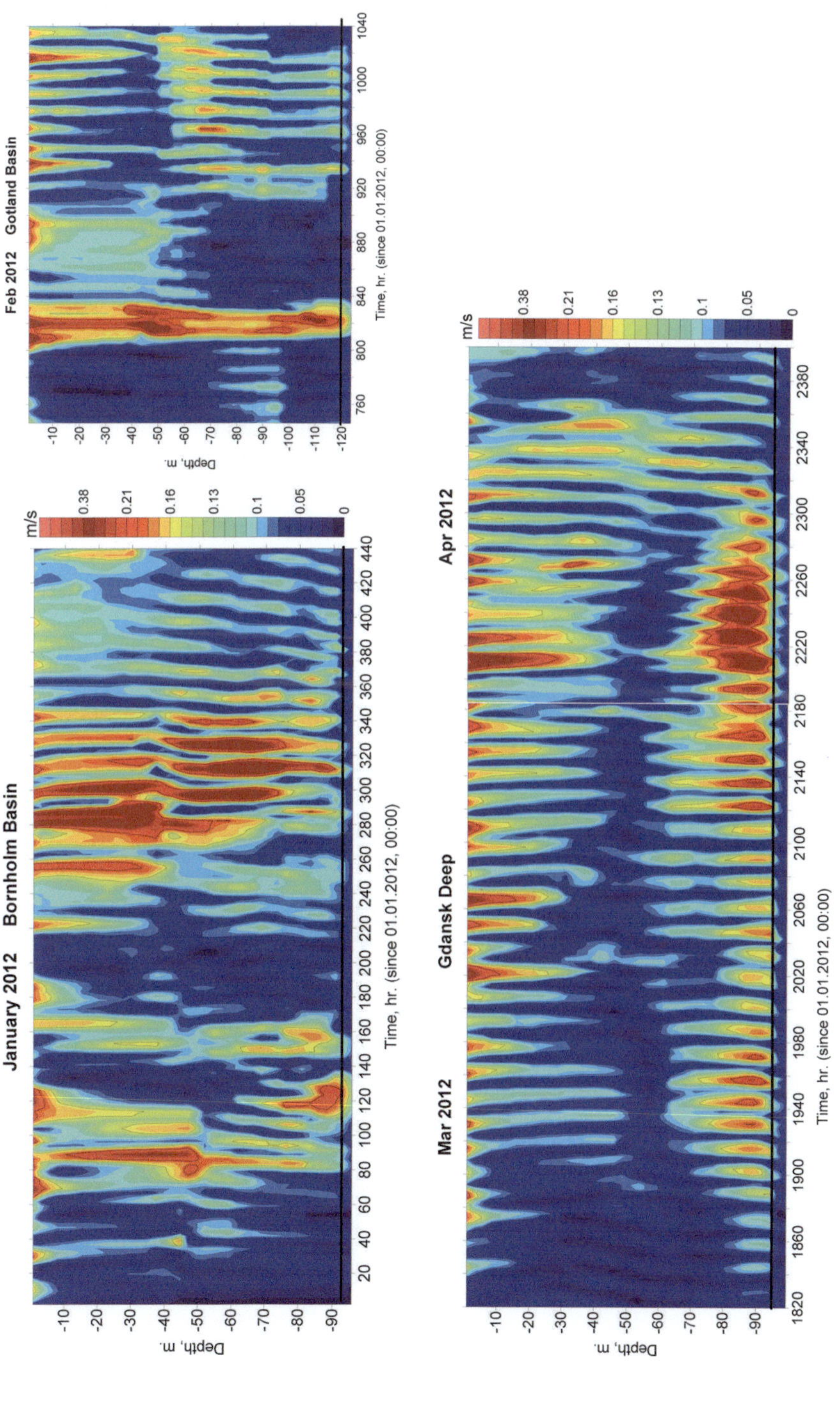

Fig. 7.5 The space-time variability of the horizontal model velocity magnitude in the points situated in the areas of dumpsites. *Black solid lines* denote the upper border of the model BBL

waves propagates directly to the bottom area in the event of wave reflection, sometimes it is possible to estimate the bottom velocity by considering the velocity at the upper border of the BBL. By accepting this hypothesis one may conclude that the re-suspension of bottom sediments due to internal waves is a possible reason of the toxic agents' transport from the dumpsites.

7.2.2.3 Verification of Random-Walk Model

Verification of model calculations of particles' trajectories has been implemented by comparative analysis with trajectories of real drifters' released from the surface and deep layers in the shelf zone of the southeastern part of the Baltic Sea in summer of 2015 and 2016. Drifters' trajectories were plotted on the basis of geographical coordinates, transferred by mobile communication channel. Temporal resolution of the data was from several minutes to half an hour (sometimes longer in the event of irregular communication). Trajectories of one of the drifters (red line) and the corresponding Lagrangian particle (magenta line) are showed in Fig. 7.6. The starting time of the trajectory is 17:08 on 02 August, 2015, the ending time is 23:08 on 09 August 2015. The starting location is marked with a black dot. The drifter's release experiment was carried out by O. Lavrova and E. Krayushkin (the Space Research Institute of the Russian Academy of Sciences). The comparison of the trajectories shows that the model rather satisfactorily reproduced all the turns of the drifter caused by changing of the direction and intensity of the wind, as well as inertial oscillations at the end of the trajectory.

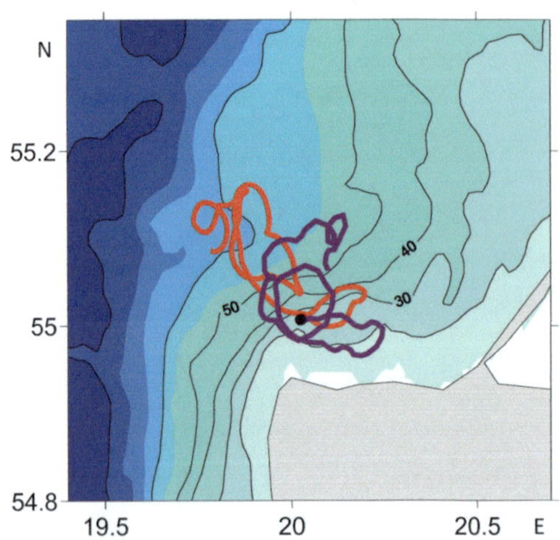

Fig. 7.6 Trajectories of the real drifter, released from the surface layer in the shelf zone of the southeastern part of the Baltic – *red line*, and the corresponding Lagrangian particle – *magenta line*. The drifter's coordinates were recorded between 17:08, 02 August, 2015 and 23:08, 09 August 2015. The start of the drifter is marked with a *black dot*

7.3 Results and Discussion

7.3.1 Integrations Based on the Passive Tracer

To obtain a comprehensive image of polluted area during potential leakage, twelve simulations for each dumpsite were performed. Each simulation began on the first day of each month and the passive tracer had been released at the beginning of each simulation. The simulations were performed for the year 2012. Although the problem connected with the chemical munitions dumped in the Baltic Sea has been known for a few decades, there are no rules on how to treat a leakage of CWA. A few assumptions were introduced to address this matter:

1. For every station the leakage material (CWA) is identical (thus, we can compare the results from the simulations, such as the pollution area, for example).
2. For such materials there is a threshold that divides contamination into two levels – dangerous and safe. The dangerous level means that it has a concentration of toxic properties and could kill, injure or incapacitate many forms of life including mammals, fishes and birds.
3. Since real concentrations are unknown, we assumed that this threshold is equal to 0.1 $\mu g/cm^3$. Those assumptions were introduced because it will help to present the estimation of the contaminated area. However, the values do not have any relationship with actual situations.
4. Based on the first two points, we can say that the following questions are of importance:

   (a) For how long is the level of pollution higher than the assumed (dangerous) threshold (AT)?
   (b) What is the greatest distance from the source of leakage to a place where concentration exceeds AT?

As it is impossible to have proper information about physical conditions in the location of potential leakage, we can say that two main parameters could describe dangerous area and time when contamination is greater than AT. The dangerous area at a specific time is the circle of a certain radius from the centre of leakage, and the length of the radius could be estimated as the greatest distance from the leakage source to the place where concentration exceeds AT.

For the purpose of point (3a) and (3b), for all the analysed dumpsite locations the following graphs will be presented:

(a) Distance from the source to the place of maximum tracer concentration
(b) Trajectories of the maximum concentration
(c) Maximum concentration vs time
(d) Maximal range to the place where concentration is higher than the assumed threshold level

Additionally, in the figures showing the distance from the source to the maximum concentration vs time, the dashed line which is an extension of the solid line denotes time dependence, but maximum concentration does not exceed AT. Also, the location of leakage is in the centre of the ellipse plotted on each b) subplot. It is well known that the two main processes that have influence on the dangerous area are advection and diffusion. In our case diffusion is the result of all the mixing processes that have influence on it (for example turbulences). Figures 7.7, 7.8, 7.9 and 7.10 show how the decrease of concentration of pollution can vary in different seasons of the year. In most cases there is no specific direction of the advection process (we could try to say that in the Słupsk Furrow (Fig. 7.8) there is a stable current, but in the beginning of the simulations directions of the movement are different). The distance from the place of leakage to the place of maximum concentration vs time provides information which processes dominate. If the distance from the leakage to the place of maximum concentration grows fast (Fig. 7.8), it means that the process of advection has greatest influence on the contaminated area. On the other hand, if the distance grows slowly, the diffusion process is the main influencing factor (for example the Bornholm Deep or the Gdansk Deep – Figs. 7.7 and 7.10, respectively). Based on the simulations, for 1-year scale we can summarise the results as the

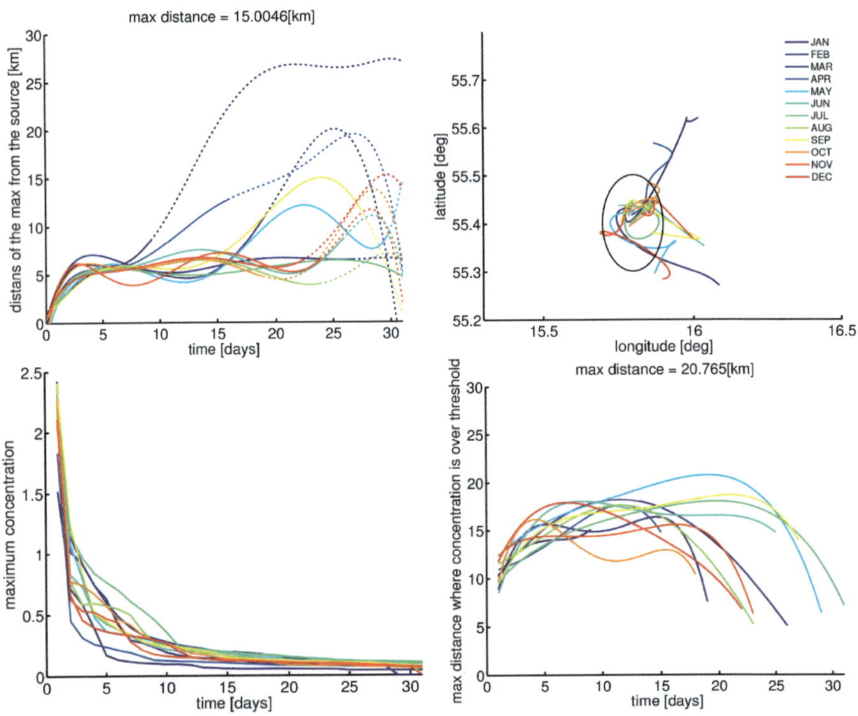

Fig. 7.7 The Bornholm deep: (**a**) distance from the source to the place of maximum tracer concentration, (**b**) trajectories of the maximum concentration, (**c**) maximum concentration vs time, (**d**) maximal range to the place where concentration is higher than the assumed threshold level

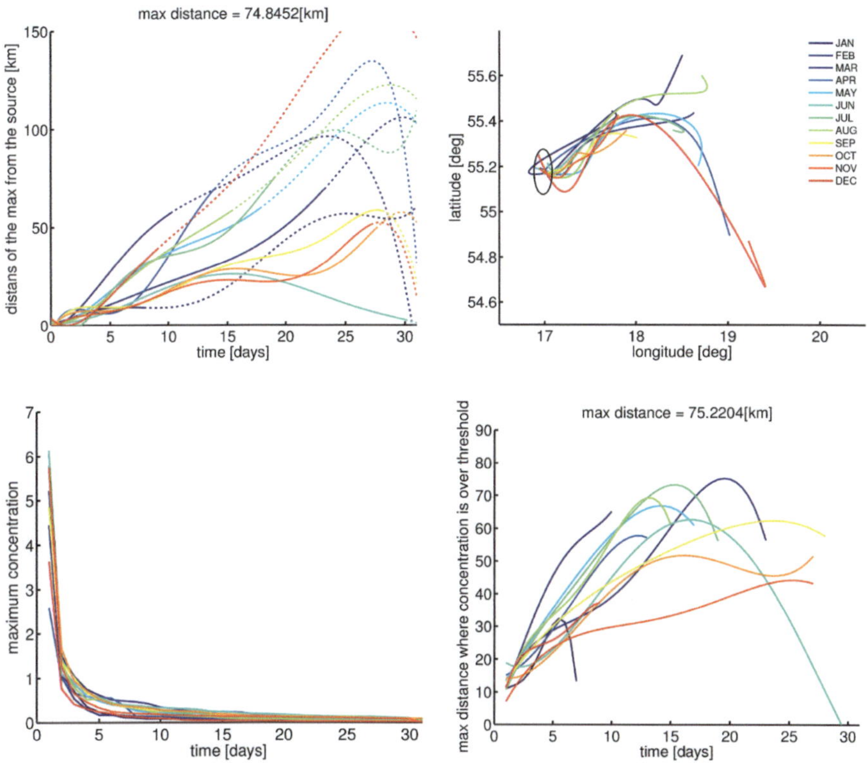

Fig. 7.8 The Słupsk Furrow: (**a**) distance from the source to the place of maximum tracer concentration, (**b**) trajectories of the maximum concentration, (**c**) maximum concentration vs time, (**d**) maximal range to the place where concentration is higher than the assumed threshold level

maximal time when concentration exceeds AT and as the maximal distance from the leakage location to the place where concentration is higher than AT. The summary is shown in Table 7.2 (recalculated values for the identical initial concentrations are provided in brackets).

It is important to add that there is no influence (in short time scale) of atmospheric conditions. Fig. 7.11 shows wind roses calculated for each month. It is clearly visible that there is no influence of wind over the Baltic Sea on the trajectories of maximum concentration shown in Figs. 7.7, 7.8, 7.9 and 7.10. For example, currents in the Słupsk Furrow transport the maximum concentration into the northeast direction irrespective of month or season.

The results presented for every dumpsite do not provide a comprehensive image of the contaminated area. A clear picture of pollution could be presented as a probability of contamination calculated based on twelve simulations (for each dumpsite separately). The probabilities are shown in Fig. 7.12.

If this probability is multiplied by real initial concentration, we will get the contamination area as a result. For real analysis real data should be taken into account,

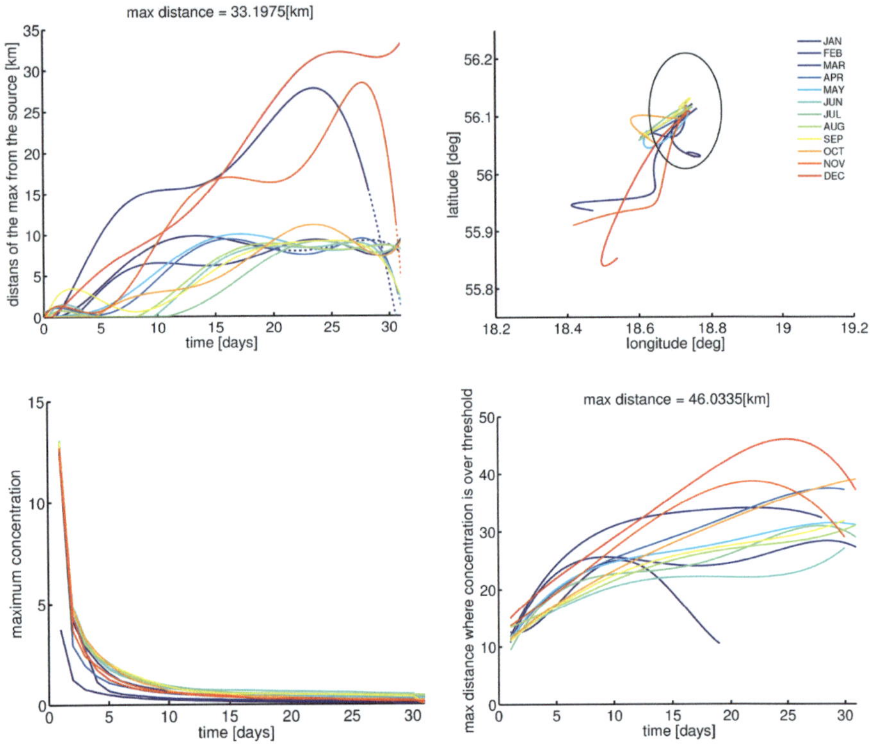

Fig. 7.9 The Gotland Dumpsite: (**a**) distance from the source to the place of maximum tracer concentration, (**b**) trajectories of the maximum concentration, (**c**) maximum concentration vs time, (**d**) maximal range to the place where concentration is higher than the assumed threshold level

these are, however, impossible to get. In the case of potential leakage we get only the location. For estimation of the contaminated area operational product should be used. But most of the operational products focus on sea surface condition, not on the lower layers. Furthermore, validation of the models is done only in specific locations and it also is provided mostly for surface. In that case it is possible to calculate the probability based on hindcast scenarios and then to multiply it by the initial concentration. For example seasonal variability of the bottom currents are shown in Fig. 7.13. The wind roses and probabilities for the Gotland dumpsite are in accordance.

7.3.2 Calculations of Trajectories

Trajectories of Lagrangian particles released from the near bottom sources at a distance of 1 cm from the bottom in the Bornholm Basin, Gdansk Deep and Eastern Gotland Basin were calculated (Fig. 7.14). Results of the calculations showed that particles released in the Bornholm Basin either remained in the vicinity of the

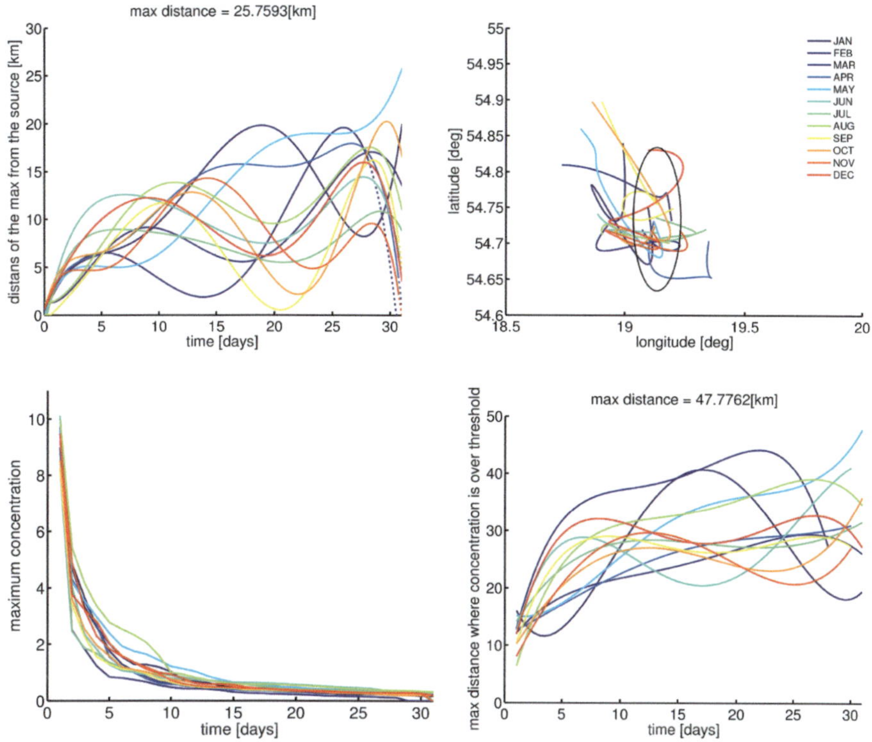

Fig. 7.10 The Gdansk Deep: (**a**) distance from the source to the place of maximum tracer concentration, (**b**) trajectories of the maximum concentration, (**c**) maximum concentration vs time, (**d**) maximal range to the place where concentration is higher than the assumed threshold level

Table 7.2 Calculated maximal time when concentration exceeds AT and maximal distance from the leakage to the place where concentration is higher than assumed

	Station	Initial concentration	Max time [days]	Max dist [km]
1	Bornholm deep	30 (10)	30 (10)	20 (6.7)
2	Slupsk_Furrow	50 (10)	30 (6)	75 (15)
3	Gotland_Deep	100 (10)	30 (3)	46 (4.6)
4	Gdansk_Deep	130 (10)	30 (2)	47 (3.6)

Recalculated values for the same initial concentrations are given in brackets

source (green lines), were trapped within the Deep, or propagated into the Arkona Basin in a counterclockwise direction (red lines) (Fig. 7.14a).

A similar situation was observed in the Gdańsk Deep. Particles released in the Gdańsk Deep either remained in the Gdańsk Bay (green lines), or moved into the Eastern Gotland Basin (red lines) (Fig. 7.14a). All the trajectories headed northward. For the particles released in the Eastern Gotland Basin the prevailing directions of propagation are the ones along the main axis of the Basin – southward and northward (Fig. 7.14b). Note that the particles released from close points propagated

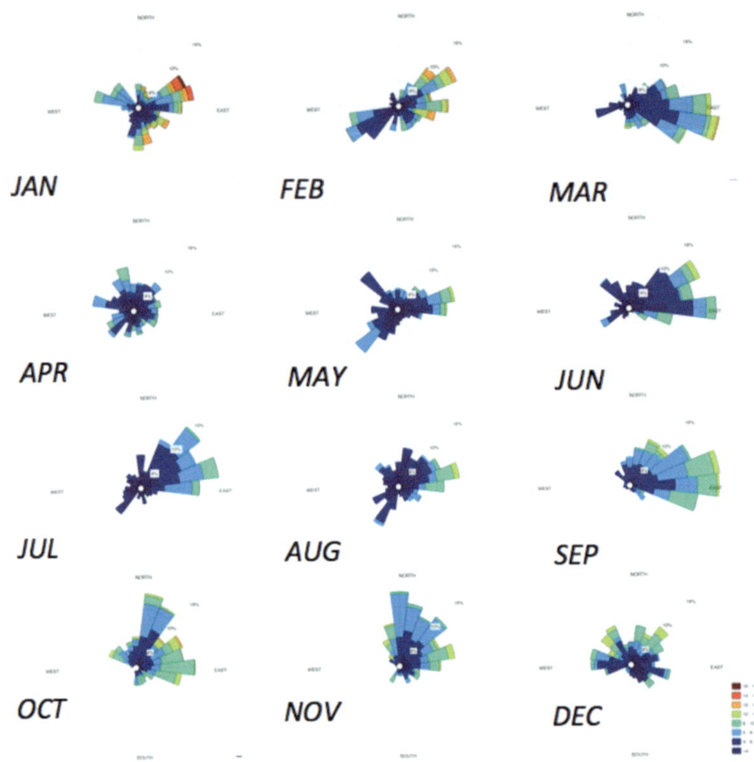

Fig. 7.11 Wind roses for each month of 2012 calculated for the Baltic Proper area (the map is shown in Fig. 7.1)

mainly in the opposite directions. Thus, the North-East direction of particles' propagation is more probable since the bottom stress distributions showed that the re-suspension threshold most probably can be exceeded in the South-Eastern border of the dumpsite.

Some trajectories show that the influence of near inertial internal waves on particles' propagation can be significant. Some particles released from the dumpsites under consideration were trapped by near inertial waves in some locations and remained within a distance of several kilometers for periods of several days (Fig. 7.14c–e).

Having analyzed the vertical displacements of particles, it can be seen that in general particles follow the bottom topography permanently oscillating with 5 m magnitude with an inertial frequency of ~14 h. (Fig. 7.15). When particles reach the crest of the sill they frequently remain at the depth of the sill or even go upward. But after some time particles return to the bottom. Apparently, it happens during periods of calm dynamics because of settling with the prescribed velocity of 2 m/day. Some particles released from the Bornholm dumpsite propagated into the surface layer and reached the shoreline.

The propagation of particles during the inflow event in the Baltic Sea was also analysed, and the results were compared to the results obtained under the non-inflow

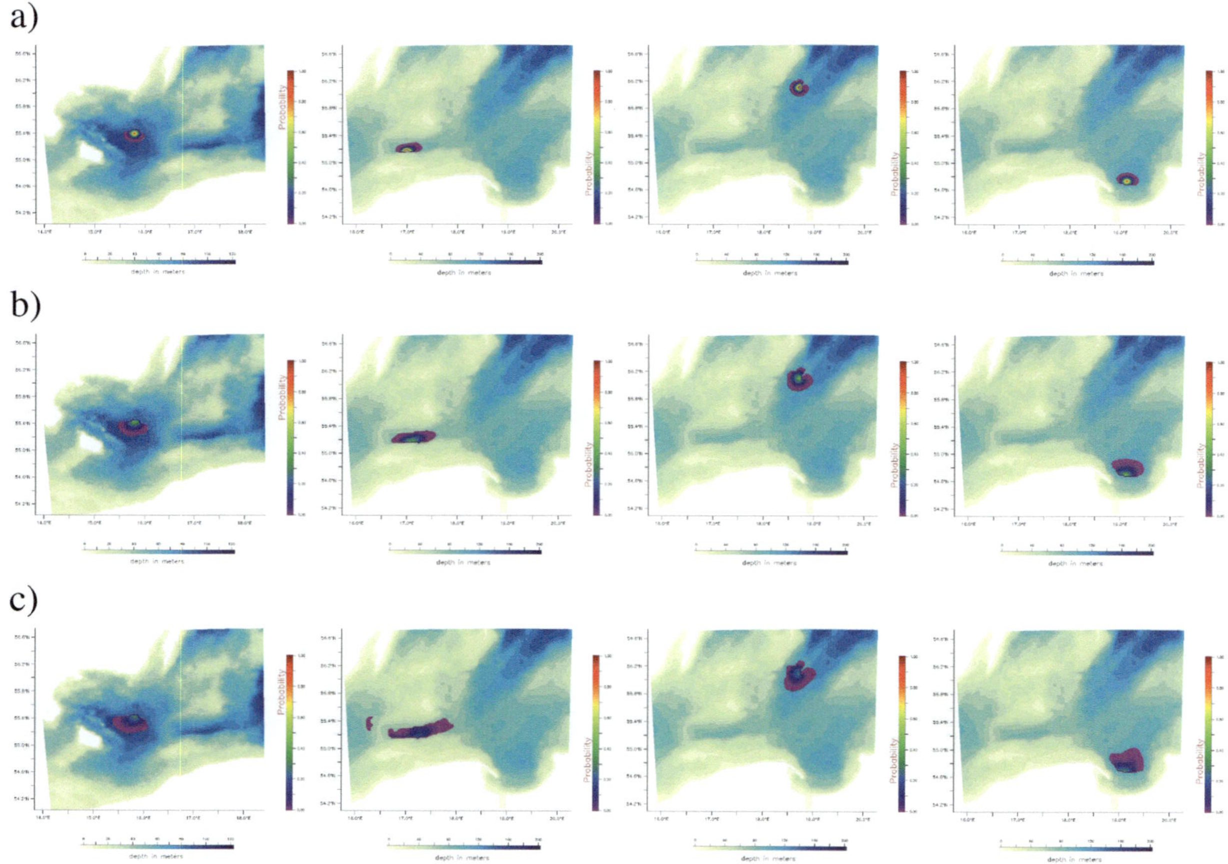

Fig. 7.12 Probability of contamination after 1, 3 and 5 days from releasing (**a–c**). Each column represents a different dumpsite – from the left are: BM, SF, GT and GD

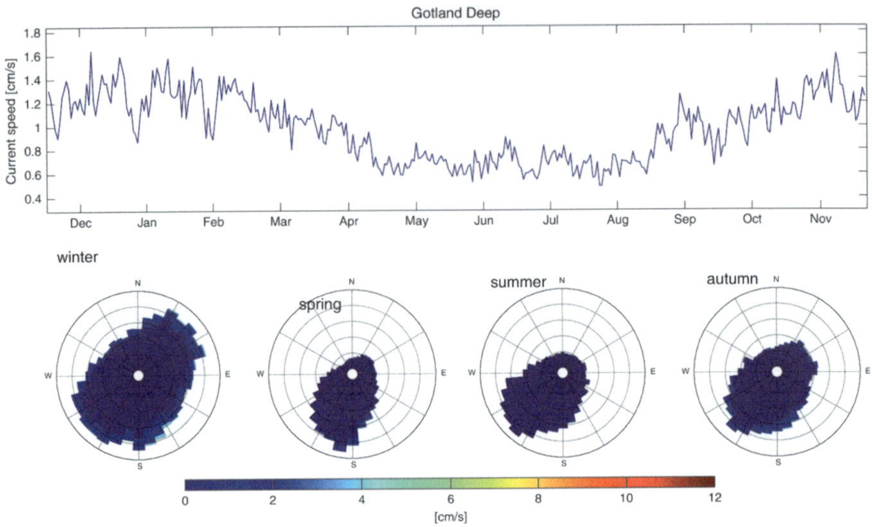

Fig. 7.13 Mean annual cycle (*upper panel*) and angle histograms (current roses) for seasonally averaged (winter, spring, summer and autumn, *lower panel*) current speed in the near-bottom layer in the Gotland dumpsite

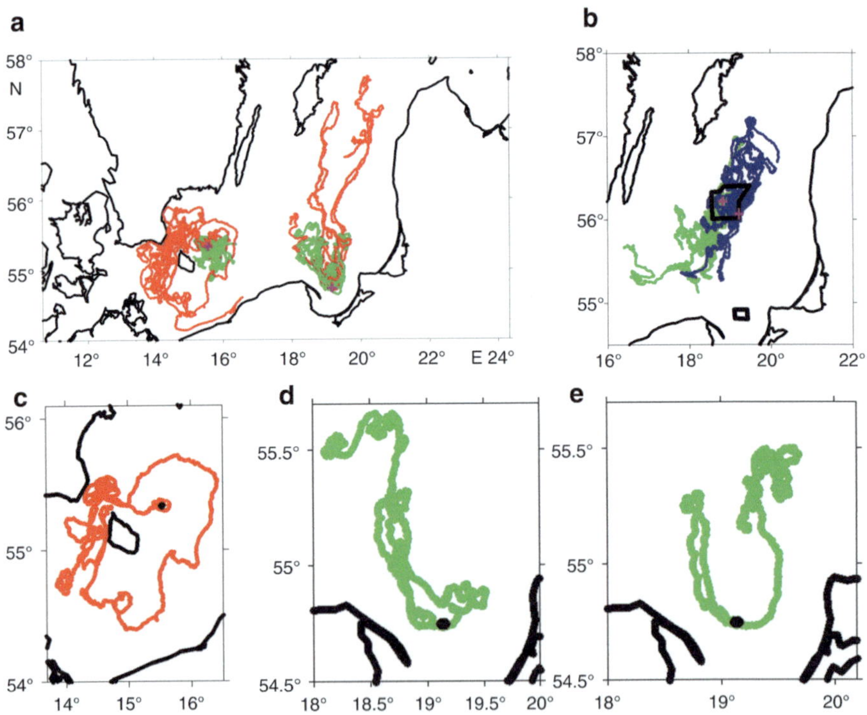

Fig. 7.14 Trajectories of Lagrangian particles released from the near bottom sources at a distance of 1 cm from the bottom in the Bornholm Basin, Gdańsk Deep (**a**) and Eastern Gotland Basin (**b**). (**c–e**) – some separate trajectories from (**a**) demonstrating the influence of the quasi inertial internal waves on the particles propagation

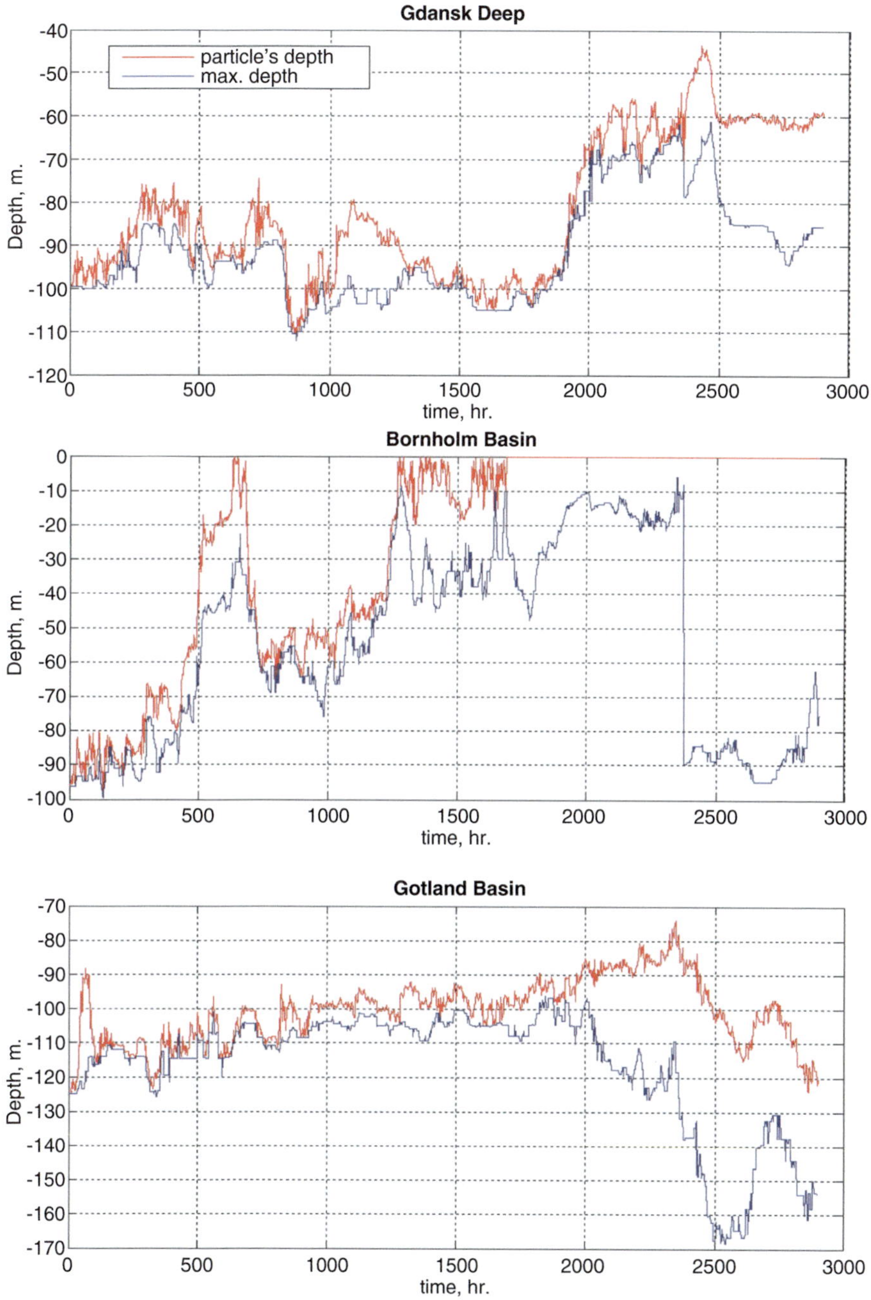

Fig. 7.15 Vertical displacements of Lagrangian particles released from the Gdansk, Bornholm and Eastern Gotland dumpsites. Note that after ~ 1700 hours particle released from the Bornholm Deep reached the shore

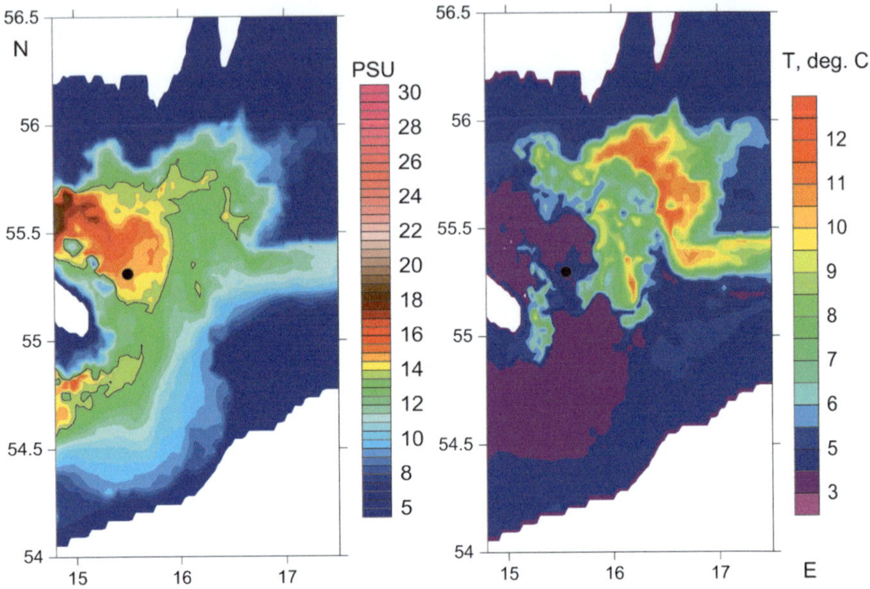

Fig. 7.16 Salinity (**a**) and Temperature (**b**) distributions in the near-bottom layer during the wind-induced inflow of salt, cold water into the Bornholm Basin. The source of trajectories is situated in the Bornholm dumpsite and denoted with a *black dot*

conditions. Both experiments were performed under constant northerly wind of 11 m/s, and total modeling time was 50 days. The inflow situation was generated due to initial filling of the extended and deepened Arkona Basin with saline and cold water. The propagation of salt, cold inflow water into the Bornholm Basin induced by constant northerly wind of 11 m/s is showed in Fig. 7.16. The source of particles was placed in the Bornholm dumpsite. It was observed that during the inflow most particles propagated directly towards the Słupsk Sill and further eastward, and after 40 days of modeling they reached the South-Eastern Gotland Basin (Fig. 7.17a). At the same time under non-inflow conditions some particles remained in the Bornholm Basin for 40 days, whereas other particles only reached the Słupsk Furrow (Fig. 7.17b).

The reason of the difference in the patterns of particles propagation is probably the strengthening of gravity current of salt dense water propagating in the Bornholm Basin and flowing down the slope of the Słupsk Sill. The gravity current facilitated the propagation of particles from the central part of the Bornholm Basin to its northeastern slope where they were "drawn" into the Słupsk Furrow. The vertical displacement of one random particle (Fig. 7.18) shows that it moves mainly along the bottom slope (at least to the end of the Słupsk Furrow), which confirms the influence of the gravity current on the particle's propagation. But at some time intervals the particle lifts up to 15–20 m above the bottom, which allows to conclude that the combination of wind-induced dynamics and gravity current determines the particle's propagation.

To investigate the influence of inflow-induced gravity current, not overlayed on the wind-induced dynamics, on particles' propagation in the near-bottom layer a

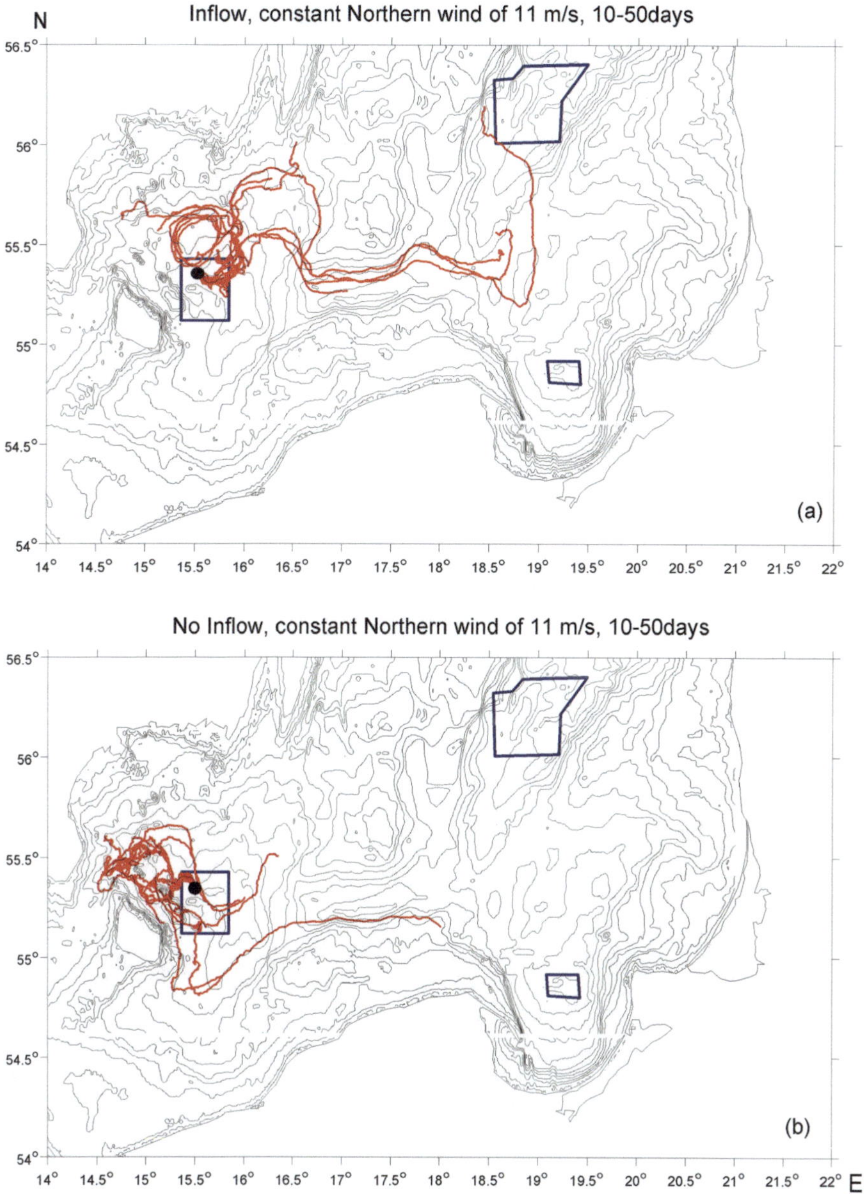

Fig. 7.17 The trajectories of Lagrangian particles (*red lines*), released from the Bornholm dumpsite under the inflow (**a**) and non-inflow (**b**) conditions. The source of particles is marked by black points. *Blue* polygons denote the dumpsites. Trajectories started after 10 days of calculations, when the inflow dense water reached the Bornholm dumpsite

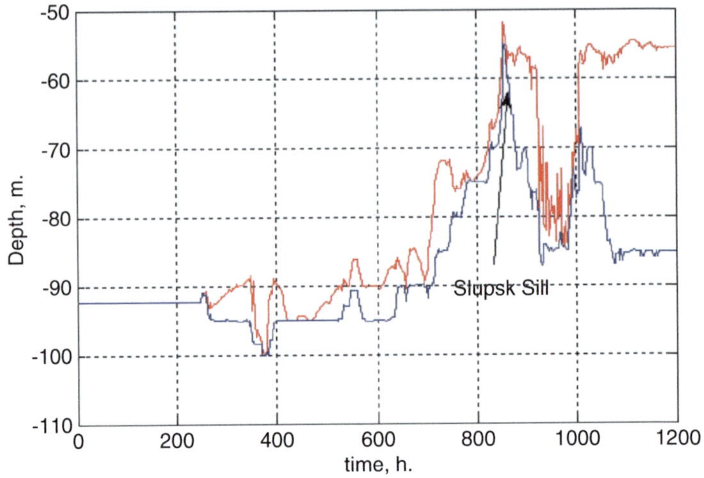

Fig. 7.18 *Red line* – vertical displacement of one particle. *Blue line* denotes topography

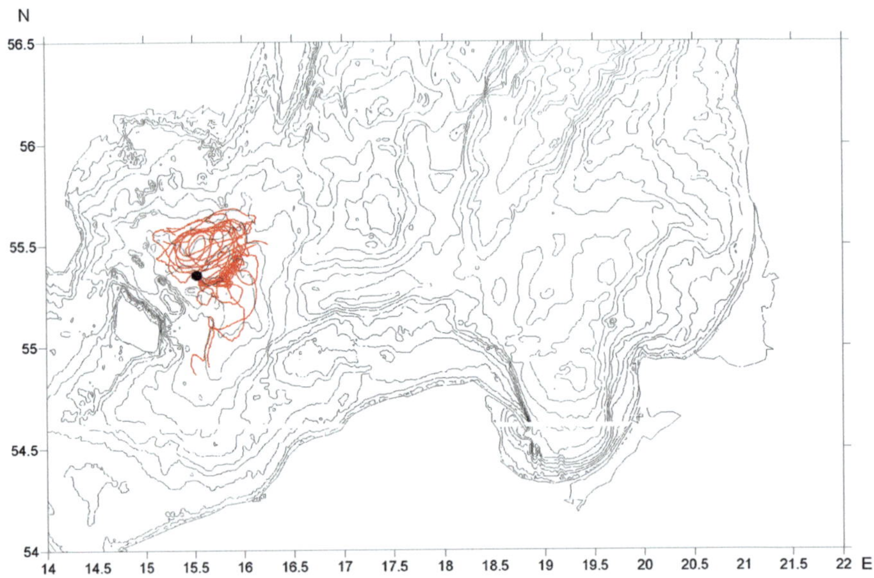

Fig. 7.19 The trajectories of Lagrangian particles (*red lines*), released from the Bornholm dumpsite under the inflow conditions for the situation when constant northerly wind of 12 m/s was stopped after 15 days since the modeling of particles' propagation began

numerical experiment was implemented in which wind stopped after 15 days since the modeling of particles' propagation began. This moment corresponds to the period of time when the inflowing dense waters reached the Bornholm Dumpsite. Results of this experiment, showed in Fig. 7.19, indicate that after the inflow water propagated in the Bornholm Basin and the wind stopped, the particles did not propagate towards the Słupsk Sill and further along the Słupsk Furrow. It confirms that only the combination of both types of currents promotes the particles' propagation in the eastward direction to the South-Eastern Gotland Basin.

7.4 Summary

Two different approaches that could be used for analysis of potential leakage have been developed. Both of them are very different and also have been embedded in different circulation models. Furthermore, both of the models have been driven by different atmospheric forces (we mean the sources – POM used data from HIRLAM and POP was forced by data form the UM model). The models have been successfully validated and at that stage we did not make a comparison of the results. It is important to add that our work is in progress and as the next point we are planning to implement both methods in the same ocean model. After that we will be able to make a detailed comparison of the presented approaches.

Bibliography

Arakawa A, Lamb VR (1977) Computational design of the basic dynamic processes of the UCLA general circulation model. Method Comput Phys 17:173–265

Bitz CM, Lipscomb WH (1999) An energy-conserving thermodynamic model of sea ice. J Geophys Res 104(C7):15669

Blumberg AF, Mellor GL (1987) A description of a three-dimensional Coastal Ocean circulation model. In: Heaps S (ed) Three dimensional coastal ocean models. American Geophysical Union, Washington, DC

Bulczak AI, Rak D, Schmidt B, Bełdowski J (2016) Observations of near-bottom currents in bornholm basin, slupsk furrow and gdansk deep. Deep Sear Res Part II 127:96–113

Craig A, Jacob R, Kauffman B, Bettge T, Larson J, Ong E, He H (2005) CPL6: the new extensible, high-performance parallel coupler for the community climate system model. Int J High Perform Comput Appl 19(3):309–328

Craig AP, Vertenstein M, Jacob R (2012) A new flexible coupler for earth system modelling developed for CCSM4 and CESM1. Int J High Perform Comput Appl 26(1):31–42

Craig AP, Mickelson S, Hunke C, Bailey DA (2014) Improved parallel performance of the CICE model in CESM1. Int J High Perform Comput Appl 29(2):154–165

Garnaga G, Stankevičius A (2005) Arsenic and other environmental parameters at the chemical munitions dumpsite in the Lithuanian economic zone of the Baltic Sea. Environ Res Eng Manag 3:24–31

Golenko MN, Golenko NN (2012) Structure of dynamic fields in the southeastern Baltic during wind forcings that cause upwelling and downwelling. Oceanology 52(5):604–616. doi:10.1134/S0001437012050086

Golenko MN, Golenko NN, Bukanova TV (2015) Investigation of structural features of coastal thermohaline frontal zones in the South-East Baltic from numerical model and satellite data. Current problems in remote sensing of the Earth from space, pp 125–135

Hu Y, Huang X, Wang X, Fu H, Xu S, Ruan H, Xue W, Yang G (2013) A scalable barotropic mode solver for the parallel ocean program, Aachen. Springer, Germany, pp 739–750
Hunke EC (2001) Viscous-plastic sea ice dynamics with the EVP model: linearization issues. J Comput Phys 170:18–38
Hunke EC, Dukowicz JK (1997) An elastic-viscous-plastic model for sea ice dynamics. J Phys Oceanogr 27:1849–1867
Jansen F, Schrum C, Backhaus JO (1999) A climatological data set of temperature and salinity for the Baltic Sea and the North Sea. Ocean Dyn 51:5–245. Deutsche Hydrographische Zeitschrift
Kay JE, Deser C, Phillips A, Mai A, Hannay C, Strand G, Arblaster J, Bates S, Danabasoglu G, Edwards J, Holland M, Kushner P, Lamarque JF, Lawrence D, Lindsay K, Middleton A, Munoz E, Neale R, Oleson K, Polvani L, Vertenste M (2015) The community earth system model (CESM) large ensemble project: a community resource for studying climate change in the presence of internal climate variability. Bull Am Meteorol Soc, August
Killworth PD, Stainforth D, Webb DJ, Paterson SM (1991) The development of a free-surface Bryan-Cox-Semtner ocean model. J Phys Oceanogr 21:1333–1348
Knobloch T, Bełdowski J, Söderström N, Rühl P, Sternheim J (2013) Chemical munitions dumped in the Baltic Sea, Baltic Marine Environment Protection Commission. HELCOM, Helsinki
Large WG, McWilliams JC, Doney SC (1994) Oceanic vertical mixing: a review and a model with a nonlocal boundary layer parameterization. Rev Geophys 32(4):363–403
Lipscomb WH, Hunke EC, Maslowski W, Jakacki J (2007) Ridging, strength, and stability in high-resolution sea ice models. J Geophys Res 112(C3):1–18
Männik A, Merilain M (2007) Verification of different precipitation forecasts during extended winter-season in Estonia. Newsletter, Estonian Meteorological and Hydrological Institute, Tallin
Matthäus W (2006) The history of investigation of salt water inflows into the Baltic Sea from the early beginning to recent results. Mar Sci Rep 65:1–73
McDougall TJ, Jackett DR, Wright DG, Feistel R (2003) Accurate and computationally efficient algorithms for potential temperature and density of seawater. J Atmos Ocean Technol 20:730–741
Meier M, Kauker F (2003) Modeling decadal variability of the Baltic Sea: 2. Role of freshwater inflow and large-scale atmospheric circulation for salinity. J Geophys Res 108(C11)
Mellor GL (1984) User's guide for a three-dimensional, primitive equation, numerical model, The revision. Program in Atmospheric and Oceanic Sciences. Princeton University, Princeton
Mellor GL, Yamada T (1982) Development of a turbulence closure model for geophysical fluid problems. Rev Geophys Space Phys 20(4):851–875
Missiaen T, Söderström M, Popescu I, Vanninen P (2010) Evaluation of a chemical munition dumpsite in the Baltic Sea based on geophysical and chemical investigations. Sci Total Environ 408(17):3536–3553
Mohrholz V, Naumann M, Nausch G, Krüger S, Gräwe U (2015) Fresh oxygen for the Baltic Sea—an exceptional saline inflow after a decade of stagnation. J Mar Syst 148:152–166
Osinski R (2007) Symulacja procesów dynamicznych w Morzu Bałtyckim zintegrowanym modelem ocean-lód; PhD. thesis. Institute of Oceanology Polish Academy of Sciences, Sopot
Paka V, Spiridonov M (2002) Research of dumped chemical weapons made by R/V 'Professor Shtokman' in the Gotland, Bornholm & Skagerrak dump sites. Belgian Ministry of Social Affairs, Public Health and the Environment
Rak D (2016) The inflow in the Baltic Proper as recorded in January–February 2015. Oceanologia 58(3):241–247
Seifert T, Kayser B (1995) A high resolution spherical grid topography of the Baltic Sea. Meereswissenschaftliche Berichte, Institute für Ostseeforschung Warnemünde, Warnemünde. Marine Science Reports
Seifert T, Tauber F, Kayser B (2001) A high resolution spherical grid topography of the Baltic Sea – 2nd ed, p 147 Stockholm.

Semtner AJ (1974) A general circulation model for the World Ocean. UCLA Dept of Meteorol Tech. Rep, August, pp 99

Smith R, Gent P (2004) Reference manual for the parallel ocean Program (POP). Los Alamos National Lab, New Mexico

Stevens DP (1990) On open boundary conditions for three dimensional primitive equation ocean circulation models. Geophys Astrophys Fluid Dyn 51:103–133

Turner JS (1973) Buoyancy effects in fluids. Cambridge Monographs on Mechanics ed

Zhurbas VM, Stipa T, Mälkki P, Paka VT, Kuzmina NP, Sklyarov EV (2004) Mesoscale variability of the upwelling in the southeastern Baltic Sea: IR images and numerical modeling. Oceanology 44(5):619–628

Zhurbas VM, Elken J, Väli G, Kuzmina NP, Paka VT (2010) Pathways of suspended particles transport in the bottom layer of the southern Baltic Sea depending on the wind forcing (Numerical simulation). Oceanology 50(6):841–854

Chapter 8
Weight-of-Evidence Environmental Risk Assessment

Patrik Fauser, Erik Amos Pedersen, and Ilias Christensen

Abstract In this chapter, the data generated within MODUM and other related projects, i.e. CHEMSEA, MERCW and NordStream, contributing to the knowledge and data on occurrence, toxicity and effects of chemical warfare agents (CWAs) and their metabolites in the Baltic Sea, are aggregated. The data are evaluated and assessed in terms of risk quotients, and whether these point to risk or no-risk towards effects on lower tier organisms (e.g. algae and daphnia), and higher tier organisms (fish) from exposure to dumped CWAs and their metabolites in the study areas. To perform a semi-quantitative assessment of the environmental risk, a number of Lines of Evidences (LoEs) are set up, that take all aspects of the performed investigations into account. Each LoE is assigned +3, +2, +1, 0, −1, −2 or −3, indicating if the LoE is found to be for (+), against (−) or neutral (0) to an increased environmental risk of dumped CWAs. The weight also reflects the predictive power, or significance, of the individual LoEs. From nine LoEs a resulting summed score of +9 indicates that there is a weak to moderate potential for confirming the hypothesis. In order to qualify and increase the precision of the weights recommendations are given that can be addressed in future investigations regarding compounds, sites and species that could be in focus. Furthermore, recommendations for activities that will improve the exposure and toxicity data, that are inherent in the environmental risk assessment, are stated.

P. Fauser (✉)
Department of Environmental Science, Aarhus University, Roskilde, Denmark
e-mail: paf@envs.au.dk

E.A. Pedersen
Department of Environmental Science, Aarhus University, Roskilde, Denmark

Department of Environmental Engineering, University of Southern Denmark, Odense, Denmark

I. Christensen
Department of Environmental Science, Aarhus University, Roskilde, Denmark

Department of Environmental Engineering, Technical University of Denmark, Lyngby, Denmark

J. Bełdowski et al. (eds.), *Towards the Monitoring of Dumped Munitions Threat (MODUM)*, NATO Science for Peace and Security Series C: Environmental Security, DOI 10.1007/978-94-024-1153-9_8

8.1 Introduction

The spatial scale of the marine environment in the Baltic Sea is large compared to the extent of dump sites, and the possibilities for dispersion of CWAs and their metabolites in sediment and bulk water are therefore significant. Given that the exact locations of all dumpsites, i.e. hot-spots, are not fully known, a complete risk assessment covering the Baltic Sea is impeded. Consequently, focus will be on the three target areas; Bornholm deep, Gotland deep and Gdansk deep where historic records indicate dumping and where the bulk of monitoring has taken place.

One way to assess the risk based on a large number of data that include multiple parameters is within the frame of a semi-quantitative Weight-of-Evidence (WoE) analysis. This builds on a number of Lines-of-Evidences (LoE) that are set-up for each relevant parameter or topic, so that they in combination cover the entire scope of the project. Although the method uses empirical data the outcome is subjective as it is founded on evaluation of data in relation to a predefined set of claims and hypothesis.

In the following the risk assessment, WoE and LoE methodologies will be briefly outlined, the data on sediment risk quotients of CWAs and metabolites, toxicity values, fish disease and meio fauna data aggregated from MODUM and other projects and studies will be presented in maps and summarizing graphs and tables. Finally, the LoEs, the resulting WoE analysis and conclusions with respect to future studies, will be presented. It is important to note that the occurrence of chemicals in this chapter will only be referred to in terms of risk quotients. With respect to measured and modelled concentration values please see elsewhere in this report.

8.2 Methods

8.2.1 Risk Assessment

The risk assessment follows the guidelines of ECHA's "Guidance on Information Requirements and Chemical Safety Assessment" (https://echa.europa.eu/guidance-documents/guidance-on-information-requirements-and-chemical-safety-assessment). The guidance is a framework consisting of an initial information gathering (hazard identification) followed by an exposure assessment and hazard assessment leading up to the risk characterization. In short the outcome is a predicted no effect concentration (PNEC), predicted environmental concentration (PEC) and a risk quotient (RQ).

8.2.1.1 Predicted No Effect Concentration (PNEC)

The aim of the hazard assessment is to determine a dose-response value for the compound concerned below which a harmful effect is not expected to occur. Exactly which value to use can be discussed; often a measured EC10 or the no observed effect concentration (NOEC) has been used. NOEC is the highest concentration in a toxicity test at which no effect is seen and is therefore very dependent on the test design and concentrations used and does not have the statistical power as a model based value. Therefore, OECD recommends the use of EC-values which are more comparable and for which confidence intervals can be calculated.

Consequently, the toxicity values used to derive the acute predicted no effect concentration (PNEC) are the measured EC50 values or QSAR derived LC50 values if there are no measured data available. Individual PNECs are derived for two trophic levels; a higher tier that includes toxicity values from fish and a lower tier that includes toxicity values from bacteria, algae or daphnia. When more toxicity values are available, then the lowest value is always chosen for conservatism.

PNEC in this report relates to the marine environment. Normally assessment factors (AF) in the range of 1–10,000 are applied to PNEC, see Eq. (8.1), depending of the amount and quality of the toxicity data. In this report we do not apply assessment factors, instead we discuss the toxicity values and how assessment factors would influence on the derived PNEC's.

$$PNEC = \frac{EC50\,or\,LC50}{AF} \tag{8.1}$$

8.2.1.2 Predicted Environmental Concentration (PEC)

In this project the exposure concentrations towards the marine organisms are derived from measured concentrations of CWAs and metabolites in the sediment. This reduces the uncertainties of the PEC and RQ estimates and avoids conservatism. The exposure model proposed by ECHA, i.e. the European Union System for the Evaluation of Substances (EUSES) (https://ec.europa.eu/jrc/en/scientific-tool/european-union-system-evaluation-substances) to estimate dispersion and concentration propagation in the Baltic Sea water and sediment, is not appropriate for the emission and exposure scenarios relevant for this project; instead a reviewed site-specific model has been developed, which is described elsewhere in this report.

When characterizing risk for aquatic organisms the exposure pathway is through the bioavailable concentration, defined as the dissolved chemicals in the water phase or the sediment pore water concentration (e.g. NRC 2003). Sampling and analytical quantification in pore water does not give adequate data due to low concentrations compared to the limits of detection of the chemical analysis methods. Accordingly, pore water concentrations (Cw) are estimated based on quasi steady-state equilibrium partitioning with the measured sediment concentration (Cs) (DiToro 1991; Sanderson et al. 2008):

$$C_{pw} = \frac{C_s}{R_s} = \frac{C_s}{\theta + K_d \times X_s} = \frac{C_s}{\theta + f_{oc} \times K_{oc} \times X_s} \tag{8.2}$$

where Xs is the sediment density of 1.2 kg dry matter per L (Forster et al. 2003), the retention factor of sediment is Rs = θ + Kd·Xs where θ is the pore volume fraction of 0.55 (Forster et al. 2003), Kd = f_{OC}·K_{OC} is the partitioning coefficient, f_{OC} is the fraction of organic carbon in sediment equal to 0.0775 (Emelyanov 1996) and Koc is the sorption coefficient between organic carbon and water which is found by means of QSAR (KocWIN v.2.0). The parameters f_{OC}, Xs and θ are site specific to the (Southern) Baltic Sea.

8.2.1.3 Risk Quotient (RQ)

Finally, the risk quotients (RQ) are calculated for each site as the ratio between the PEC and the PNEC:

$$RQ = \frac{PEC}{PNEC} \tag{8.3}$$

According to ECHA, if the RQ is below 1 there is no risk as the PEC is below PNEC. On the other hand, RQ ≥ 1 indicates there is a risk and the PEC exceeds the PNEC. A RQ is calculated for each detected compound. If more than one compound are found together or expected to form a "mixture" the RQs of these compounds are summed.

8.2.2 *Weight of Evidence (WoE)*

Assessing the environmental risk of CWAs in the Baltic Sea requires reviewing many data and studies of toxicity and exposure of varying quality and magnitude. This brings the need for organizing and weighing the information as lines-of-evidences (LoEs). The process of weighing LoEs includes assembling evidence (literature search, toxicity testing, exposure assessment, and risk assessment), identifying categories of evidence, assigning weights, and weighing of bodies of evidence. The process of organizing the WoE is carried out according the framework of Suter and Cormier (2011), see Fig. 8.1.

Weighing is in practice done by assigning scores to each LoE with for example a +,- or 0. The estimate is subjective and given according to whether the LoE is for, against or neutral to the claim/hypothesis. Categories are assigned to matching

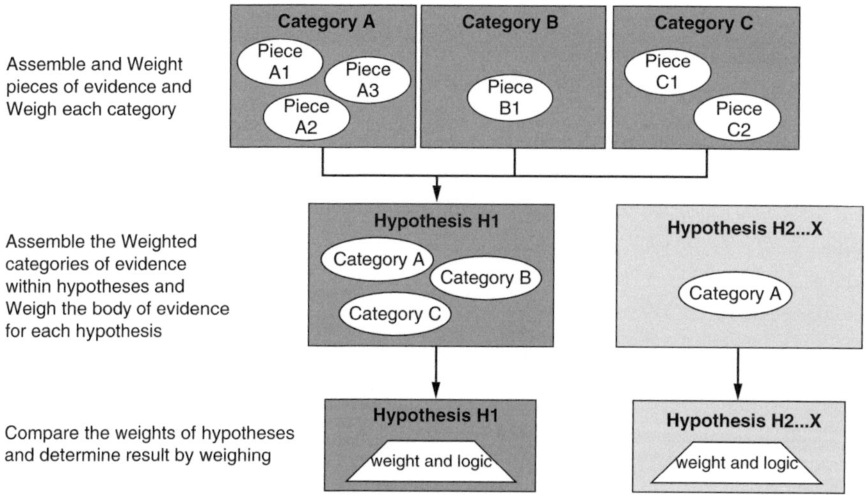

Fig. 8.1 Framework for tiered weighing of evidence in environmental assessments (Suter and Cormier 2011)

claims and the scores are summed. The concluding result is subjective and not only determined by whether the sum of scores is for or against the claims.

8.3 Materials and Data

8.3.1 Sediment Samples

The sediment samples that are used in this assessment are from the projects stated in Table 8.1. It shows the number of sediment samples analysed and also the chemical species that were detected, see Table 8.2 for full names of chemicals. A total of 627 sediment samples are included in this assessment. Measured concentrations can be found elsewhere in this report and reports in the Nord Stream, CHEMSEA and MERCW projects.

8.3.2 Toxicity Data

In Table 8.2 Toxicity values representing PNEC are shown for lower tier (bacteria, algae and daphnia) and higher tier organisms (fish). Measured data within this project, e.g. EC50 microtox for lower tier organisms, are preferred. When such a value is not available the lowest measured value found in the scientific literature is used, and a reference is stated. When no measured values are available a QSAR value is

Table 8.1 Sediment samples, sampling area, number of samples and chemical species detected used in this assessment

Project	Sampling area	Number of sediment samples	Chemical species detected
CHEMSEA (2011–2014)	Dumpsites at Bornholm, Gotland and Gdansk deep	171	1, 1_1o, 1_2, 1_3, 1_4, 1_5, 1_7, 2o, 3o, 3t, 4, 4o, 5o, 5t, 6, 7o, 8o
MERCW (2006–2008)	Dumpsite at Bornholm deep	66	1_1o, 2o, 3a, 3o, 4, 5o, 10t, cb
Nord stream 2008	Pipeline route west of Bornholm deep	49	2o, 3o, 4, 5o
Nord stream 2010	Pipeline route west of Bornholm deep	98	2o, 3o, 5o, 8o, 10t
Nord stream 2011	Pipeline route west of Bornholm deep	98	2o, 3o, 5o, 8o, 10t
Nord stream 2012	Pipeline route west of Bornholm deep	98	2o, 3o, 5o, 8o, 10t
MODUM (2014–2016)	Dumpsites at Bornholm, Gotland and Gdansk deep	47	1, 1_2, 1_3, 1_4, 1_5, 2o, 3o, 3t, 4, 4o, 5o, 5t, 6
Total		627	

derived and used. Missing measured values can be due to restrictions towards production and purchase according to the Convention on the Prohibition of the Development, Production, Stockpiling and Use of Chemical Weapons and on their Destruction (Chemical Weapons Convention, CWC) from 1997. Only values used in this project are shown in the table. No assessment factors are used in the conversion to PNEC values according to Eq. (8.1).

The fish 14d value of 1.53 mg/L is found for the Sulphur mustard cyclic degradation product 1,4,5-oxadithiepane (Storgaard et al. 2016). This value is used for all six cyclic degradation products of sulphur mustard found in the projects for two reasons; Firstly, it has a high frequency of detection and has the highest measured concentration of the degradation products. Secondly, the compound has been characterized as toxic, according to the Globally Harmonized System of Classification and Labelling (GHS).

8.3.3 *Other Data*

Fish disease data and meio fauna data are described and assessed in the Weight of Evidence section.

CWAs and metabolites will disperse with bulk water currents and degrade and thus the exposure concentrations towards organisms decrease on a spatial and temporal scale. Special circumstances may lead to dispersion to vulnerable areas, which

Table 8.2 Toxicity values for detected parent CWAs and metabolites representing PNEC without use of assessment factors. Koc is found using (KOCWIN v.2.0) with molecular connectivity index (MCI) as descriptor or from the HSDB Toxnet database (https://toxnet.nlm.nih.gov/newtoxnet/hsdb.htm)

				QSAR derived		Measured lower tier		Measured higher tier	
Compound	Short name	CAS#	Koc (L/kgOM)	LC50 daphnia (mg/L)	LC50 algae (mg/L)	EC50 Microtox (mg/L)	EC50 daphnia (mg/L)	Fish 14d (mg/L)	Fish SSD (mg/L)
Sulphur mustard and metabolites:									
Sulphur mustard	1	505-60-2	240				0.06[g]		0.69[f]
Thiodiglycol sulfoxide (TDG[OX])	1_1o	3085-45-8	1			74250[a]		1.53[e]	
1,4-Dithiane	1_2	505-29-3	1			9.97[a]		1.53[e]	
1,4-Oxathiane	1_3	15980-15-1	2.2			47.4[a]		1.53[e]	
1,4,5-Oxadithiepane	1_4	3886-40-6	36			1.7[a]		1.53[e]	
1,2,5-Trithiepane	1_5	6576-93-8	265			1.17[a]		1.53[e]	
Thiodiglycolic acid	1_7	123-93-3	13			22.5[a]		1.53[e]	
Adamsite metabolites:									
Phenarzasinic acid (DM[ox])	2o	4733-19-1	335,583			5.33[a]			0.29[b]
Clark I and metabolites:									
Clark I (DA)	3a	712-48-1	15,304	0.165					0.29[b]
Diphenylarsenic acid (DA[ox])	3o	4656-80-8	269			124[a]			0.29[b]
Diphenylpropylthioarsine (DPTA)	3t	17544–92-2	29,188	0.196					0.29[b]
Triphenyl arsine and metabolites:									
Triphenyl arsine (TPA)	4	603-32-7	335,583			200[a]			0.29[b]
Triphenyl arsine oxide	4o	1153-05-5	520,715			155[a]			0.29[b]
Parent compounds:									
α-Chloroacetophenone (CAP)	6	532-27-4	99			0.0112[a]			0.5[c]
Trichloroarsine (TCA)	10t	7784-34-1	25		12.8				0.29[b]
Chlorobenzene	CB	108-90-7	233.9				2.5[d]		0.1[c]

(continued)

Table 8.2 (continued)

Compound	Short name	CAS#	Koc (L/kgOM)	QSAR derived		Measured lower tier		Measured higher tier	
				LC50 daphnia (mg/L)	LC50 algae (mg/L)	EC50 Microtox (mg/L)	EC50 daphnia (mg/L)	Fish 14d (mg/L)	Fish SSD (mg/L)
Organo arsenicals:									
Phenylarsonic acid (pdca_ox)	5o	98-05-5	1001			97.1[a]			0.29[b]
Dipropyl phenylarsonodithioite	5t	1776-69-8	2017	0.69					0.29[b]
Lewisite metabolites:									
2-Chlorovinylarsonic acid (L1ox)	7o	64038-44-4	157			0.0312[a]			0.29[b]
Bis(2-chlorovinyl)arsinic acid (L2ox)	8o	157184-21-9	560	55.553					0.29[b]

[a](Christensen et al. 2016)
[b](Sanderson et al. 2012)
[c](Sanderson et al. 2008)
[d](Calamari et al. 1983)
[e](Storgaard et al. 2016)
[f](Sanderson and Fauser 2016)
[g](Muribi 1997)

can be investigated with a high resolution model developed within MODUM to predict dissolved and particulate CWAs and their metabolites. Calculations performed with this model will however not be included in this chapter.

8.4 Results and Discussion

8.4.1 Risk Assessment

In Fig. 8.2 the risk quotients (RQs) calculated from Eq. (8.3) summed for all chemicals at any given site are shown in logarithmic intervals. This is done for lower tier (upper figure) and higher tier (lower figure) organisms for all sediment samples/sites stated in Table 8.1, and for a reference site chosen to represent an area not influenced by dumping activities. The blue and red lines indicate designated

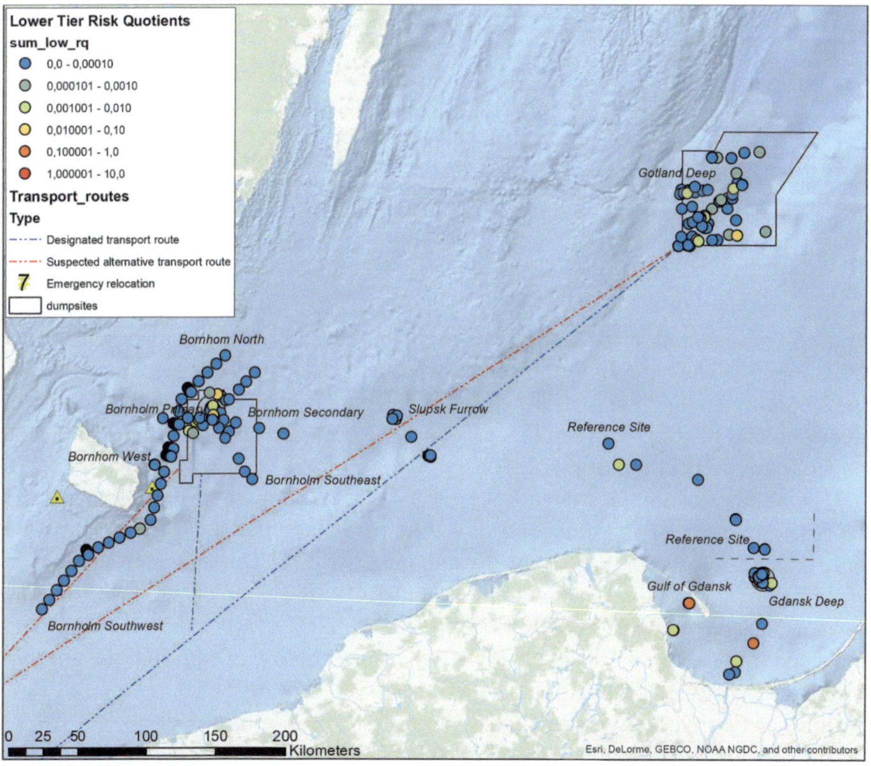

Fig. 8.2 Sum of risk quotients (RQs) divided in logarithmic intervals for lower tier (*upper figure*) and higher tier (*lower figure*) organisms. Figures comprise all 627 sediment samples for the three study areas; Bornholm deep, Gotland deep and Gdansk deep and also Slupsk furrow and a reference site. In Appendix 1 the sum RQ maps are shown individually for the three dumping areas

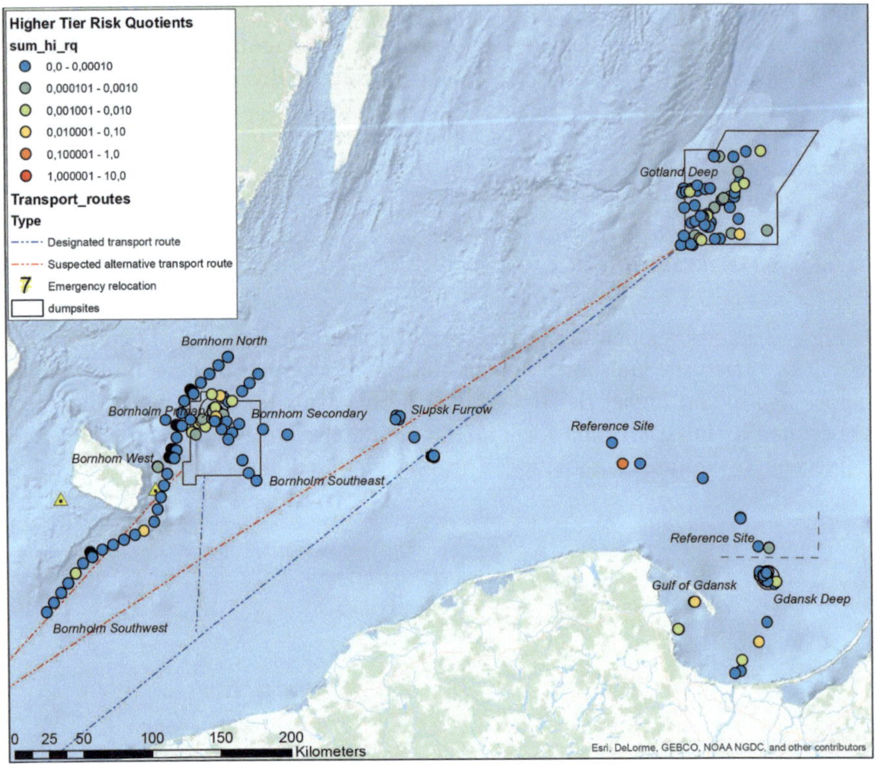

Fig. 8.2 (continued)

transport routes and suspected transport routes, respectively, of ships performing the dumping. The sampling trajectory for investigating the environmental risk for the Nord Stream pipeline is clearly discernible going from the north of the Bornholm dumpsite and passing between the east side of Bornholm and the primary and secondary dumpsites at the Bornholm deep.

The number of samples in each sum RQ interval is shown in Fig. 8.3. Sum RQ ranges from zero, i.e. no detection of CWAs or metabolites, to 1.45, which is the only sample/site that has a sum RQ value larger than 1. The abundant number of samples (> 95% of total samples) have sum RQ < 0.01, which indicate low risk towards the analyzed CWAs and metabolites. A sum RQ = 0.01 indicate no risk with an uncertainty factor of 100 on the PNECs. The distribution of sum RQs are approximately equal for lower and higher tier organisms. Discrepancies are due to variabilities of PNECs between chemicals.

Broken down in individual chemicals (CWAs and metabolites) the maximum RQ and mean RQ ± st.dev. Are shown in Table 8.3 and Fig. 8.4 together with detection frequencies (DF) and areas where max RQs occur. The DF for a chemical is calculated as the number of detections divided with total number of samples where the chemical was analyzed for.

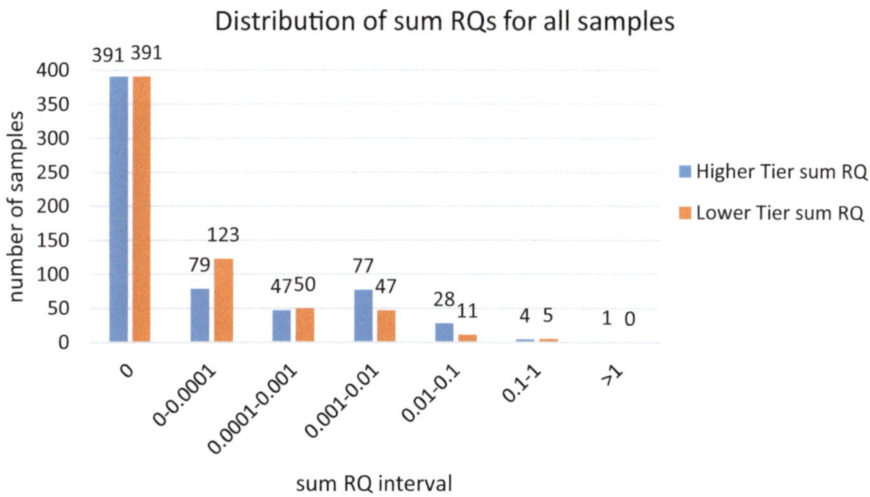

Fig. 8.3 Distribution of sum RQ values for all 627 sediment samples. The majority of samples (> 95% of all samples) indicate no risk with a assessment factor of 100. One sample for higher tier organisms show a sum RQ > 1. The number of samples in intervals is shown on top of the bars

The chemical with the highest max RQ (0.28) for lower tier organisms is the Lewisite metabolite 7o. 7o was only detected in the CHEMSEA project and has a relatively low DF of 0.8%. The highest value is found in the gulf of Gdansk, see Fig. 8.5. A relatively low EC50 value, see Table 8.2, contribute to the high RQ. It is also 7o that has the highest mean RQ. Clark I (3a), which was detected in the MERCW project, has a max RQ of 0.17 for lower tier organisms found in the primary dumpsite in the Bornholm deep. For higher tier organisms trichloroarsine (10t), detected in the MERCW and Nord Stream projects, has the highest max RQ (0.84) and mean RQ (0.20) values found in the primary dumpsite in the Bornholm deep with a DF = 1.4%. The same site has the highest max RQ (0.48) for phenylarsonic acid (5o), which was detected in all projects. These highest max RQs occur at the site with the overall highest calculated max RQ (1.46) in the primary dumpsite of Bornholm deep. For higher tier organisms the max RQ at this sampling site is a factor of four higher than the second highest (0.36) also found at the primary dumpsite.

From Fig. 8.5 it is seen that for lower tier organisms the sum RQ reaches highest values in the gulf of Gdansk and the Gdansk deep and that 7o is the main contributor to the high risk. The main contributors to the sum RQ in the Bornholm primary dumpsite are more diverse with 3a, 1_2 and 1_4 for lower tier organisms and 10t, 5o and 1_2 for higher tier organisms. Max sum RQ for Gotland deep is more than a factor of 10 lower than the max sum RQs for Bornholm and Gdansk sites for both tiers. The max sum RQ for the reference area is more than a factor of 25 lower than Bornholm and Gdansk sites for lower tier. For higher tier organisms the max sum RQ is relatively high for the reference site, due to significant detection of a Sulphur mustard degradation product (1_3).

Table 8.3 Maximum and mean ± st.dev. Risk quotients (RQs) for individual chemicals (CWAs or metabolites) for all 627 sediment samples

Chemical	Lower Tier			Higher Tier			DF (%)	Area max RQ
	RQ max	RQ mean	St. dev.	RQ max	RQ mean	St. dev.		
1	0.040	0.020	0.029	0.0035	0.0018	0.0025	0.4% (n = 533)	Bornholm_primary
1_1o	1.5E-05	8.1E-06	6.3E-06	0.00049	0.00026	0.00020	0.9% (n = 580)	Reference_area
1_2	0.042	0.0037	0.0084	0.27	0.024	0.055	15.6% (n = 218)	Bornholm_primary
1_3	0.0040	0.0011	0.0017	0.12	0.033	0.052	2.8% (n = 218)	Reference_area
1_4	0.045	0.0024	0.0060	0.050	0.0027	0.0067	32.1% (n = 218)	Bornholm_primary
1_5	0.015	0.0012	0.0031	0.011	0.00089	0.0023	16.5% (n = 218)	Bornholm_primary
1_7	0.016	0.016	0.016	0.043	0.043	0.0011	1.2% (n = 171)	Gdansk_deep
2o	4.3E-05	1.5E-06	6.6E-06	0.00079	2.8E-05	0.00012	10.7% (n = 599)	Bornholm_secondary
3a	0.17	0.011	0.035	0.099	0.0063	0.020	6.4% (n = 409)	Bornholm_primary
3o	0.0094	0.00011	0.0010	0.0060	0.00018	0.00086	14.5% (n = 599)	Bornholm_primary
3t	0.28	0.0058	0.030	0.014	0.00074	0.0027	13.7% (n = 190)	Bornholm_primary
4	1.6E-05	3.5E-07	2.1E-06	0.011	0.00024	0.0014	9.4% (n = 627)	Bornholm_primary

4o	9.4E-08	2.3E-08	2.8E-08	0.00095	0.00012	0.00031	4.7% (n = 190)	Bornholm_primary
5o	0.0014	2.1E-05	0.00013	0.48	0.0069	0.045	9.4% (n = 627)	Bornholm_primary
5t	0.0044	0.00040	0.00086	0.011	0.00095	0.0020	19.5% (n = 218)	Bornholm_primary
6	0.082	0.040	0.046	0.0019	0.0012	0.00072	14.2% (n = 533)	Bornholm_primary
7o	0.28	0.22	0.059	0.030	0.023	0.0063	0.8% (n = 514)	Gulf_of_Gdansk
8o	2.6E-05	1.3E-05	1.1E-05	0.0049	0.0025	0.0021	0.6% (n = 514)	Bornholm_west
10t	0.019	0.0046	0.0081	0.84	0.20	0.36	1.4% (n = 360)	Bornholm_primary
Cb	0.00081	0.00023	0.00019	0.020	0.0058	0.0047	18.2% (n = 66)	Bornholm_primary

The detection frequency (DF) is number of findings larger than the limit of detection relative to the number of samples where the chemical was analyzed for (Table 8.1). The sample area with maximum RQ is also stated. See full name of chemicals in Table 8.2

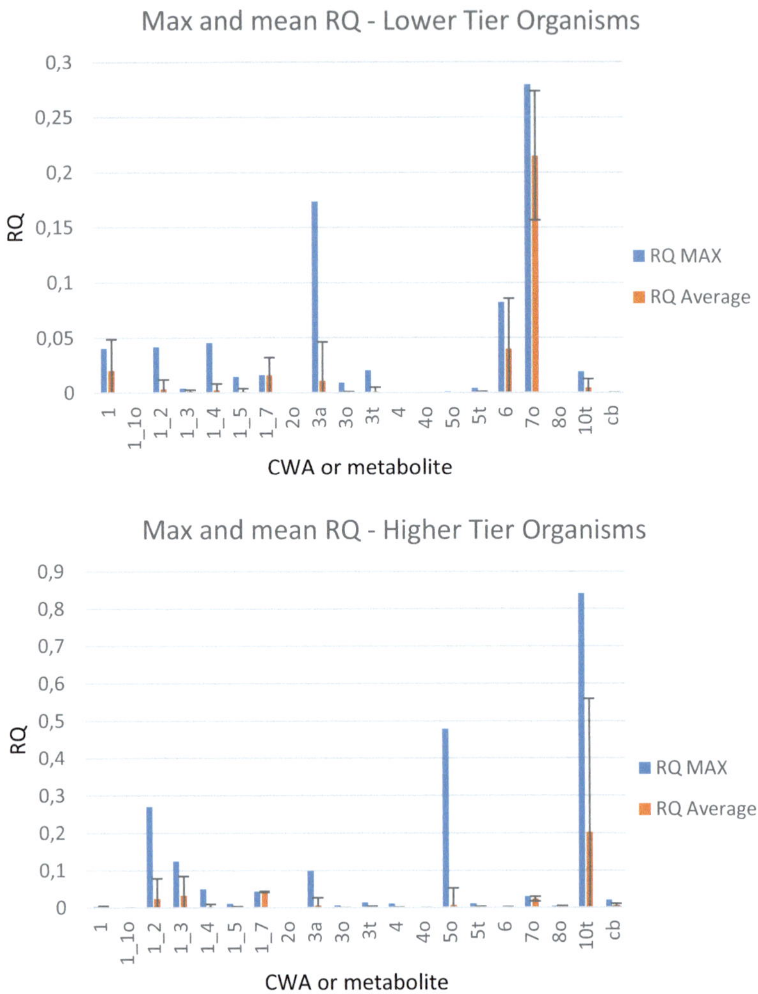

Fig. 8.4 Maximum and mean RQ ± st.dev. for single chemicals (CWAs or metabolites) across all sediment samples for lower tier organisms (*upper figure*) and higher tier organisms (*lower figure*)

8.5 Weight of Evidence (WoE)

8.5.1 Lines of Evidence (LoE)

To perform a semi-quantitative assessment of the environmental risk in the study areas from occurrence of CWAs and their metabolites, a number of LoEs are set up, that take all aspects of the performed investigations into account. Each LoE is assigned +3, +2, +1, 0, −1, −2 or −3, indicating if the LoE is found to be for (+), against (−) or neutral (0) to an increased environmental risk of dumped CWAs.

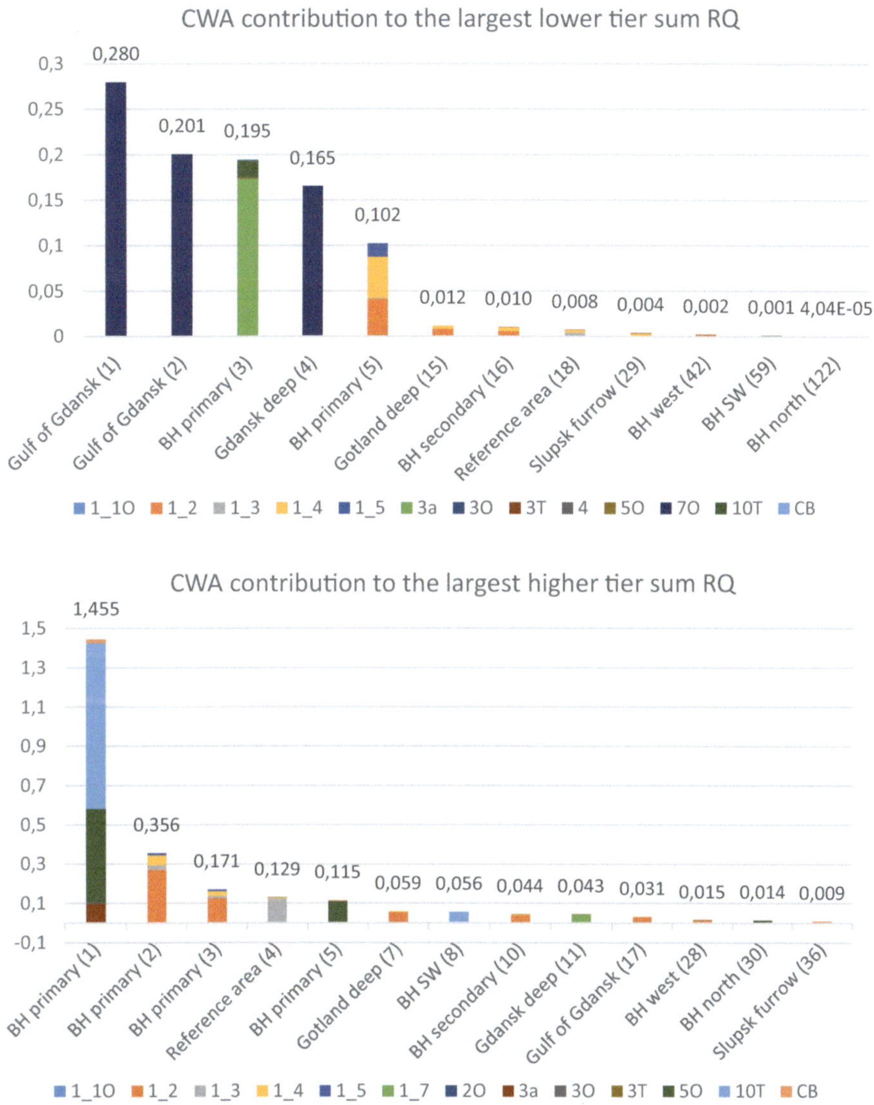

Fig. 8.5 Sediment samples with highest sum RQs for all study areas including reference area and Slupsk furrow, broken down in individual CWAs and metabolites. The number in parenthesis indicates the rank of sum RQ compared to all 629 samples

Each LoE is based on information and data that have been derived from the projects on the study areas in Table 8.1. This means that all references, e.g. literature references, used to derive results within the projects will not be weighted individually, but rather weighted as aggregates in the overall results from the projects. It can be argued that the individual studies should have one weight score each, but in the derivation of the results within the projects covered in this assessment we include previous significant conclusions and data, which justifies the use of overall results from the projects.

Some LoEs have more predictive power than others. For example the calculated sum RQs (LoE1) comprise important answers to the hypothesis; "Dumped CWAs constitute an environmental risk in the Baltic Sea". Accordingly, there is the possibility for weighing the parameters from 1, 2 or 3, with 3 representing the most influential LoEs.

LoE1: Environmental risk quotients (sum RQs) indicate risk.

This LoE has arguably the highest predictive power in relation to environmental risk assessment of dumped CWAs. It combines the knowledge and data on measured and modelled occurrence of CWAs and metabolites together with measured and modelled toxicity values. Both types of parameters have been derived in the studies included in this assessment with addition of best available toxicity data for a couple of compounds.

Based on the 627 sediment samples that have been taken during the last 10 years and analysed for known dumped parent CWAs and a wide range of their metabolites, only one sediment sample is found to have a sum RQs exceeding 1, being the limit value of risk, see Fig. 8.5. No assessment factors (AFs) have been assigned to the toxicity values.

The issue of assigning AFs is crucial in risk assessment. Uncertainties may apply both to the toxicity estimate PNEC and the exposure estimate PEC. Here we will briefly illustrate the consequences of applying AFs to the toxicity value PNEC, as outlined in ECHA (2008). The general principle is that the result from a laboratory test is divided by an appropriate AF. The sparser the available data, the higher is the AF which is applied. PNECs are estimated by division of the lowest value for the toxicity with the relevant AF. We will tentatively apply an AF of 10,000 for the EC50 lower tier tests and Fish 14d higher tier test. Higher AFs could be applied to the QSAR derived values and lower AFs could be assigned to the Fish SSD. If we assign AFs of 10,000 to all PNECs then 25% and 18% of all samples will show a risk (sum RQ > 1) for higher and lower tier organisms, respectively. If more knowledge on the toxicity values are obtained the AFs will be lowered, and as an example an AF of 100 will result in 5.3% and 2.6% of the samples showing a risk (sum RQ > 1) for higher and lower tier organisms, respectively. So, AFs are clearly instrumental in risk assessment.

Since we have a situation ranging from only one sample with sum RQ > 1 (AF = 1) to significant risk where approx. 20% samples have sum RQ > 1 (AF = 10,000) there is an indication with considerable uncertainty that the calculated sum RQs show risk, and we will assign the value +1 to LoE1.

How to act on a result that reveals risk for a given number of samples is a management decision, which can be based on the precautionary principle that takes all kinds of factors into consideration. Our scientific recommendations are to enhance the PNECs for the lower and higher tier organisms by performing toxicity tests for species that represent the actual ecosystem more completely and thus ensuring that the extrapolation from laboratory to the Baltic Sea becomes more realistic. This will reduce the applied AFs and make the RQs more precise and reliable.

LoE2: CWAs and metabolite detections are in areas that can be associated with dumping of CWAs.

The study areas cover the suspected and recorded areas of dumping, designated and suspected transport routes of ships performing the dumping as well as sites with no suspected dumping activities, see Fig. 8.2.

From Table 8.4 it is seen that the areas with high expectancy of finding traces of CWAs and metabolites have relatively high DFs, e.g. 91% and 85% for Bornholm primary and secondary dumpsites. Lower DFs are found for the dumpsites Gdansk deep and Gotland deep with 34% and 37%, and for designated and suspected ship routes; Bornholm SW, Bornholm W, Gulf of Gdansk and Slupsk furrow with 31%, 21%, 56%, 25%, respectively. Sampling areas with no or low expectancy to detect CWAs and metabolites are Bornholm North and Bornholm SE with DFs of 11% and 0%. Approximate CWAs dumped amounts in the Bornholm deep, Gotland deep and Gdansk deep are 1200, 900 and 60 tonnes (HELCOM 1994), so apart from the reference area where the DF is relatively high (44%), the found DFs are in accordance with what would be expected on a relative scale.

Table 8.4 Sampling area, area description, number of samples in study areas and detection frequency (DF)

Sampling area	Description in relation to dumping activities	Number of samples	Detection frequency (DF)%
Bornholm_North	Nord stream transect, MERCW transect	63	11
Bornholm_Primary	Primary Bornholm hot spot	88	91
Bornholm_SE	MERCW transects	3	0
Bornholm_Secondary	Secondary Bornholm hot spot	20	85
Bornholm_SW	Nord stream transect and designated ship route	118	31
Bornholm_West	Nord stream transect and partly designated ship route	173	21
Gdansk_Deep	Gdansk hot spot	32	34
Gotland_Deep	Gotland hot spot	100	37
Gulf_of_Gdansk	Transect to shore	9	56
Reference_Area	No suspected dumping	9	44
Slupsk_Furrow	Designated and suspected ship routes to Gotland deep	12	25
Sum		627	

Given the large spatial scale of the dumping areas and the uncertainties pertaining to the exact location of dumping there is an indication that CWAs and metabolite detections are in areas that can be associated with dumping of CWAs, and therefore LoE2 is assigned +2.

LoE3: CWA lumps, munitions, intact or fragments have been found or caught by fishermen.

From Fig. 8.6 it is seen that there are no reported encounters in the Bornholm primary dumpsite, due to the trawling prohibition. The two yellow triangles in Fig. 8.6 mark emergency relocation sites for caught CWAs. The number of reported encounters by fishermen has decreased significantly since the early 1990ies (Fig. 8.7), which indicates that the problem of encountering CWA lumps or munition by activities near or in dumpsites is decreasing. In conclusion, all other things being equal, such as trawling frequency or efficiency of equipment, there is evidence that CWA lumps, munitions, intact or fragments have been found or caught by fishermen, and LoE3 is assigned +2.

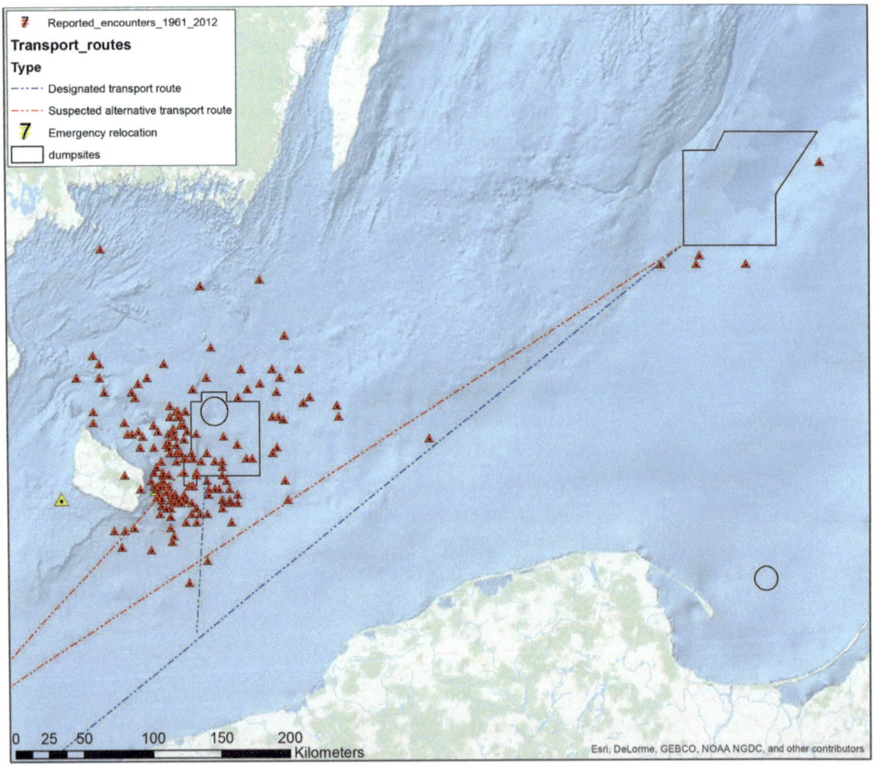

Fig. 8.6 Spatial representation of reported encounters with lumps of CWAs or munition parts caught by fishermen in the period from 1961 to 2012 (HELCOM 2013)

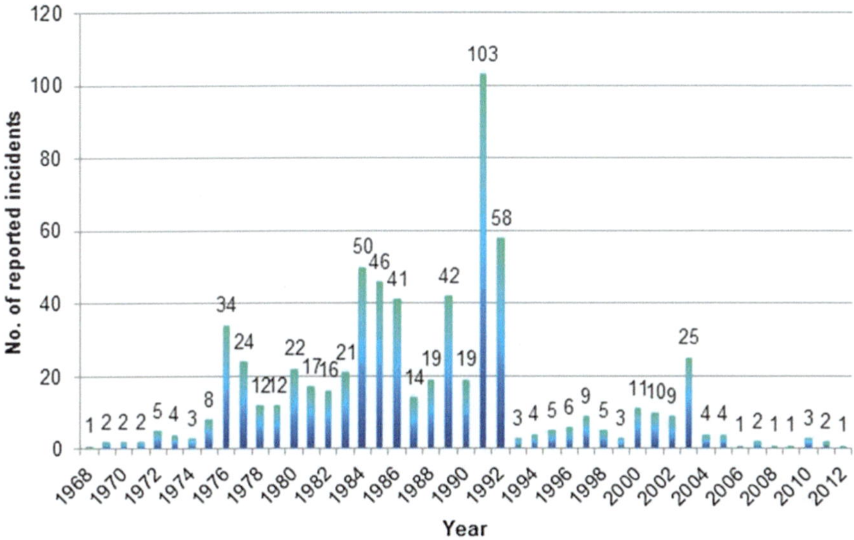

Fig. 8.7 Temporal representation of reported encounters with lumps of CWAs or munition parts caught by fishermen in the period from 1961 to 2012 (HELCOM 2013)

LoE4: Found lumps contain parent compounds and/or metabolites.

No chemical analysis has been performed on the caught lumps, but based on the registered health effects and visual inspection it is assumed that they are lumps of mustard gas and occasional other CWA remnants. LoE4 is assigned +1.

LoE5: Intact munitions with CWAs are buried in sediment.

This can potentially be a considerable risk source. Although the performed sonar and magnetic measurements indicate several submerged magnetic objects it is associated with high uncertainty to determine the type and origin of the objects. LoE5 is assigned 0.

LoE6: Meio fauna are exposed and affected by dumped CWAs.

According to Czub et al. (2016) the only known benthic fauna inhabiting the areas of CW dumpsites in the Baltic Sea is meio fauna, represented by Nematoda. Furthermore, detailed taxonomical analyses showed statistically valid differences between infauna communities from dumpsites in Bornholm, Gdansk and Gotland Deeps most likely linked with differences of type, source and quantities of suspending organic matter.

Figure 8.8 shows the meio fauna (nematode) counts performed in 61 samples taken in all study areas including the reference area and Slupsk furrow. At each sampling point sediment samples were analysed for CWAs and metabolites. Two sites had mean counts >1000, and 15 sites had mean counts >100 and <1000. Forty five stations have counts different from zero (Fig. 8.9).

The detection frequency of any CWA or metabolite is relatively low for the 61 sampling sites (37%) with meio fauna counts, and the sum RQs are relatively low

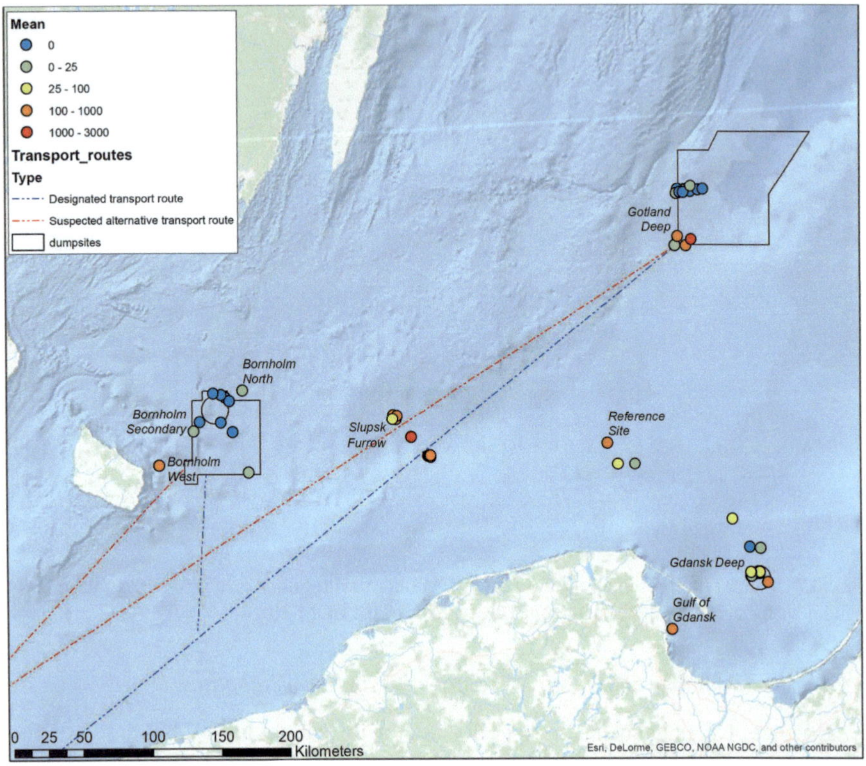

Fig. 8.8 Meio fauna (nematode) counts at 61 sampling stations covering all study areas, Slupsk furrow and reference area

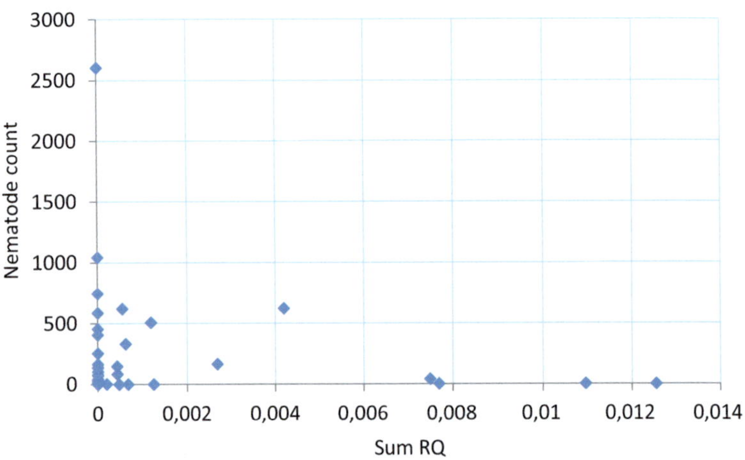

Fig. 8.9 Nematoda count vs. sum RQ at 61 sampling stations. Forty five stations have counts different from zero. Twenty three samples have sum RQs different from zero

with six samples having sum RQ > 0.002 and two samples with sum RQ > 0.01. Mean and standard deviation sum RQs is 8E-4 ± 2E-3.

The frequency of meio fauna counts larger than zero is relatively high (73%), ranging from 1 to 2603, with a mean and standard deviation of 156 ± 384 counts.

A multivariate analysis of site specific parameters such as dissolved oxygen (DO), depth, sediment characteristics, pH, and occurrence of CWAs or metabolites and biota was performed by Sanderson et al. (2012) for data from the Bornholm deep. Similar to that study the present study found that CWA metabolites are more frequently found and at concentrations, which do not constitute a significant risk at the sites with meio fauna samples. In Sanderson et al. (2012) depth, salinity and temperature were co-variants that were strongly inversely correlated with the macro zoobenthos data, and oxygen (DO) was strongly positively correlated with the measured biota. The total CWA risk was moderately to strongly negatively correlated with biota.

In other words, based on the statistical relationships the observed biota is primarily affected by the availability of DO, which is a function of water depth, which again govern water temperature; i.e. greater depth, less DO and less biota. Likewise, according to the multivariate analysis, there is a probability of observing reductions in biota where the total CWA risk is predicted to be elevated. Elevated DO levels suggest lower CWA concentration, e.g. due to more effective oxidation and degradation of the CWA and less dumped munition in the first place.

The conclusions that can be drawn in this analysis are; A quantitative assessment of the observed meio fauna counts indicate a moderate to strong negative correlation with occurrence of dumped CWAs and metabolites. The calculated sum RQs are low, i.e. all sum RQs are below unity with a factor of 100. To enhance the predictive power of data, the number of samples with corresponding concentration measurements and meio fauna counts should be increased, and more detailed information on parameters such as DO, species richness and biomass should be included.

LoE6 (meio *fauna*) is assigned +1 regarding exposure and 0 regarding effects.

LoE7: Fish are exposed and affected by dumped CWAs.

Disease data from 19.171 Atlantic cod (*Gadus morhua*) caught in the Baltic between 2011 and 2015 are used to evaluate the toxicological effect on fish from dumped CWAs. The standardized fish disease index (sFDI) is used to quantify the disease status of the cod and it is based on seven externally visible fish diseases, which is described in detail elsewhere in the report.

A hotspot analysis is used to visualise the regional differences in the disease status of the cod. The hotspot analysis is a statistical tool, that identifies significant spatial clusters of high sFDI values (hot spots) and low sFDI values (cold spots). Each fish that has been caught is plotted as a feature on a map, using the geographical location (latitude and longitude) and the calculated sFDI value. From this plot each feature is analysed within the context of neighbouring features. Neighbouring features inside the specified critical distance receive a weight of one and exert influence on computations for the target feature. Neighbouring features outside the critical distance receive a weight of zero and have no influence on a target feature's computations.

A critical distance of 40 km is used in the analysis. It is based on the geographical size of the areas where the fish are caught. The critical distance of 40 km makes it possible only to identify regional differences in fish disease, and not local differences. But since the fish are not static objects in the sea, it is more useful to look on regional differences than local.

In the Baltic Sea there are two genetically distinct cod populations; the western and the eastern stock (Schaber et al. 2011). They are therefore analysed individually in the hotspot analysis. Consequently, it is possible to see relative differences in fish disease between the two populations as well as between the different regions (Figs. 8.10, 8.11 and 8.12).

The eastern population is generally healthier than the western population in all geographical regions. Among the western populations, Mecklenburg Bight has the highest mean sFDI value of 4.3, Kiel Bight 3.6, Arkona Sea east 2.26 and Arkona Sea west 2.22. In the eastern population Gdansk Bay Reference has the highest mean sFDI value of 2.15, Gdansk Deep Dumpsite 2.11, Bornholm Dumpsite 1.87, Gotland Deep Dumpsite 1.61, Hanö Bight 1,17 and Gotland Reference 0.16. The

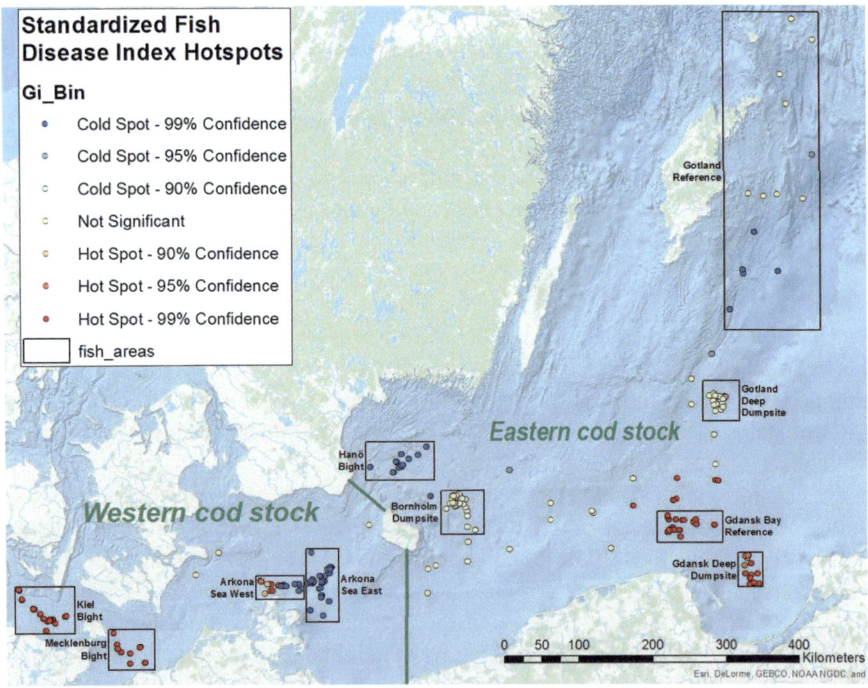

Fig. 8.10 Hotspot analysis of the disease status of western and eastern population of Atlantic cod in the Baltic Sea

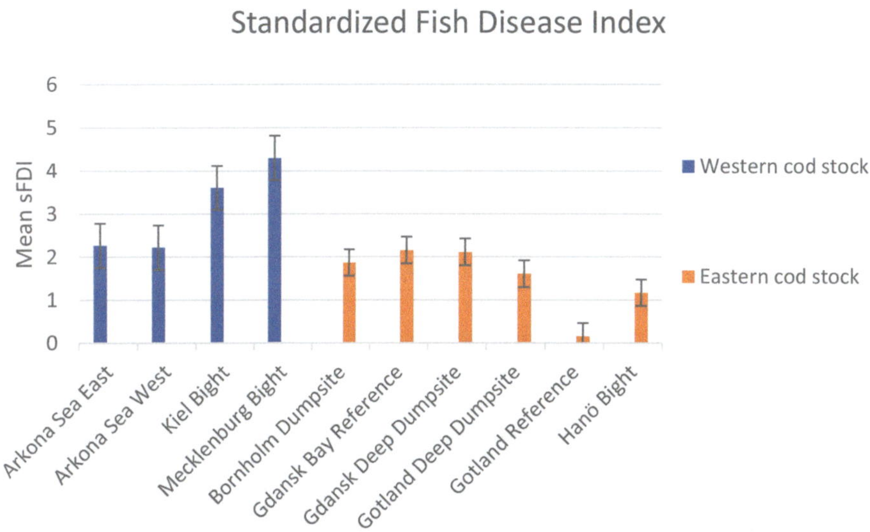

Fig. 8.11 Mean standardised Fish Disease Index values of the cod caught in the difference geographical regions in the Baltic Sea

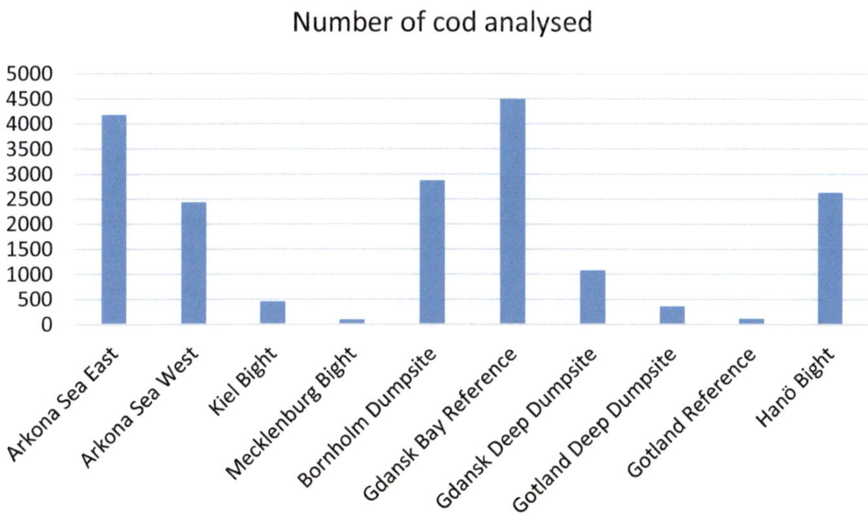

Fig. 8.12 The number of cod analysed in each area

low numbers of cod analysed in Mecklenburg Bight (100) and Gotland Reference (115), cause some statistical uncertainty on those areas.

Comparing the regional differences between Hanö Bight (1.17) and Bornholm Deep (1.87) which is separated by 75–115 km, shows a significant difference in fish disease. The same difference can also be found between Gotland Deep Dumpsite

(1.61) and Gotland reference (0.16) which is separated by 100–500 km of water, but it is based on a very small number of caught fish (115), making it more uncertain. The regional difference between Gdansk Reference (2.15) and Gdansk Deep Dumpsite (2.11), which is separated by 65–150 km of water, is almost insignificant.

The conclusion is that the fish disease data do not give a clear correlation between fish disease and dumped CWA's. The marine environment in the Baltic Sea is contaminated with various xenobiotic compounds, including hydrocarbons, radionuclides, heavy metals, pesticides, that are all causing environmental stress making it hard to see if and how much the CWAs are contributing (Baršiene et al. 2014), the dynamic nature of fish is also adding to that. The relative large difference in fish disease status between Hanö Bight and Bornholm Deep indicates that the fish health could be affected by the dumped CWA. Therefore, more reference data is needed in other areas that are in close proximity to the dump sites, in order to establish a clear correlation.

Studies of the horizontal and vertical distribution pattern of cod in the Baltic Sea, shows that the cod can be found at the bottom in the Bornholm Deep Dumpsite (Schaber et al. 2011) and therefore the fish can be exposed to CWA from the sediment.

LoE7 (fish) is *assigned* +1 regarding exposure and 0 regarding effects.

LoE8: Detected parent CWAs and degradation products are toxic.

The acute and chronic toxicity of chemicals is classified according to the Globally Harmonized system of classification and labelling of chemicals (UN 2010). Measured data for fish, crustaceans or algae are preferred as test species for the aquatic environment, and when these data are missing other species or QSAR data can be used. In this toxicity assessment the measured EC50 values, fish 14d only for sulphur mustard metabolite (1_4), fish SSD for sulphur mustard (1), and QSAR LC50 values in Table 8.2 are used. The fish SSD of 0.29 mg/L is derived from toxicity data on inorganic As from a conservative point of view (Sanderson et al. 2008). This value is not chemical specific and will not be used in this toxicity assessment. An acute toxicity of a compound can be assigned a label of very toxic, toxic, harmful or not toxic according to the following:

- Very toxic: EC50 $\leq$ 1 mg/L
- Toxic: 1 mg/L < EC50 $\leq$ 10 mg/L
- Harmful: 10 mg/L < EC50 $\leq$ 100 mg/L
- Not toxic: 100 mg/L < EC50

Of the detected compounds in Table 8.2, six chemicals are classified as very toxic to aquatic life: 1, 6, 7o, 3a, 3t and 5t. Five chemicals are classified as toxic: 1_2, 1_4, 1_5, 2o and cb, five as harmful: 1_3, 1_7, 10t, 5o and 8o, and four as not toxic: 1_1o, 3o, 4 and 4o.

Among the parent CWAs and metabolites that have highest DFs and max and mean RQs the following are classified as very toxic or toxic: 1_2, 1_4, 1_5, 3a, 3t and 7o. Based on this assessment LoE 8 is assigned +1.

LoE9: Detected parent CWAs and degradation products cause harmful gene responses.

Basically, chemicals that damage DNA causing lesions that result in cell death or mutations are genotoxins. To evaluate the genotoxicity of chemicals the Hazardous Substances Databank (HSDB) (https://toxnet.nlm.nih.gov/cgi-bin/sis/search2) is used. HSDB is a toxicology database, developed by the National Institutes of Health (NIH) that focuses on the toxicology of potentially hazardous chemicals.

The majority of chemicals are not listed in the database, which is probably due to the fact that no toxicological studies have been performed. The chemicals that do have an entry show the following results:

- Sulphur mustard (1) has evidence of genotoxicity.
- 1,4-dithiane (1_2) is found to have low mutagenicity. Mutagenicity refers to its capacity to cause mutations, i.e. genetic alterations. All mutagens are genotoxic, but not all genotoxins are mutagens as they may not cause retained alterations in DNA sequence.
- α-chloroacetophenone (CAP) (6) is not genotoxic.
- Trichloroarsine (TCA) (10t) shows limited evidence of animal carcinogenicity, which can be categorized as either genotoxic or non-genotoxic.
- Chlorobenzene (cb) shows limited evidence of genotoxicity.

1, and to a lesser extent 1_2 and cb and possibly 10t are the only chemicals which are found to show evidence of genotoxicity. One has only been detected in two samples (DF = 0.04%), so the potential impact is limited. For 1 the max RQ is 0.04 and 0.004 for lower and higher tier organisms, respectively. The Sulphur mustard metabolite 1_2 has a relatively high DF (16%) and is one of the main contributors to the sum RQ for both tiers in the Bornholm primary dumpsite, which is the study area with the highest DF (91%). Cb has a relatively high DF (18%) but is an industrial chemical and thus not a significant indicator of dumped CWAs. 10t has the highest max and mean RQs for higher tier organisms although with a relatively low DF (1.4%).

It is important to bear in mind that most of the chemicals (parent CWAs and metabolites) that have been detected in MODUM and the other related studies have no data with respect to genotoxicity. Therefore the full genotoxic potential of CWAs and their metabolites is not known. It is recommended to perform tests with the following parent CWAs and metabolites based on their DFs and RQs, to evaluate their genotoxicity; Sulphur mustard metabolites (1_2, 1_4, 1_5), parent compounds 10t, 6 and 3a, and metabolites 7o, 5o and 3t. Based on this assessment LoE9 is assigned 0.

8.6 Conclusions

Based on the listed Lines of Evidences (LoEs) a Weight of Evidence (WoE) assessment has been made by assigning them with a weight of +3, +2, +1, 0, −1, −2 or −3, indicating if they are for, neutral or against the hypothesis: "Dumped CWAs constitute an environmental risk in the Baltic Sea". The weight of each LoE is stated in parenthesis.

LoE1: Environmental risk quotients (sum RQs) indicate risk (+1).
LoE2: CWAs and metabolite detections are in areas that can be associated with dumping of CWAs (+2).
LoE3: CWA lumps, munitions, intact or fragments have been found or caught by fishermen (+2).
LoE4: Found lumps contain parent compounds and/or metabolites (+1).
LoE5: Intact munitions with CWAs are buried in sediment (0).
LoE6: Meio fauna are exposed (+1) and affected (0) by dumped CWAs.
LoE7: Fish are exposed (+1) and affected (0) by dumped CWAs.
LoE8: Detected parent CWAs and degradation products are toxic (+1).
LoE9: Detected parent CWAs and degradation products cause harmful gene responses (0).

The weight reflects the claim of being for, neutral or against but also the predictive power, or significance, of the individual LoEs. The summed score of +9 indicates that there is a weak to moderate potential for confirming the hypothesis. In order to qualify and increase the precision of the weights there are a number of issues that can be addressed in future investigations:

- Focus can be on parent CWAs and metabolites that have highest DFs and max and mean RQs, namely; Sulphur mustard metabolites (1_2, 1_4, 1_5), parent compounds 10t, 6 and 3a, and metabolites 7o, 5o and 3t.
- The dump sites and designated and suspected ship routes that have been investigated in the projects have relatively high DFs, ranging from 21% to 91%. However, max RQs and DFs have been found at Bornholm primary dumpsite and Gulf of Gdansk and emphasis could be on these areas with main focus on Bornholm primary dumpsite.
- Performing toxicity tests for species that represent the actual ecosystem more completely will reduce the assessment factor applied to PNEC and thus reduce the uncertainties on the RQs.
- Perform tests to evaluate the genotoxicity of prioritized chemicals.
- Chemical analysis of found lumps of mustard gas will reveal the composition of parent compound and metabolites and thus their potential hazard.
- Investigate hot spots of suspected submerged intact munitions as these may be a potential source of instantaneous or continuous release in case of disturbance or corrosion.
- Investigate occurrence of meio fauna in more sites where there are detected CWAs or metabolites. More detailed information on parameters such as DO, species richness and biomass should be included.
- Obtain more data on fish disease of cod caught in reference areas that are in close proximity (40–100 km) to the dumpsites.

Appendix 1 Risk Maps for Study Areas

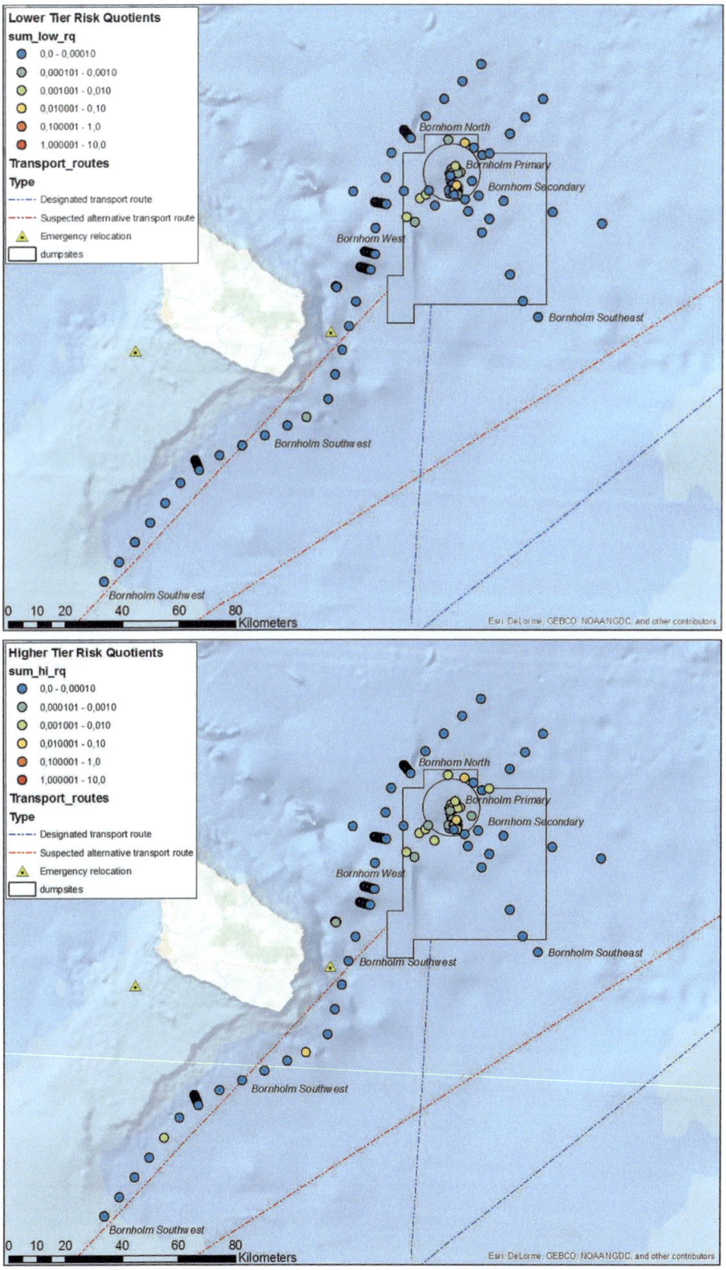

Fig. 8.13 Sum of risk quotients (RQs) for Bornholm deep with primary and secondary dumpsites and Nord Stream pipeline trajectories. The *blue* and *red* lines indicate designated transport routes and suspected transport routes, respectively, of ships performing the dumping. Lower tier (*upper figure*) and higher tier (*lower figure*) organisms

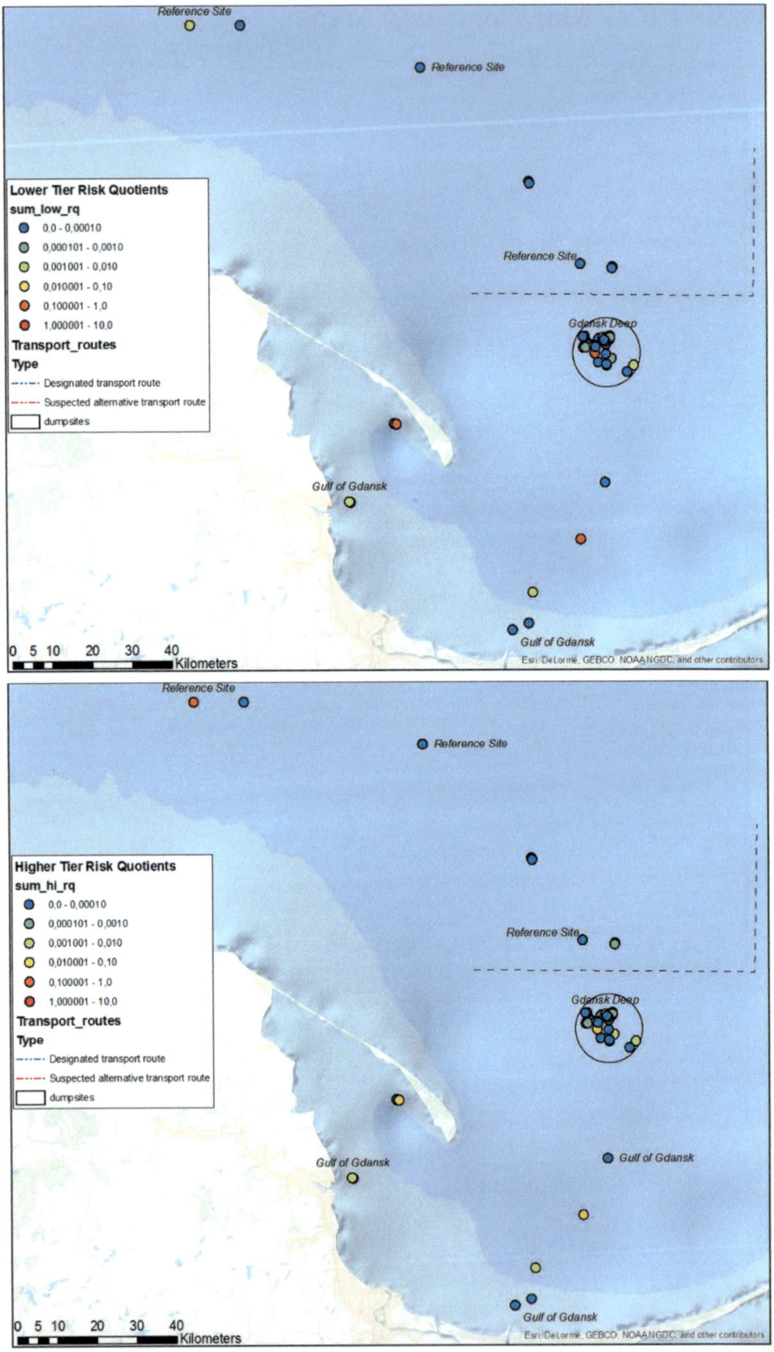

Fig. 8.14 Sum of risk quotients (RQs) for Gdansk deep, gulf of Gdansk and reference site with samples concentrated around suspected dump site. Lower tier (*upper figure*) and higher tier (*lower figure*) organisms

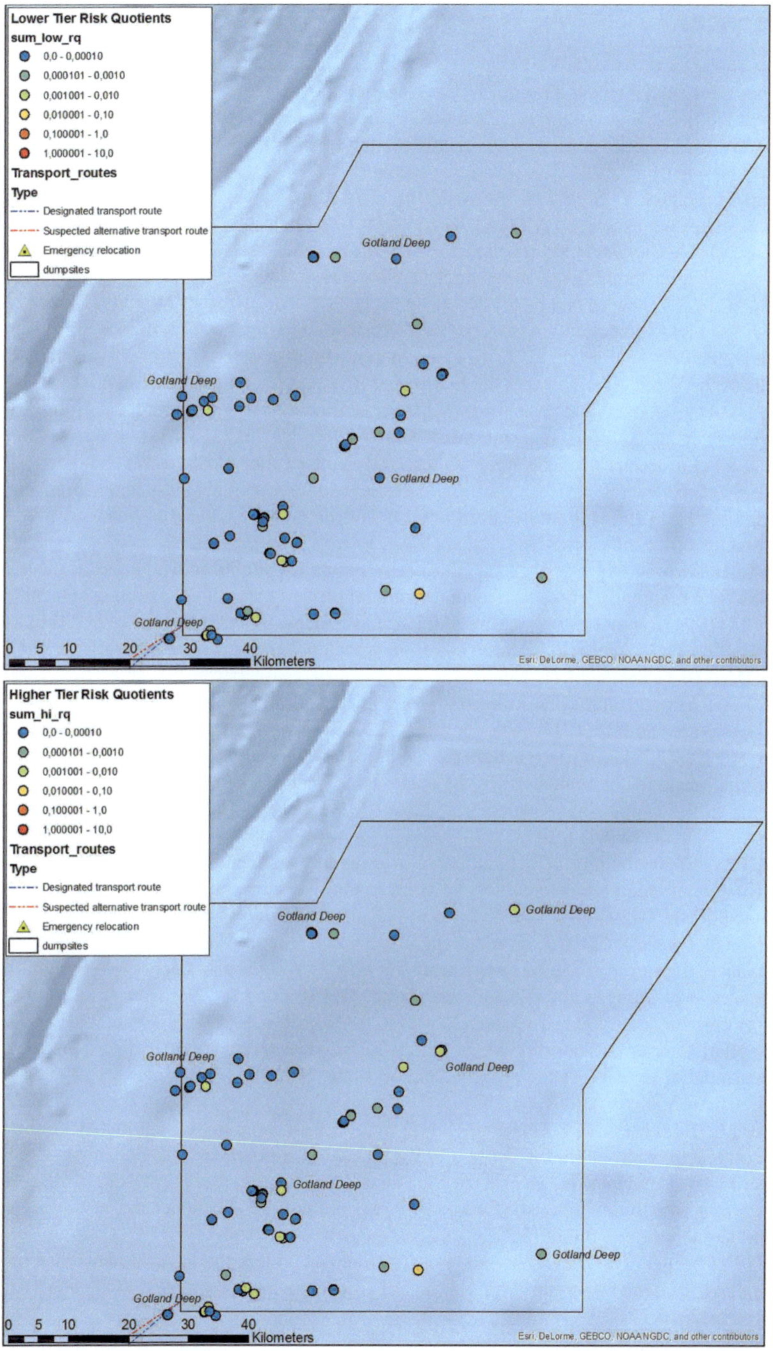

Fig. 8.15 Sum of risk quotients (RQs) for Gotland deep for lower tier (*upper figure*) and higher tier (*lower figure*) organisms

References

Baršiene J, Butrimavičiene L, Grygiel W, Lang T, Michailovas A, Jackunas T (2014) Environmental genotoxicity and cytotoxicity in flounder (Platichthys flesus), herring (Clupea harengus) and Atlantic cod (Gadus morhua) from chemical munitions dumping zones in the southern Baltic Sea. Mar Environ Res 96:56–67

Calamari D, Galassi S, Setti F, Vighi M (1983) Toxicity of selected chlorobenzenes to aquatic organisms. Chemosphere 12(2):253–262. doi:10.1016/0045-6535(83)90168-6

Christensen IMA, Storgaard MS, Hansen SF, Baatrup E, Sanderson H (2016) Acute toxicity of sea-dumped chemical munitions : luminating the environmental toxicity of legacy compounds, global security: health, science and policy. Routledge 1(1):39–50. doi:10.1080/23779497.2016.1219962

Czub M, Kotwicki L, Jąkacki J, Andrzejewski J, Grabowski M, Szubska M, Bełdowski J, Klusek Z, Rak D, Sanderson H, Lang T (2016) Deep Marine Habitats in the Chemical Warfare Dumping areas of the Baltic Sea: Towards the Ecological Risk Assessment (draft Nov 2016)

DiToro DM (1991) Technical basis for establishing sediment quality cri-teria for non-ionic organic chemicals using equilibrium partitioning. Environ Toxicol Chem 10:1541–1583

ECHA (2008) European Chemicals Agency. Guidance on information requirements and chemical safety assessment. Chapter R.10: Characterisation of dose [concentration]-response for environment

Emelyanov EM (1996) Chemical components and elements in the Sus-pend-ed matter and sediments of the western Baltic. Baltica 9:5–15

Forster S, Bobertz B, Bohling B (2003) Permeability of sands in the coastal areas of the southern Baltic Sea: mapping a grain-size related sediment property. Aquat Geochem 9:171–190

HELCOM (1994) Report on chemical munitions dumped in the Baltic Sea. Dan Environ Prot Agency. Available at: http://www.helcom.fi/helcom-at-work/publications/other-helcom-reports/

HELCOM (2013) Knobloch T, Bełdowski J, Böttcher C, Söderström M, Rühl N-P, Sternheim J (2013) Chemical munitions dumped in the Baltic Sea, Baltic Sea environment proceedings No. 142. Available at: http://www.helcom.fi/helcom-at-work/publications/other-helcom-reports/

Muribi M (1997) Toxicity of mustard gas and two arsenic based chemical warfare agents on Daphnia magna : for evaluation of the ecotoxicological risk of the dumped chemical warfare agents in the Baltic Sea. Available at: https://www.foi.se/rapportsammanfattning?reportNo=FOA-R--97-00430-222-SE#.

NRC (2003) Bioavailability of contaminants in soils and sediments: processes, tools, and applications committee on bioavailability of contaminants in soils and sediments, national research council. ISBN: 0–309-50578-X, 432 pages, available at: http://www.nap.edu/catalog/10523.html

Sanderson H, Fauser P (2016) Confidential Nordstream 2 Report

Sanderson H, Fauser P, Thomsen M, Larsen JB (2012) Weight-of-evidence environmental risk assessment of dumped chemical weapons after WWII along the Nord-stream gas pipeline in the Bornholm deep. J Hazard Mater 215–216:217–226. doi:10.1016/j.jhazmat.2012.02.057

Sanderson H, Fauser P, Thomsen M, Sørensen PB (2008) Screening level fish community risk assessment of chemical warfare agents in the Baltic Sea. J Hazard Mater 154(1–3):846–857. doi:10.1016/j.jhazmat.2007.10.117

Schaber M, Hinrichsen H-H, Gröger J (2011) Seasonal changes in vertical distribution patterns of cod (Gadus morhua) in the Bornholm Basin, central Baltic Sea. Fish Oceanogr 21:33–43

Storgaard MS, Sanderson H, Henriksen PG, Fauser P, Östin A, Baatrup E (2016). Suppressed swimming activity in Zebrafish (Danio rerio) exposed to 1,4,5-oxadithiepane, a sulphur mustard degradation product. Submitted

Suter GW, Cormier SM (2011) Why and how to combine evidence in environmental assessments: weighing evidence and building cases. Sci Total Environ 409:1406–1417. doi:10.1016/j.scitotenv.2010.12.029

UN (2010) Globally harmonized system of classification and labelling of chemicals (GHS)

Chapter 9
Best Practices in Monitoring

Jacek Bełdowski, Jaromir Jakacki, Miłosz Grabowski, Thomas Lang, Kela Weber, Lech Kotwicki, Vadim Paka, Daniel Rak, Maria Golenko, Michał Czub, and Martin Söderström

Abstract This chapter summarizes the methods used within the MODUM project for monitoring chemical munition dumpsites. It includes general introduction to monitoring process, listing the requirements that are a basis for the establishment of full scale monitoring programme. It describes survey procedures, for locating dumped munitions, Sampling and analytical procedures for the detection of Chemical Warfare Agents, as well as the usage of fish as bioindicators are described. Modelling of pollutants originating from dumped munitions is presented and discussed. Only methods, which were proven to be most effective within the MODUM projects were selected, also data interpretation methods providing coherent information regarding the environmental risk are explained in details.

J. Bełdowski (✉) • J. Jakacki • M. Grabowski • L. Kotwicki • D. Rak •
M. Golenko • M. Czub
Institute of Oceanology, Polish Academy of Sciences,
Powstancow Warszawy 55, 81-712 Sopot, Poland
e-mail: hyron@iopan.gda.pl

T. Lang
Thünen-Institut für Fischereiökologie, Deichstraße 12, 27472 Cuxhaven, Germany

K. Weber
Environmental Sciences Group, Department of Chemistry and Chemical Engineering, Royal Military College of Canada, Kingston, Ontario K7K 7B4, Canada

V. Paka
Atlantic Branch of the P.P. Shirshov Institute of Oceanology, Russian Academy of Sciences, Kaliningrad, Russian Federation

M. Söderström
Finnish Institute for Verification of the Chemical Weapons Convention, University of Helsinki, A. I. Virtasen aukio 1, P.O. Box 55, FI-00014 Helsinki, Finland

J. Bełdowski et al. (eds.), *Towards the Monitoring of Dumped Munitions Threat (MODUM)*, NATO Science for Peace and Security Series C: Environmental Security, DOI 10.1007/978-94-024-1153-9_9

9.1 Introduction

Chemical warfare agents are by design toxic versus humans. Their presence on the bottom of the sea raises concerns in decision makers and environmental agencies (Tobias Knobloch et al. 2013). Since corrosion estimates are difficult, release of toxic agents into environment vary from total release (Korotenko 2000) to slow leaching over long time period (Sanderson et al. 2008). Both intact munitions and completely corroded pieces were found in the same dumpsites, therefore it is difficult to make any predictions regarding release and pollution of the marine environment (Beldowski et al. 2016a). According to models developed within CHEMSEA and MODUM projects, munition constitutents may be transported as far as 30 km away from dumpsites, and can stay in the water column for weeks, although according to some models used within this study individual particles may travel significantly longer distances, reaching northern Gotland (Jakacki et al., Chap. 7 this issue). Therefore the risk may be considered to be larger than purely local. Studies in the sediments show that detectable CWA degradation product concentration can be detected as far as 30 m from the object for mustard gas derivatives, and 50 m from the object for arsenic based CWA's (Beldowski et al. 2016b). This suggests, that time trends of CWA contamination in dumspites and their vicinity is advisable, but it does not necessarily justify regular monitoring programme. This chapter summarizes best practices for evaluating time trends connected to the presence of chemical munitions on the sea bottom, starting from locating objects in question, monitoring physical parameters that may affect spreading of those substances, most cost and time effective methods for analysis of the degradation products of chemical warfare agents, establishing their effects on marine organisms, and modelling their environmental fate.

9.2 When to Monitor

Monitoring can be described as the process of observing and tracking the progress, quality, and quantities within a process, system, or item. Environmental monitoring is more specific to an environmental system whereby data is collected to characterize and track changes over time. When a system is fully defined and understood, the impact of inputs, such as stressors, can be better connected to system outputs and internal system variables. With respect to contaminated sites, monitoring can be undertaken for many reasons. Most often though monitoring is undertaken in some way to ensure minimal risk to environmental or human health.

The end goal of contaminated site management is to close the site, which can also be described as reaching a point where the site poses no risk to human or environmental health for the foreseeable future. The management of contaminated sites, in and of itself, can be challenging. It requires long-term vision and understanding of the different steps required to reach eventual site closure. Most jurisdictions have adopted specific procedures for handling contaminated sites. For example, In Canada the Federal Contaminated Sites Action Plan (FCSAP) uses a specified 10 step framework (Chapman 2010). Although there will be jurisdictional differences, contaminated sites management often proceeds in the following phases: (A) Identify potentially contaminated site, (B) historical review, (C) site assessment (usually consisting of several iterative phases), (D) risk assessment (where required), (E) remediation and/or risk management, (F) confirmatory sampling and/or long term monitoring. In this context, for the management of contaminated sites, monitoring is usually used to measure the return of ecosystem function in response to remediation or long term risk management/ecosystem recovery strategies.

The expectation is that remediation activities at a contaminated site will assist in the recovery of ecosystem function. The United States Environmental Protection Agency has defined a six-step process for the development and implementation of monitoring when used to assess the success of remediation activities (USEPA 2004). These steps are: (1) Identify monitoring plan objectives, (2) Develop monitoring plan hypotheses (which is assisted by a conceptual site model), (3) Formulate monitoring decision rules, (4) Design the monitoring plan, (5) Conduct monitoring analyses and characterize results, and (6) Establish the management decision. Although these steps are specific to assisting remediation projects the concepts are important to the general field of environmental, or contaminated site monitoring. Developing a conceptual site model (CSM) is important for understanding, studying, modifying, or remediating any environmental system. It will usually be completed for any site assessment or risk assessments prefacing monitoring programs. The CSM serves as the basis point for formulating hypotheses on how the site/ecosystem will respond to remediation strategies, but also serves as a basis point for re-evaluating site management strategies throughout the monitoring process (when required). Ideally monitoring will occur before remediation activities occur in order to provide baseline data. Later temporal trend analysis can then inform whether ecosystem recovery is occurring throughout the monitoring period. Creating decision points, and rules for action throughout the monitoring period should be completed with input from all major stakeholders. Many rules of action may exist for a site, especially if partial economic function or usage can potentially be restored before full site closure, however the most important decision rules are those defining what constitutes full site closure (and therefore no further action or monitoring). Monitoring with iterative data assessment, and comparison to site specific goals, should allow for the ongoing evaluation of remedial actions or risk mitigation which in turn informs site owners and stakeholders of any program oriented, practical, or economic investment changes required throughout the site management program (Hooper et al. 2016). Monitoring plan design and data analysis methodologies are site and case specific, but will relay heavily on the CSM as a basis. The USEPA has some recommended ways to ensure and track data quality throughout monitoring programs (USEPA 2006), however to the best knowledge of the authors there currently exists no guidance documents on monitoring of underwater munition sites. The MODUM and CHEMSEA underwater munition projects are unique in their encompassing nature. Best practices for underwater munition sites from the MODUM and CHEMSEA projects will be covered in detail later.

Apart from supporting and evaluating remediation activities, there exists other situations where monitoring programs may be employed at contaminated sites. Demonstrating due diligence is an important aspect of contaminated site management and is directly tied to continually understanding any potential risk to human or environmental health originating from a contaminated site. Risk exists when there is a chemical hazard, an exposure route, and a receptor. Therefore, risk mitigation strategies aim to remove the hazard, route, or receptor, thereby removing risk. Where risk management strategies are required or chosen over remediation activities/actions, monitoring is continually required to ensure that the original assumptions, actions and understanding behind risk mitigation are still valid, operational, and effective. For example, risk mitigation strategies that rely on containment or fixation of a contaminant, thereby removing exposure routes, require ongoing monitoring to ensure that the containment stays intact. Another common situation where monitoring is employed is at site boundaries between property owners. Migration of contaminants across property or jurisdiction borders is a source of litigation. Tracking the source zone, release rates/flux, and span of migration encompass due diligence requirements. Similar monitoring is also required for natural attenuation projects where understanding contaminant dynamics in relation to original risk assessment assumptions is important. Other contaminated site monitoring situations exist, however the concepts behind the above examples are the basis for applying monitoring methods in most situations.

Environmental science and engineering are now well developed disciplines, yet emerging contaminants and emerging issues pose significant challenges to our current conventional scientific understanding and associated technologies. Although underwater munition sites have existed for quite some time, attention to the issue itself is quite new. Most underwater munition sites have been left without monitoring for many decades. The original reasons and assumptions behind dumping in-tact munitions in water bodies was related to risk mitigation, mostly a mitigation of risk to human health and safety in the terrestrial setting (i.e. munitions were simply removed from the terrestrial environment). However, it could also be said that there was another aspect of risk mitigation at play as well. When dumping in water bodies, the majority of munition items themselves could be said to have been "contained" at the time. The munition fillings were most often well sealed within metal casings, therefore there was no route for the fillings to interact with receptors. Therefore, the process could be described as a form of risk mitigation or risk management. The major failing however was in a lack of monitoring afterwards to check the original risk mitigation/management assumptions.

The management of underwater munition sites poses unique challenges. They are often found in highly dynamic areas. Ocean currents, sedimentation, burying/uncovering cycles, and other stressors expose munition items to varying environmental conditions (e.g. redox, physical impacts). As the munition items change over time via corrosion or physical movement and impacts, they will release chemical fillings at unpredictable rates starting at unknown/unpredictable times. Given the dynamic nature of underwater munition sites, long term data sets via monitoring are required to understand some of the most basic scientific questions: What constitutes natural variation? How quickly do munition item fillings release once casings are penetrated/corroded? How far do the fillings move once released? How quickly do the fillings degrade? Are fillings taken up by biota, and if so do the bioaccumulate? The MODUM project is the first large scale research endeavour to truly focus on monitoring, and begin building comprehensive data sets to allow for true management and understanding of underwater munition sites in the future.

9.3 Survey

To get the best results of monitoring, it is essential to conduct a precise survey with maintaining of the best hydrographic practice. Also, an ability to replay an observation on detected targets is crucial for monitoring. Hence, there was a strong need of establishing a survey procedure in the MODUM project. The procedure was divided into two stages: Area Wide Assessment – focused on selecting areas of interest and performing preliminary surveys on them, and Detail survey and Investigation – immersed on detail survey and sampling campaigns. Applying of developed procedure gave good, repeatable results. This subchapter will give a detail description of the survey procedure. Scheme of the workflow (Fig. 9.1) shows step-by-step activities to achieve desirable outcomes.

9.3.1 Stage I – Area Wide Assessment

In this phase, a side scan survey has been conducted with a usage of traditional, towed side scan system (EdgeTech DF – 1000) in areas selected on the basics of historical sources. Recorded acoustic data were processed with SonarWiz 6 software, and non-detailed seafloor maps were prepared. This activity gave knowledge about obstacle located on the seafloor, which could be a potential threat for the AUV during its missions, for example, shipwrecks, big rocks, and ghost nets.

9.3.2 Stage II Detail Survey and Investigation

Based on non-detailed seafloor maps created from data recorded by towed side scan sonar, AUV missions were planned. The autonomous underwater vehicle was equipped with hull-mounted side scan sonar, towed magnetometer, and set of environmental sensors. This vehicle was conducting a survey with low altitude above the bottom, which gave high-resolution data as a result. Acoustic data were also processed with SonarWiz 6 software. To detect targets immersed into muds and

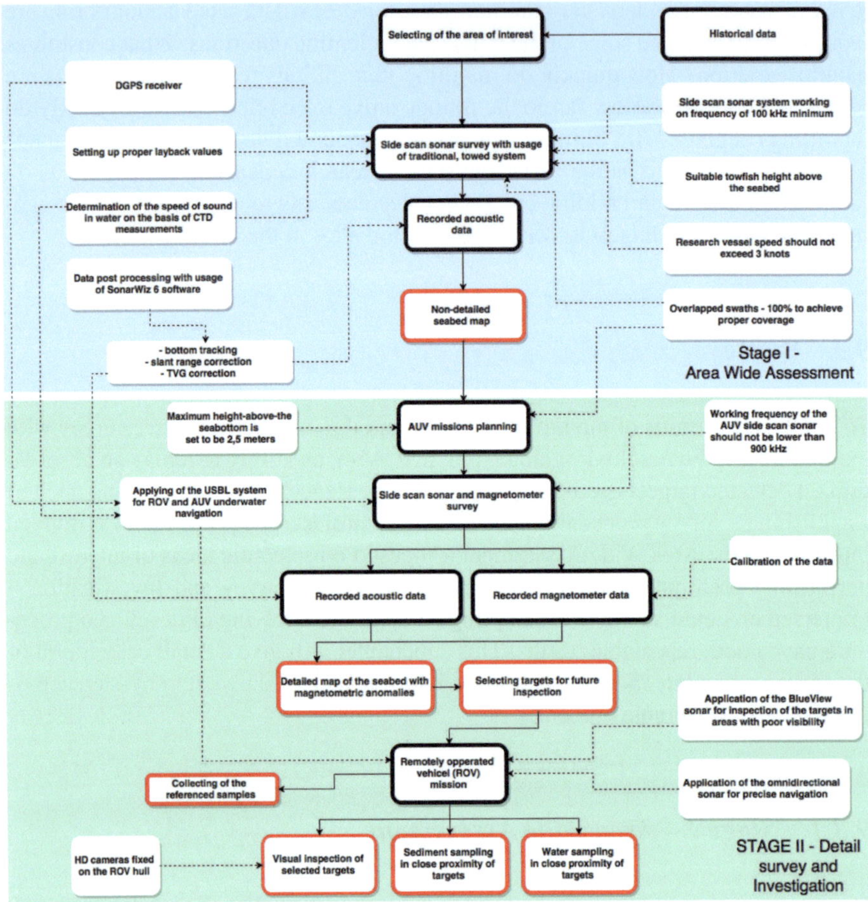

Fig. 9.1 Scheme of Chemical Warfare targets detection. Following steps to obtain results (*red envelope*) are marked with a *black envelope*. Boxes without envelopes describe certain actions/equipment, which should be applied to complete the two stages of the procedure

invisible for sonars, a magnetometric survey must go along with the side scan survey. The main task is to find accumulations of magnetic targets (not the single targets!) by means of magnetometers towed at high altitude above the bottom, or at low altitude behind the AUV then to correlate results of the two independent records. Acoustic targets found within the magnetic anomalies should be chosen as the foremost for identification.

A detailed map, overlaid with a grid of magnetometer data provides a location of the targets with accuracy up to 5 m. Next, visual inspection of selected targets was carried on with employment of the remotely operated vehicle ROV, together with actions of water and sediment sampling in close proximity of contacts. ROV can dive up to 300 m and is equipped with HD cameras, LED lights, and BlueView sonar, which allow performing an inspection of the targets, even when visibility is low. Underwater navigation was conducted with help of USBL system. Also, omnidirectional sonar mounted on the hull of the ROV or installed as independent device increases the precision of navigation.

9.3.3 Detail Description of Activities and Equipment Essential for Complete the Procedure

Stage I

(a) DGPS receiver – *(Differential Global Positioning System)* A GPS receiver enhanced with difference correction for satellites. The difference correction improves location accuracy from 0.5 to 2 m. The output signal from DGPS receiver is used for positioning of the research vessel and as an input for USBL system. Data recording is carried out in WGS84 coordinate system.

(b) Setting up a proper layback values – layback is a real distance between research vessel and towed side scan sonar. It is essential to set a proper value of the layback to get real, geo-referenced data. Layback can be calculated with following formula:

$$d = \sqrt{L^2 - (H - h)^2}$$

where d is a real distance between survey vessel and towed sss system, L is a length of cable connecting ship and towfish, H is total depth of water under the ship, and h is an altitude of towfish above the seafloor (Blondel and Blondel 2009). During the post processing of the data, layback value (in meters) is added to the ship position. Corrected data represents a real position of the towfish.

(c) Determination of the speed of sound in water on the basis of CTD measurements – knowledge about the speed of sound in water is necessary to obtaining good results in sea bottom scanning. It is an input parameter during sonar and USBL system calibration. The speed of sound is a function of temperature, salinity (conductivity), and pressure. Hence, determination of the speed of sound can be done only on the basis of *in situ* measurements. To calculate a speed of sound in water the following formula has been applyied (Medwin 1975)

$$c = 1449{,}2 + 4{,}6T - 0{,}055T^2 + 0{,}00029T^2 + (1{,}34 - 0{,}01T)(S - 35) + 0{,}016Z$$

where c is a speed of sound in water, m/s, T is a temperature of the water, °C, S is a salinity of the water, PSU, and Z is a pressure represented as a depth, m.

(d) Data post-processing with a usage of SonarWiz 6 software – collected raw acoustic data requires it's processing in order to create sea bottom map. For this purpose special software SonarWiz 6 was employed. Processing of the data consists of three actions: bottom tracking, slant range correction, and usage of the TVG (time-varied gain) gain.

(e) Side scan sonar working on a frequency of 100 kHz minimum – side scan sonar is a device that allows performing seafloor mapping. It generates acoustic pulses, which are traveling through the water column and gain the sea bottom. Next, pulses are on-going phenomena of reflection and refraction on the bottom. Side scan sonar record reflected pulses from the bottom and base on them, a two-dimensional image of the seafloor is generated. The frequency of 100 kHz is directly connected with a resolution of collected data and should be no less than this.

(f) Suitable towfish height above the seabed - during the survey side scan sonar towfish should be driven on certain altitude above the bottom to get good quality data. This altitude can be calculated with formula:

$$h = K^{-1}$$

where h is an altitude of towfish above the seabed, and K is maximum swath range of acoustic pulses generated by the sonar. Swath range is connected with a frequency selected for the survey.

(g) Research vessel speed should not exceed 3 knots speed – it is important to preserve a speed of the vessel less than 3 knots during a survey to obtain good quality data.

(h) Overlapped swaths – 100% to achieve proper coverage – due to phenomenon of acoustic pulses refraction and undulating of the towfish it is necessary to design survey grid with overlapped swaths. Typical cover rate is 200%.

(i) Special configuration of the towing system is recommended to stabilize the towfish depth (Fig. 9.2). In general, the fish is attached to the long cable, its depths variations are proportional to the cable length and strongly affected by the vessel's speed variations. To decrease the towfish depths instability, a heavy depressor should be attached to the towing cable at the distance about 30 ± 10 m from the fish. A depth of the heavy depressor is almost independent from the vessel speed variations. Some variations of the depths of the heavy depressor could be due to the roll. The 30 m long horizontal cable section plays a role of a low-pass filter for these disturbances as well as a barrier for vibrations coming from the tight quasi vertical cable section.

Stage II

(a) Maximum height above the bottom is set to be 2.5 m – AUV magnetometric missions should be designed with a certain parameter of altitude above the bottom, which is 2.5 m. This value is associated with maximum detecting range of magnetometer. Additionally, this preset allows collecting high-resolution data with sonar slant range from 25 to 30 m.

(b) Applying of the USBL system for ROV and AUV underwater navigation – USBL (Ultra Short Base Line) system is employed for tracking of the AUV when it is under the water surface and for ROV homing.

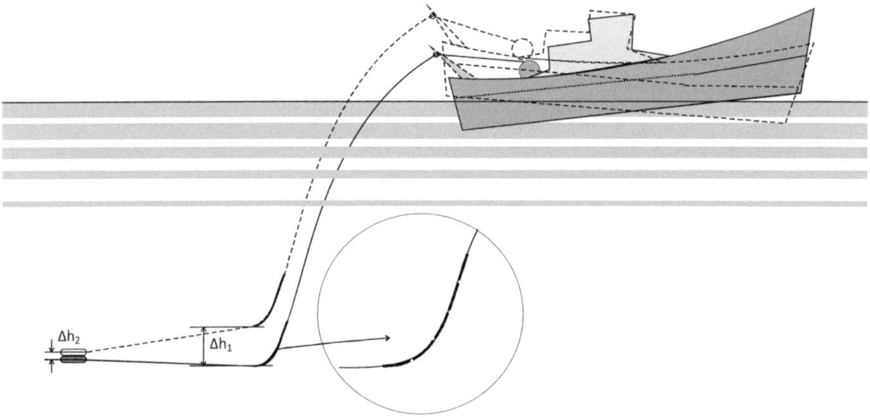

Fig. 9.2 Configuration of a towing system, which decreases displacement of the towfish caused by the vessel rolling. In circle: short section of the towing cable loaded with heavy lead cylinders. Two positions of the towline correspond to its displacement due to rolling with magnitude Δh1. Vertical displacement of the tow fish, Δh2, is significantly smaller than Δh1

(c) HD cameras fixed on the ROV hull – with a usage of high-definition cameras, it is possible to perform visual inspection and identification of detected targets. Video footage from the HD cameras can be captured and is treated as one of the inputs for the database.
(d) Working frequency of the AUV side scan sonar should not be lower than 900 kHz – autonomous underwater vehicle (AUV) performs bottom scanning along the pre-designed route. It is equipped with navigation sensors, which allows completing missions according to its assumptions. Also, it has several environment sensors for collecting information about water column and hull-mounted side scan sonar. The side scan sonar works on the same principles as towed side scan systems, yet its frequency is higher. To obtain photo-like images of the bottom it is necessary to use frequency of at least 900 kHz.
(e) Calibration of the data – magnetometric data should be calibrated to separate magnetic anomalies (targets) from the environmental magnetic background. It is connected with a resolution of the magnetometer, which collects the data in a wide range.
(f) Application of the BlueView sonar for inspection of the targets in areas with poor visibility – the BlueView sonar is a type of an acoustic camera fixed on the hull of ROV. It works with a principle of traditional video camera, but it uses high-frequency acoustic impulses to generate the image. It is useful in areas where visibility is poor due to prevailing conditions. Optical cameras have a limited range there, however, BlueView system provides visibility up to several meters.
(g) Application of the omnidirectional sonar for precise navigation – this kind of sonar is also fixed on the ROV hull and it generates acoustic impulses with directivity of 360°. With this device it is possible to imaging area around the

vehicle. Employment of the omnidirectional sonar simplifies reaching preselected targets and increase precision of navigation.

(h) When the vessel is gone by wind, operations with ROV are impossible. If a dynamic positioning system is impossible, anchoring of small boats would be an alternative.

9.4 Physics

A specific feature of the Baltic Sea is its salinity stratification. The greatest interest represents the saltiest bottom layer within which interaction of bottom sediments with bottom current is concentrated. There are many areas in the Southern Baltic where the thickness of this layer is small. That creates difficulties for its research. Thickness of the saltwater layer over the Slupsk Sill is especially small – around several meters, and sometimes it absolutely disappears. To obtain the nearbottom stratification, the probe has to reach at each sounding to the bottom, and at the same time there is a risk of its damage. To minimize risk, it is offered to carry out sounding in the mode of quasifree fall. In Fig. 9.3 the composition of freely falling probe is proposed.

The element which is slowing down the falling speed is the cylindrical float made of a plastic pipe with caps. Such float maintains external pressure to 50 bars that allows to use it in any point of the Baltic Sea. The probe can fall on the cord or on a thin flexible cable, but it doesn't influence quality of data. The mode of free fall is reached thanks to what a cord (cable) is issued loosely. A cord or a cable can be reeled up on

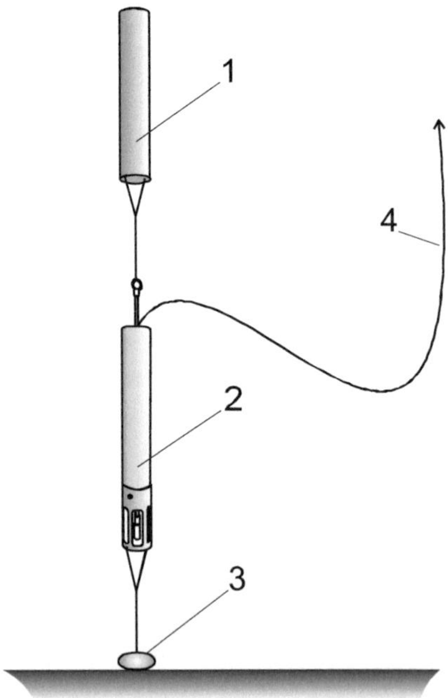

Fig. 9.3 A sketch of the high resolution profiling system. The float 1 attached to the probe 2 along with the lead load 3 allows the probe to move steadily with the constant speed, the cable 4 is not tight that time, and to stay vertical on the bottom for some time before recovery

the winch if its speed easily changes in necessary limits, or it is laid on the deck by the running end up, then the probe pulls a cable without experiencing any influence. Lowering speed is regulated by selection of weight which ties under the probe. In such option it is possible to use Idronaut's Sea&Sun's and other probes. Lowering speed is chosen within 0.5–0.7 m/s taking into account requirements to spatial resolution. Buoyancy of the float has to be rather big – about 2–3 kg that at the moment when the weight reaches the bottom, the lifting force could shut down the probe before sensors reach the bottom. In practice, it is enough to have length of a cord of 0.5 m. If the probe stops insufficiently close by the bottom, the cord can be truncated, having at the same time smaller weight to slow down speed at the initial stage of braking.

The attached float allows the probe to stop at the given small distance from a bottom that gives the chance to inert sensors to finish measurements. If the recovery begins without pause, then the value of gaged parameter F related to the bottom differs from the true value at a rate $\Delta F = \tau \cdot dF/dt$, где τ – a response time, dF/dt – a rate of change of parameter at the end of profiling. Leaving the probe hanged over the bottom for a while near 3τ, it is possible to get more precise assessment of parameter characterizing its value at the bottom. To understand as far as it happens important, we will address Fig. 9.4 in which the T, S, oxygen, and pressure profiles are shown at lowering and recovery of the probe. Temperature profile at the same time changed slightly; salinity profile possesses the hysteresis, but both T and S profiles at one value come to an end and begin. On a profile of oxygen big change of a signal during the exposition is found. Such is the effect of the large response time of the Idronaut's oxygen sensor ($\tau > 3$ s). Let's pay an attention that from the beginning of rising value

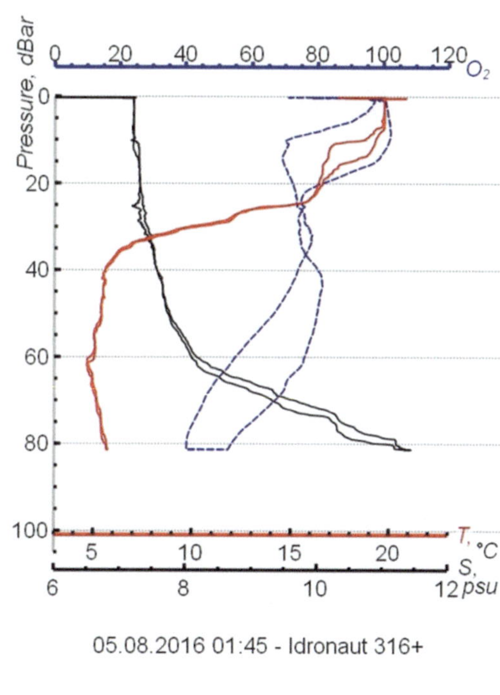

Fig. 9.4 An example of T, S, and O_2 profiles measured during the downward and upward motion of the Idronaut OS 316 probe with a half of minute stay on the bottom. The strongest difference between the profiles is seen for the oxygen sensor due to large response time (3 s or more)

of a signal O_2 does not change that testifies to uniformity of the oxygen distribution in close proximity to the bottom. Therefore, O_2 value at the time of the beginning of recovery is the most reliable assessment of concentration of oxygen at the bottom. Difference of this value from the indication of the sensor at the time of achievement of the bottom makes about 13% on a saturation measure. Such larger error needs to be taken into consideration and to define whenever possible in the recommended way. The error on other channels with fast sensors is not so big.

In Fig. 9.5 distribution of oxygen in the Southern Baltic from Bornholm to Gdansk Basins is presented. Under a 2D distribution of oxygen two curves are shown: the dotted curve is a value of a signal of the sensor at the time of touching

a

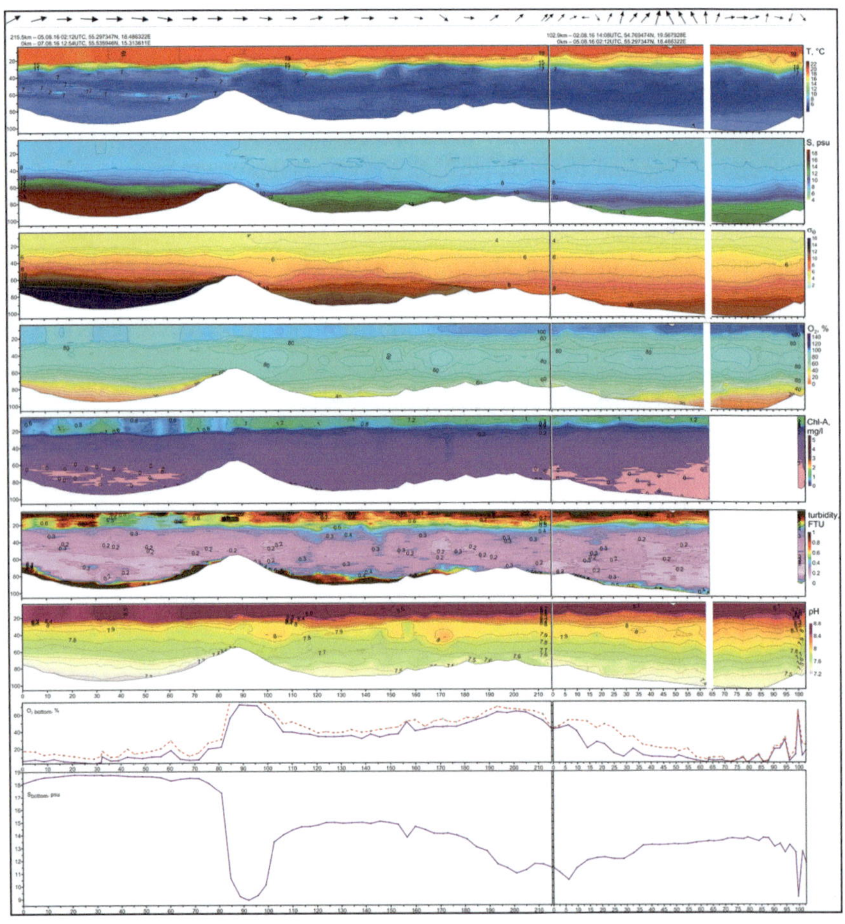

Fig. 9.5 (**a**) a spatial distribution of oxygen (saturation, %) on a section in the Southern Baltic. (**b**) the concentration of oxygen at the bottom measured at the time of reaching the bottom (*dotted line*) and at the time of a separation from the bottom (*solid line*)

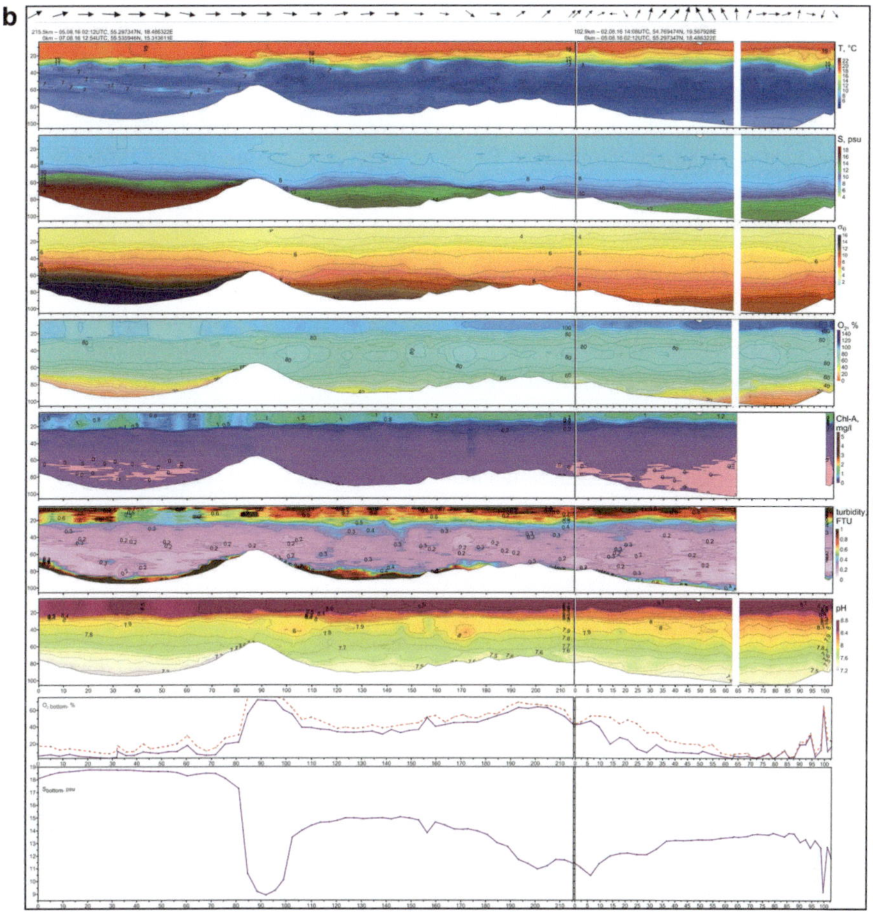

Fig. 9.5 (continued)

the bottom, and a solid line - the same at the time when recovery begins, i.e. the corrected values of oxygen at the bottom.

As we see, without amendment, the mistake in determination of content of oxygen averages about 10% of saturation, it is minimum over sills where the content of oxygen is close to the full saturation, and the gradient of the oxygen distribution is small; the maximal value of an error exceeding 20% was found in the section where Slupsk waters inflow in the Gdansk Deep.

9.5 Sampling

Underwater unexploded ordnance is recognised to pose serious threats to the marine environments and their users (AMATO et al. 2006). This also includes researchers studying the munitions dumpsites. From anchoring to sampling of sediments and biota (Benthos, fishes), there is an elevated risk in almost every operation taking place in those areas. To exclude the danger of vessel contamination, underwater explosions or any accidents including workers getting exposed to the toxic agents, several precautions need to be implemented. Levels of risk vary between the types of measurements or applied methods, however, benthic studies (including bottom trawling for fish) are among the most difficult in terms of risk management.

The typical equipment used for the wide spectrum of sediment sampling represents simple, but smart design. Grabs and corers depend on vertical freefall, while bottom trawls and dredges simply scrap the sediment surface while being towed by the vessel. Simplicity of those mechanisms results in their high efficiency and receiving comparable datasets, however it also limits their accuracy. Unfortunately, the last feature is the most important for working in an underwater munitions dumpsite.

To assess the ecological effects of potentially exposed CWA on biota and their habitats, multiple sediment and biota samples had to be collected. The sampling strategy depended on the precision of sonar and magnetometer surveys. Thanks to the acoustic image of the sea-bottom, the following recognition of suspicious objects served for the targets selection.

Whenever the anchored vessel was positioned in a safe distance from targets, sediment samples were collected by Box Corer, Van Veen grab or Niemisto bottom corer. The use of the first one is limited by sea conditions, however, it leaves the collected sediment undisturbed. Box Corers cut and enclose sediment profiles from specific areas, protecting the sample from vertical mixing. It works perfectly as a tool for collecting samples that are later used in analyses requiring information about natural sediment layering. It is a widely used in benthic research for analyses of sediment, chemistry of pore-water, meiofauna assemblages. If the sea conditions do not favour the use of BoxCorer, the VanVeen grab comes handy. The simplicity of its design makes a perfect tool for fast collection of sediments, even during harsh sea condition. What is problematic about the samples collected by VanVeen grab, is that it only preserves the specific area of sediment, mixing the whole sediment. On the other hand, it is a perfect tool for the investigation of macrofaunal assemblages. Niemisto corer also produces undisturbed structure, but its small area of sampling limits its usability for biological sampling.

From the CW dumpsites investigation point of view, both tools worked perfectly during collection of data used later for background information about habitat's natural characteristics. On the other hand, they were both completely useless for the precise sampling in the vicinity of the sea-dumped bombs and mines. Such operations required precision and visual recognition that was possible thanks to the use of ROV – Remotely Operated Vehicles.

Underwater robots are equipped with propellers, sonars and positioning systems that provide safe underwater navigation. In overall, the obtained precision, together with set of cameras mounted and additional scientific equipment, perfectly fits the requirements for samples collection and visual recognition of sea bottom with all its features. The ROV setup used during the MODUM project allowed to collect sediment samples that were later used for chemical, biological and geological analyses. As the consequence of the ROV construction and navigation, the amount of collected sediment is limited and it takes relatively long time to perform the whole operation. It may not be the perfect tool for classical benthic research, however, without the visual image from the ROV mounted cameras, we would not be able to detect the presence of active benthic Amphipoda in the Bornholm Deep area, that were absent in sediment samples.

The problem of precisely coordinated sampling of sediments in close proximity to the underwater objects which are presumably sources of chemical pollution was solved in the project MODUM by development of cassette sampler. The cassette sampler with 6 grabs sequentially put in action by latex rubber bands operated on a cable under monitoring of the video camera was constructed. The device is placed on the motor-boat which can be held over an object and to move according to the given scheme. Positioning at the time of sampling is conducted by fixing of coordinates of the motor-boat by means of GPS/GLONASS, or the provision of a sampler concerning an object by means of an imaging sonar. Composition of the cassette 6-grab sampler of bottom sediments is presented in Fig. 9.6. The main parts are

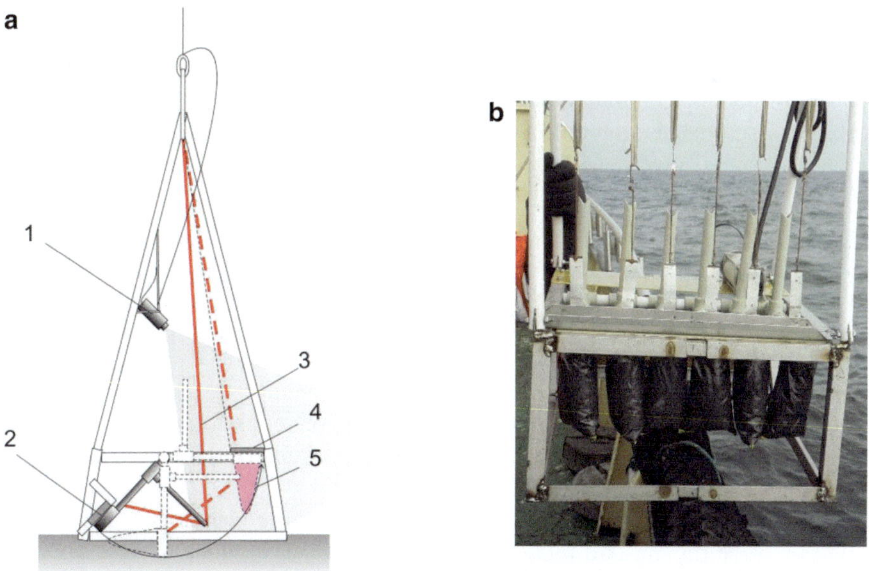

Fig. 9.6 (**a**) Sediment sampler: principle of work. *1* – video camera, *2* – a grab in the launching site, *3* – rubber band, *4* – horizontal plate at the end of combustion stroke, *5* – a bag filled with a sampled soil.; (**b**) In the photo the 6-grab sediment sampler after recovery on the deck with full grabs is shown

placed in a framework. In an upper section the video camera with lighting is located. The operator can observe achievement of the bottom and work of grabs. In the lower section grabs with trigger mechanisms and the electric motor putting them in action are located. Grabs are fixed on the arms established in a row on the common horizontal axis. Each grab consists of two parts: a metal ring and a disposable elastic plastic bag which stretches on a ring and fastens by means of a disposable collar. At the launch site the bag is placed in the ring closed on the one hand by a folding cover, and other party pressed to the starting panel. The cover excludes casual hit of sediments in a package at the repeating contacts with ground during capture of a series of samples. Latex rubber bands by effort ca 3 kg forces a grab to move an arch of a circle. At the same time it plunges into a soil on depth about 70 mm, scooping about 200 ml of a mud. The filled grab rather densely nestles on the horizontal plate interfering the sample washing away. For inspection of an object a light boat capable to be anchored and at the same time to move in any direction is chosen. The boat is equipped with the GPS, a source of power supply and a simplified electric winch. With such means, it was possible to receive dozen of samples in a radius 15–20m around an object with locating accuracy within 0.5 m. To assess the impact of sea-dumped CW on natural environment, researchers in MODUM decided to test the suitability of taxonomical analyses for this specific monitoring case. Meio- and macrofaunal assemblages may serve as pollution bioindicators, therefore they are widely used in many monitoring programmes.

Due to severe environmental disturbances occurring in the Baltic Sea deeps, especially creation of "benthic deserts" resulting in a temporary absence of macrofauna in studied areas, taxonomical studies were limited to Nematodes, the only representatives of meiofauna. Meiofauna is an ecological group of small-sized organisms (<5 mm) that is well-recognised as an abundant and cosmopolitan component of benthic communities. Statistical analyses of the structure and composition of assemblages combined with the knowledge about taxa environmental preferences are a precise tool used for ecological status description.

9.6 Best Practices for Chemical Analysis of CW Contaminated Sediments

Chemical analysis can be used at two stages of monitoring chemical munition dumpsites. It can be applied on board the research vessel performing the sediment sampling. Typically, analysis is performed in a well-equipped reach back laboratory on land.

At the moment, the methods for on board analysis are mature enough only for analysis of sulfur mustard-related chemicals. This can be done using a portable headspace gas chromatography-mass spectrometer. Still, this capability could be useful during a sampling campaign to redirect the efforts into correct direction e.g. by quickly classifying found objects based on sediment samples.

In the reach back laboratory, all CW-related chemicals can be identified and quantified. The list of possible target chemicals for the tandem mass spectrometric analysis typically contains 20+ analytes. The methods should be optimized for the most probable target chemicals. In Baltic Sea, realistic target limit of quantitation is at the level from 1 to 10 μg/kg dry weight.

Before actual analysis, pore water and sediment are separated by centrifugation. This way they can be analyzed separately and it can be determined if any of the target chemicals are present in pore water, which would mean that they could bioavailable. For determination of the sediment concentration a portion of the sediment should be dried separately to determine the dry weight of the sediment. It is not recommended to use freeze drying as cross-contamination of samples has been observed.

The sediment should be extracted using both non-polar (e.g. dichloromethane) and polar solvents (e.g. acetonitrile) and the extracts should be concentrated. One part of each extract should be analyzed as such and one part derivatized. Typically, propane-1-thiol has been used for gas chromatography-based analyses of arsenic-containing chemicals and hydrogen peroxide for oxidation of arsenic-containing chemicals for liquid chromatography-based analyses. Optionally, silylation can be applied for gas chromatography-based analyses of some sulfur mustard-related chemicals.

For the analysis, both gas chromatography-tandem mass spectrometry (GC–MS/MS) and liquid chromatography-tandem mass spectrometry (LC–MS/MS) should be used. Typically, the mass spectrometric analysis without tandem option will not produce satisfactory results. In order to guarantee reliable results the analysis should be carried out using a dilution series of reference chemicals and strict identification criteria should be applied to verify identity of the chemicals.

As sediment samples contain large amount of extractable chemicals, the analysis of large number of sample will results in dirty ion source and chromatographic system. The instrumentation will require regular cleaning to overcome these issues. Therefore, quality control samples and blank samples have to be included in the analysis scheme to guarantee the reliability of the analyses and to avoid false negative results.

During the analysis of larger amount samples, the samples are typically divided into batches. The analysis of the sample should be done immediately after the sample preparation as the samples will deteriorate during storage. One batch of 20–30 sample can usually be processed in 1 week.

9.7 Studies Using Fish as Bioindicators

As part of the biological component of the MODUM project, studies on the health status of fish in CWA dumpsites and reference areas in the Baltic Sea were carried out (for results, see Lang et al. Chap. 6 this issue). These constituted a continuation of efforts already initiated during the previous CHEMSEA project (2011–2013) (Beldowski et al. 2016a), Lang et al. Chap. 6 this issue).

The main rational and advantages of studying fish in a monitoring framework related to ecological risks associated with dumped munitions (both conventional and chemical) can be summarised as follows:

- Fish can easily be collected by means of adequate gears (see below) and can ideally be used for integrated chemical and biological analyses, aiming at quantitative measurements of (a) the uptake and metabolism of hazardous substances originating from dumped muntions and their degradation products in fish tissues and (b) biological effects of exposure to hazardous substances and their degradation products at different levels of biological organisation (from molecular to individual and population effects).
- Since fish are non-sessile and mobile organisms, monitoring of their health status is best suited for area-wide monitoring and assessment of effects.
- Commercial fish species are usually subject of stock assessment programmes, which offer valuable background data on stock dynamics and performance that can be utilised in the assessment of data from monitoring eco-risks due to dumped munitions.
- If commercial fish species are used, monitoring results may provide information relevant for human food safety issues.

9.7.1 Rational for Monitoring Fish Health Effects of Dumped Munitions

Diseases of wild marine fish have been studied in European seas (North Sea, Baltic Sea and adjacent areas) on a regular basis for more than three decades. Fish disease surveys are often integrated with other types of biological and chemical investigations, as part of national monitoring programmes aiming at an assessment of the health of the marine environment, in particular in relation to the impact of anthropogenic pollution (Lang 2002).

Fish diseases are considered as an ecosystem health indicators, reflecting ecologically relevant effects of environmental stressors at the individual and population levels. At the individual level, fish diseases may act by adversely affecting behaviour, growth, reproduction and survival of affected specimens. Individual effects may lead to ecologically relevant population effects (especially in epidemic situations) and ultimately to biodiversity effects at the community level. The occurrence of most infectious and non-infectious common diseases in wild fish stocks is a natural phenomenon which, per se, cannot be regarded as an indicator of ecosystem deterioration. However, when disease prevalence increases above a natural threshold, this can be regarded as a sign of ecosystem stress, leading to an imbalance in the interactions between the impact of physical, chemical and biological environmental factors, disease agents (pathogens in case of infectious diseases and parasitoses) and the immune functions of fishes.

The multifactorial aetiology of diseases, in particular of infectious diseases, is generally accepted ((Vethaak and Rheinallt 1992; Lang 2002). Most common wild fish diseases are caused by pathogens (viruses, bacteria, parasites). However, other endogenous or exogenous factors may be required before the disease develops. One of these factors can be environmental pollution (e.g. originating from dumped munitions), which may either affect the immune system of the fish in a way that increases its susceptibility to disease or may alter the number and virulence of pathogens. In addition, contaminants may also cause specific and/or non-specific toxic changes at various levels of biological organisation (molecule, sub-cellular units, cells, tissues and organs), leading to disease without involving pathogens.

Because of the role of diseases as ecosystem health indicators, it is recommended that fish disease surveys should be part of monitoring programmes aiming at an assessment of ecological risks associated with dumped chemical and conventional munitions in the Baltic Sea and elsewhere. Results obtained may act as an "alarm bell", initiating further and more specific studies on cause-effect relationships.

The strengths of fish disease monitoring can be summarized as follows (Lang 2002, with modifications):

- Diseases are an overt and integrative biological endpoint of physiological changes at different levels of biological organization, affecting the organism's homeostasis, that are associated with environmental change.
- In concert with more contaminant-specific early-warning biological effects techniques, e.g., biomarkers of contaminant exposure and contaminant-induced damage, fish diseases can be used more specifically as indicators of effects of contaminants released from dumped chemical and conventional munitions.
- Significant changes in disease prevalence are a biologically and ecologically relevant warning sign ("alarm bell") for adverse environmental changes, since diseases may affect growth, reproduction, and survival of affected individuals and may, therefore, have implications at the population level.
- Data on the prevalence and spatial distribution of diseases (including parasites) of commercial fish species are of direct use for quality controls of fish as a food resource for human consumption.
- Fish disease monitoring is cost-effective since it can be carried out directly on board research or even commercial vessels, possibly in combination with stock assessment surveys, without involving subsequent laboratory work (except histopathology).
- Externally visible target diseases identified are, with a certain degree of training, easy to recognize.
- A large number of fish and large geographical areas can be screened and results are immediately available.
- Methodologies for fish disease surveys have been established and repeatedly intercalibrated largely through activities of the International Council for the Exploration of the Sea (ICES).

9.7.2 How to Sample Fish

For adequate sampling of fish in a monitoring framework, a number of prerequisites have to be met: An appropriate fish species has to be selected that is abundant in the areas of interest and that is at risk to get in contact with dumped munitions. Consequently, the species should predominantly inhabit bottom-near waters. It is advisable to select a species for which data from previous studies or already running monitoring programmes (e.g., stock assessment or fish disease surveys) are available for comparative purposes (see above).

For the MODUM project and under the conditions present in the CWA dumpsites in the Baltic Sea, the Atlantic cod (*Gadus morhua*) was the species of choice, because it is abundant and widely distributed in the Baltic Sea. Furthermore, cod is a demersal species, with a preference to stay close to the sea bottom if environmental conditions (i.e., oxygen content of the water) permit (see Lang et al. Chap. 6 this issue). Cod in the Baltic Sea has been subject of many scientific studies and partly internationally standardised monitoring and assessment programmes, e.g., addressing stock performance and dynamics, migration patterns, contamination with anthropogenic hazardous substances as well as the occurrence of diseases and parasites. Data generated through such activities in general are valuable tools for the interpretation of results obtained in project or monitoring programmes focussing on effects of dumped munitions.

If monitoring is done in shallow regions with sufficient oxygen concentrations in the bottom water, other demersal species such as flatfishes (in the Baltic Sea, e.g., flounder (*Platichthys flesus*), plaice (*Pleuronectes platessa*) or, in the western Baltic Sea, dab (*Limanda limanda*) are suitable bioindicator candidates. In coastal areas, the eelpout (*Zoarces viviparus*) may be an appropriate species due to its marked stationary behaviour.

The sampling design applied in MODUM was largely according to standard guidelines for fish disease surveys (ICES 1989; Bucke et al. 1996) and comprised of multiple hauls in each study area to reduce sampling variation. Sampling of cod in the Baltic Sea (and elsewhere) usually is done by means of bottom or pelagic (midwater) trawling. In MODUM, fish were sampled using either a 180 ft. bottom trawl with cod end (20 mm mesh width) or a pelagic trawl (PSN 205), also with cod end (20 mm mesh width). The latter gear was used in CWA dumpsites to exclude the risk to catch up munitions or parts thereof. Towing time was between 30 and 60 Min, towing speed 3–4 knots. Prolonged trawling is not recommended, because it would lead to too much superficial damage of the fish and, thus, problems to identify externally visible diseases/parasites. It, furthermore, would increase stress in fish caught which should be avoided if, e.g., biomarker samples (e.g., for enzymatic or immunological measurements) are to be taken in addition to the disease examination.

The depth in which the pelagic trawl was operated was variable and depended on the depth of the fish signals on the echo sounder device. Alternative methods could be the use of passive gears (e.g., gill nets, fish traps, longlines), but this was not an option in the deep basins of the Baltic Sea due to practical reasons. It is recommended that – if available - standardised fishing gears are being used in order to obtain comparable results, e.g., those used in coordinated fish stock assessment surveys such as the ICES International Bottom Trawl Survey (IBTS) (ICES 2012a) or the Baltic International Trawl Survey (BITS) (ICES 2014).

For monitoring purposes it is crucial to select appropriate geographical sampling areas. These should include the target sites (i.e. munitions dumpsites) and appropriate reference sites. Study areas in MODUM were selected in a way that they covered areas representing the habitat of the western cod stock (west of the island of Bornholm) and of the eastern cod stock (east of Bornholm) which differ in their biological traits. Known ("official" sites, marked in the sea charts) or suspected CWA dumpsites studied were the major CWA dumpsite east of Bornholm (area B13), the dumpsite in the Gotland Basin (area B14) and the Gdansk Deep (area B15) (see Lang et al. Chap. 6 this issue). As references, sites unaffected by munitions dumping east (outside Bay of Gdansk) and west (Arkona Sea, Kiel Bight) of Bornholm were selected. Sampling areas were defined as boxes with nominated latitude and longitude at each corner of the box. Within the boxes, either randomised hauls were conducted or, in case of the dumpsite in the Bornholm Basin, repeated trawling was carried out through the centre of the radial dumpsite with the trawl tracks creating an asterisk-like pattern. An approach as described above is advisable for monitoring purposes in general. However, there may be other suitable strategies, e.g. if effects of a clearly defined locally restricted point sources are to be monitored. In that case, sampling sites may be arranged on a known or suspected contaminant gradient and positioned at fixed nominated geographical latitudes and longitudes.

9.7.3 How to Examine Fish for Health Effects

In the MODUM project, two basic indicators of health effects were recorded (see Lang et al. Chap. 6 this issue):

- Fulton's condition index K (also termed condition factor, CF), representing the mathematical relationship between length and weight of the fish,
- The occurrence and prevalence of a range of seven externally visible fish diseases and parasites

The condition factor was measured as indicator of general body condition or fitness of the cod and showed some marked differences between MODUM sampling areas (see Lang et al. Chap. 6 this issue). The condition factor is the ratio of total cubed body weight to body length, and is calculated as Fulton's condition factor (K), depicted in the formula below (Fulton 1904; Ricker 1990), It expresses that heavier fish of a certain length are in a better condition. The factor 100 is used to bring K close to a value of one.

$$Condition\ Factor\left(K\ or\ CF\right)=\left(total\ wet\ weight\ /\ total\ length^{3}\right)^{*}100$$

The disease conditions recorded are common ones known for a long time. Details are provided in Lang et al. (Chap. 6 this issue). There is consensus that they respond to environmental stressors rather in a non-specific than in a contaminant-specific way (Lang 2002). The following methodological description should be taken into account for monitoring purposes.

Directly after hauling, cod were sorted from the total catches and kept alive in tanks with running seawater of ambient water temperature. Fish disease examination was carried out according to guidelines generated through activities of the International Council for the Exploration of the Sea (ICES) (ICES 1989; Bucke et al. 1996; Lang and Mellergaard 1999), and through the fish disease component of the BEQUALM programme (Biological Effects Quality Assurance in Monitoring Programmes, http://www.bequalm.org/).

The method consists of a visual external examination of freshly collected fish for the occurrence of a defined set of externally visible diseases with clinical signs and of parasitoses (see Lang et al. Chap. 6 this issue). The external examination focuses on the body surface, including the spread-out fins as well as the mouth and gill cavities. In addition to information on the presence of a disease condition, its severity grade was recorded.

Disease examination on board and sampling for subsequent analyses were carried out by trained experts following a strict protocol (SOPs). The specimens selected for examination were inspected while fresh, i.e., shortly after they have been landed on the ship.

Fish were examined for externally visible diseases and parasites after rinsing in clean water. Thin gloves to protect were worn to protect the skin of the observers. Each fish was length-measured (to the nearest 1.0 cm below) and weighed by using an appropriate balance suitable for operation under instable conditions at sea (®MAREL Marine Scales). Only the fish inspected internally were sexed.

At least two people were needed for examining a large number of fish: one conducting the examination and the other one for recording the data directly onto the computer keyboard. These positions were made interchangeable, so that both workers knew how to take the measurements and how to transcribe the data.

A computer data entry software was developed as a MODUM deliverable which was used instead of hand-written protocols. Data entry was accomplished using two rugged notebooks suitable for work under wet conditions on a research vessel. The entry software is a part of the DMAR-System and based on a Python web service. An automatic storing process, transferred the data to a Debian based PostgreSQL-Database (DEMAR).

As a minimum requirement, a subsample of at least 500 specimens per sampling area and survey were examined where possible. This was, however, not possible in all cases because of a lack of fish in some of the catches.

9.7.4 How to Treat Fish Disease Data

Based on the examination of single fish for the range of target diseases, the prevalence of each disease can be calculated:

$$p = x / n$$

where p = prevalence, x = number of fish affected and n = number of fish examined.

The prevalence can either be expressed as a numeric value (p = ≤ 1.0) or as a percentage (p = ≤ 100%). The latter is more common.

For a comparison of prevalences (p) recorded (e.g., between sampling sites or years), confidence intervals (C.I.) for p can be calculated. Because of the nature of the data, C.I. (described by the limits π_{upper} and π_{lower}) for binomial distributions are calculated, e.g., the 95% C.I., according to the following formulas (for p > 0.0 and <1.0):

$$\pi_{\text{upper}} = \frac{(x+1)F}{n-x+(x+1)F} \; with \; F_{(DF1=2(x+1),DF2=2(n-x))}$$

$$\pi_{\text{lower}} = \frac{x}{x+(n-x+1)F} \; with \; F_{(DF1=2(n-x+1),DF2=2x)}$$

For p = 0.0 (none of the fish examined is affected) the following formula applies:

$$\pi_{\text{upper}} = \frac{F}{n+F} \; with \; F_{(DF1=2(x+1),DF2=2(n-x))}$$

For p = 1.0 (all of the fish examined are affected) the following formula applies:

$$\pi_{\text{lower}} = \frac{n}{n+F} \; with \; F_{(DF1=2,DF2=2n)}$$

As indicator of the general health status of cod, a Fish Disease Index (FDI) for externally visible disease conditions was calculated for individual fish, based on the presence/absence of the seven disease conditions, their intensity (3 severity grades), their suspected impact on the host (weighting by expert judgment) as well as on adjustment factors compensating for effects of length and sampling season on the disease prevalence. The (individual) FDI summarizes the disease status of individual specimen by a single number which can be in the range of 0 (no disease present) to 100 (all seven cod diseases recorded present with the highest severity stages) (for methodological details see Lang and Wosniok 2008 (Lang and Wosniok 2008; ICES 2012b; Weirup 2015; Lang et al. 2017). The aim of the FDI tool is to summarise information on the disease status of individual fish into one robust and easy-to-understand and easy-to-communicate numeric figure per fish. Its design principle allows the FDI to be applied to other species with other sets of diseases. Therefore, the FDI approach is applicable for wider geographical areas, e.g., munitions dumpsites in other regions.

9.8 Modelling

Modelling is a unique tool that could help in prediction of contaminated area. We could have two main cases: the first one is to diagnose leakage of CWA in real time; the second one is to analyse data from historical point of view (hindcast situation). In both cases it is important to have a validated, properly working model. The validation procedure usually depends on the type of model and data availability and it should also strongly depend on the analysed area. Here we assume that we are limited to the Baltic Sea, which is the main basin of the MODUM project, and the model has been successful validated. As the first step we should take into account the spinup. The Baltic Sea is under a strong influence of water exchange with the North Sea. Saline and dense waters income from Kattegat in specific atmospheric conditions. The characteristic location of the Baltic Sea implies a salinity memory of about 30 years (Omstedt and Hansson 2006a, b) Thus, in the case of the Baltic Sea, the spinup time should be longer than 30 years. Such a very long time of spinup could be decreased to 3 years, if we start with data that represent correct salinity of the Baltic Sea. Another important thing to consider is seasonal variability. The Baltic Sea is under influence of a lot of rivers and the maximum discharge occurs in spring. At this time currents obtain an additional barotropic part of flow, which is caused by seasonally increased surface elevation. It is important to be aware that barotropic flow is usually more important than baroclinic one.

A very significant element of introducing any model is to analyse time and space scale of the process of leakage. Dumped munitions are usually smaller than 1 m, thus, the horizontal resolution of the model should be at the same space scale. Currently it is impossible to cover the Baltic Sea by a grid of about 1 m and run the integration. However, the horizontal resolution plays very important role in estimating contamination and should be less than 1 km. Furthermore, the vertical resolution is not without significance. It affects the estimation of overflow. Models with sigma coordinates are better then z-levels because of better representation of the bottom turbulence.

Another important thing is the analysis of leakage location. The Baltic Sea does not have a constant circulation. It means that there is no area with stable bottom currents. Thus, in any case it is good to know the main direction of bottom current and its seasonal variability. For example Fig. 9.7 shows seasonal variability of the bottom currents calculated for the Gotland dumpsite and shows visible changes of bottom currents (Fig. 9.7).

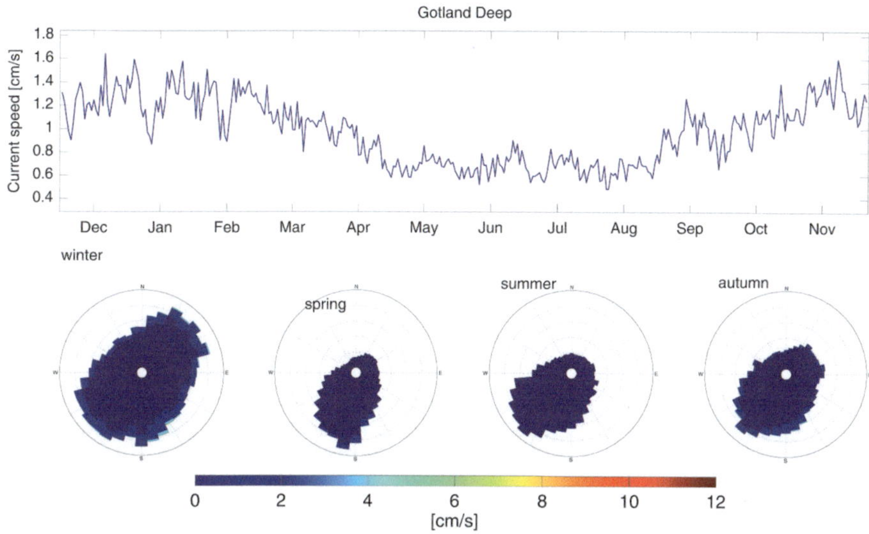

Fig. 9.7 Seasonal variability of the bottom currents for the Gotland dumpsite (based on 20 years of simulation)

Firstly, the average magnitude of the currents presents a seasonal cycle with a visible minimum in summer (strongly connected with river discharge, upper image). Secondary the direction of the currents depends on seasons and the influence of the barotropic part from sea level variability is visible mostly in spring and summer (lower image). In some areas the bottom currents could have a velocity of up to 50 cm/s (for example in the Słupsk Furrow). In other areas (the Bornholm Deep) the bottom currents are very slow and quite stable. But in autumns and winters the probability of inflows from the North Sea increases and saline and dense waters from Kattegat could disturb the stable situation in the deep. Depending on density differences of water in the Bornholm Deep and water from the North Sea, the incoming water could move in the most bottom layers (when the incoming water is heavier).

Assuming that we know the pollution level and threshold that represents the boundary between safe and unsafe concentration of dangerous material, during the first stage it is good to make a simulation of potential leakage. The number of simulations will depend on the type of model and also on the method used for estimating contamination. In the case of Lagrangian tracking with random disturbance, the number of integration should be sufficient for calculation of the density of trajectories. These densities could be transformed into probability of contamination (PC). Random disturbance in Lagrangian walking model plays the role of diffusion. Thus, if the passive tracer is used as a main tool, diffusion is a natural part of equation - it requires much less number of integrations.

In both methods switching into PC level is better then making operational simulations. It does not mean that the operational data are wrong, but PC gives a much wider image and could help to visually estimate the contaminated areas, which is impossible in the case of the operational simulations. For example when multiplying PC by real initial concentration, the result shows estimated contamination based on seasonal simulation. A contaminated area should be estimated as a circle around the source of leakage with radius equal to distance from the leakage location to place where contamination is higher than the assumed threshold. This method gives much larger area than the polluted area, but it takes into account the variability of the currents. When PC images are ready, operational prediction of contamination provides an additional value of assessment.

9.9 Conclusions

In conclusion, it seems appropriate to recommend that regular fish disease monitoring in munitions dumpsites and reference areas should be carried out, applying standardised and established methodologies (including the measurement of fitness parameters). If significant changes occur, the results may primarily serve as an "alarm bell", reflecting adverse effects of environmental stressors on fish health. Fish disease surveys should, however, be part of an integrated monitoring programme, comprising chemical measurements of pollutants derived from dumped munitions and biological effects measurements at different levels of biological organisation (from the molecular to the individual and population level). Ideally, studies on specific biomarkers indicating exposure to and effects of toxic munitions compounds should be incorporated. However, these still need to be developed and validated as a future research priority.

Whatever environmental monitoring strategies are being applied, they should always be accompanied by research activities in order to provide scientific background data required for the interpretation of monitoring data and to improve monitoring strategies and techniques, as appropriate.

In overall, the combination of presented tools successfully provided the MODUM project with all necessary data used in habitat description.

References

Amato E, Alcaro L, Corsi I, Della Torre C, Farchi C, Focardi S, Marino G, Tursi A (2006) An integrated ecotoxicological approach to assess the effects of pollutants released by unexploded chemical ordnance dumped in the southern Adriatic (Mediterranean Sea). Springer, Heidelberg

Beldowski J, Klusek Z, Szubska M, Turja R, Bulczak AI, Rak D, Brenner M, Lang T, Kotwicki L, Grzelak K, Jakacki J, Fricke N, Ostin A, Olsson U, Fabisiak J, Garnaga G, Nyholm JR, Majewski P, Broeg K, Soderstrom M, Vanninen P, Popiel S, Nawala J, Lehtonen K, Berglind R, Schmidt B (2016a) Chemical munitions search & assessment-an evaluation of the dumped munitions problem in the Baltic Sea. Deep-Sea Res Pt Ii 128:85–95. doi:10.1016/j.dsr2.2015.01.017

Beldowski J, Szubska M, Emelyanov E, Garnaga G, Drzewinska A, Beldowska M, Vanninen P, Ostin A, Fabisiak J (2016b) Arsenic concentrations in Baltic Sea sediments close to chemical munitions dumpsites. Deep-Sea Res Pt II 128:114–122. doi:10.1016/j.dsr2.2015.03.001

Blondel P, Blondel P (2009) The handbook of sidescan sonar. Springer/Praxis Pub, Berlin/New York/Chichester

Bucke D, Vethaak AD, Lang T, Mellergaard S (1996) Common diseases and parasites of fish in the North Atlantic: training guide for identification. ICES Techniques in Marine Environmental Sciences 19, 27 p

Chapman P (2010) Guidance document: framework for addressing and managing aquatic contaminated sites under the Federal Contaminated Sites Action Plan (FCSAP). Final Report, Canada

Fulton TW (1904) The rate of growth of fishes. In: 22nd Annual Report Fish Board Scotland 3:141–241

Hooper MJ, Glomb SJ, Harper DD, Hoelzle TB, McIntosh LM, Mulligan DR (2016) Integrated risk and recovery monitoring of ecosystem restorations on contaminated sites. Integr Environ Asses 12(2):284–295. doi:10.1002/ieam.1731

ICES (1989) Methodology of fish disease surveys. Report of an ICES Sea-going workshop held on RV U/F 'Argos' 16–23 April 1988. ICES Cooperative Research Report 166. ICES

ICES (2012a) Manual for the International bottom trawl surveys. Series of ICES survey protocols. SISP 1-IBTS VIII. ICES

ICES (2012b) Report of the working group on pathology and diseases of marine organisms. ICES CM 2012/SSGHIE 3. ICES

ICES (2014) Manual for the baltic international trawl surveys (BITS). ICES survey protocols SISP 7 – BITS. ICES

Korotenko KA (2000) In: John D, Catherine T (eds) Chemical weapons dumped in the baltic sea: facts, tendency and prediction of level and scales of possible ecological disaster. MacArthur Foundation, Moscow

Lang T (2002) Fish disease surveys in environmental monitoring: the role of ICES. Paper presented at the ICES Marine Science Symposia 215:202–212

Lang T, Mellergaard S (1999) The BMB/ICES sea-going workshop "fish diseases and parasites in the Baltic sea" – introduction and conclusions. ICES J Mar Sci 56(2):129–133. doi:10.1006/jmsc.1999.0468

Lang T, Wosniok W (2008) The fish disease index: a method to assess wild fish disease data in the context of marine environmental monitoring. ICES CM 2008/D:01. ICES

Lang T, Feist SW, Stentiford GD, Bignell JP, Vethaak AD, Wosniok W (2017) Diseases of dab (Limanda limanda): analysis and assessment of data on externally visible diseases., macroscopic liver neoplasms and liver histopathology in the North Sea, Baltic Sea and off Iceland. Mar Environ Res 124:61–69. doi:10.1016/j.marenvres.2015.12.009

Medwin H (1975) Speed of sound in water – simple equation for realistic parameters. J Acoust Soc Am 58(6):1318–1319. doi:10.1121/1.380790

Omstedt A, Hansson D (2006a) Erratum to: "The Baltic Sea ocean climate system memory and response to changes in the water and heat balance components". Cont Shelf Res 26(2):236–251. Cont Shelf Res 26 (14):1685–1687

Omstedt A, Hansson D (2006b) The Baltic Sea ocean climate system memory and response to changes in the water and heat balance components. Cont Shelf Res 22(2):236–251

Ricker WE (1990) 100 years of computations – a citation-classic commentary on computation and interpretation of biological statistics of fish populations, (Whole Issue) by Ricker WE. Cc/Agr Biol Environ (39):18–18

Sanderson H, Fauser P, Thomsen M, Sorensen PB (2008) Screening level fish community risk assessment of chemical warfare agents in the Baltic Sea. J Hazard Mater 154(1–3):846–857. doi:10.1016/j.jhazmat.2007.10.117

Tobias Knobloch, Jacek Beldowski, Claus Böttcher, Martin Söderström, Niels-Peter Rühl, Sternheim J (2013) Chemical munitions dumped in the Baltic Sea. Report of the ad hoc expert

group to update and review the existing information on dumped chemical munitions in the baltic sea (HELCOM MUNI), Baltic sea environmental proceedings. HELCOM
USEPA (2004) Guidance for monitoring at hazardous waste sites: framework for monitoring plan development and implementation. United States Environmental Protection Agency
USEPA (2006) Guidance on systematic planning using the data quality objectives process. Office of Environmental Information
Vethaak AD, Rheinallt T (1992) Fish disease as a monitor for marine pollution: the case of the North Sea. Rev Fish Biol Fish 2:1–32
Weirup L (2015) Diseases and parasites of Baltic cod (Gadus morhua L): Spatio-temporal patterns and host effects. University of Hamburg

MIX
Papier aus verantwortungsvollen Quellen
Paper from responsible sources
FSC® C105338
www.fsc.org

Printed by Libri Plureos GmbH
in Hamburg, Germany